T0133062

Applied Math
FOR WATER PLANT OPERATORS

JOANNE KIRKPATRICK PRICE
Training Consultant

CRC Press
Taylor & Francis Group
Boca Raton London New York

CRC Press is an imprint of the
Taylor & Francis Group, an **informa** business

Published in 1991 by
CRC Press
Taylor & Francis Group
6000 Broken Sound Parkway NW, Suite 300
Boca Raton, FL 33487-2742

© 1991 by Taylor & Francis Group, LLC
CRC Press is an imprint of Taylor & Francis Group, an Informa business

No claim to original U.S. Government works

ISBN 13: 978-0-87762-874-3 (hbk)

Library of Congress catalog number: 91-65983
Cover art adapted from photograph of Fallbrook Sanitary District Reclaimed Water System

Library of Congress Cataloging-in-Publication Data

Catalog record is available from the Library of Congress

Visit the Taylor & Francis Web site at
http://www.taylorandfrancis.com

and the CRC Press Web site at
http://www.crcpress.com

Dedication

This book is dedicated to my family:

To my husband Benton C. Price who was patient and
supportive during the two years it took to write these texts, and
who not only had to carry extra responsibilities at home during
this time, but also, as a sanitary engineer, provided frequent
technical critique and suggestions.

To our children Lisa, Derek, Kimberly, and Corinne,
who so many times had to pitch in while I was busy writing,
and who frequently had to wait for my attention.

To my mother who has always been so encouraging and who
helped in so many ways throughout the writing process.

To my father, who passed away since the writing of the first
edition, but who, I know, would have had just as instrumental
a role in these books.

To the other members of my family, who have had to put up
with this and many other projects, but who maintain a sense
of humor about it.

Thank you for your love in allowing me to do something
that was important to me.

J.K.P.

Contents

Contents—Cont'd

Contents—Cont'd

Contents—Cont'd

Preface to the Second Edition

The first edition of these texts was written at the conclusion of three and a half years of instruction at Orange Coast College, Costa Mesa, California, for two different water and wastewater technology courses. The fundamental philosophy that governed the writing of these texts was that those who have difficulty in math often do not lack the ability for mathematical calculation, they merely have not learned, or have not been taught, the "language of math." The books, therefore, represent an attempt to bridge the gap between the reasoning processes and the language of math that exists for students who have difficulty in mathematics.

In the years since the first edition, I have continued to consider ways in which the texts could be improved. In this regard, I researched several topics including how people learn (learning styles, etc.), how the brain functions in storing and retrieving information, and the fundamentals of memory systems. Many of the changes incorporated in this second edition are a result of this research.

Two features of this second edition are of particular importance:

- the **skills check section** provided at the beginning of every basic math chapter

- a **grouping of similar types of calculations** in the applied math texts

The skills check feature of the basic math text enables the student to pinpoint the areas of math weakness, and thereby customizes the instruction to the needs of the individual student.

The first six chapters of each applied math text include calculations grouped by type of problem. These chapters have been included so that students could see the common thread in a variety of seemingly different calculations.

The changes incorporated in this second edition were field-tested during a three-year period in which I taught a water and wastewater mathematics course for Palomar Community College, San Marcos, California.

Written comments or suggestions regarding the improvement of any section of these texts or workbooks will be greatly appreciated by the author.

Joanne Kirkpatrick Price

Acknowledgments

"From the original planning of a book to its completion, the continued encouragement and support that the author receives is instrumental to the success of the book." This quote from the acknowledgments page of the first edition of these texts is even more true of the second edition.

First Edition

Those who assisted during the development of the first edition are: Walter S. Johnson and Benton C. Price, who reviewed both texts for content and made valuable suggestions for improvements; Silas Bruce, with whom the author team-taught for two and a half years, and who has a down-to-earth way of presenting wastewater concepts; Mariann Pape, Samuel R. Peterson and Robert B. Moore of Orange Coast College, Costa Mesa, California, and Jim Catania and Wayne Rodgers of the California State Water Resources Control Board, all of whom provided much needed support during the writing of the first edition.

The first edition was typed by Margaret Dionis, who completed the typing task with grace and style. Adele B. Reese, my mother, proofed both books from cover to cover and Robert V. Reese, my father, drew all diagrams (by hand) shown in both books.

Second Edition

The second edition was an even greater undertaking due to many additional calculations and because of the complex layout required. I would first like to acknowledge and thank Laurie Pilz, who did the computer work for all three texts and the two workbooks. Her skill, patience, and most of all perseverance has been instrumental in providing this new format for the texts. Her husband, Herb Pilz, helped in the original format design and he assisted frequently regarding questions of graphics design and computer software.

Those who provided technical review or assistance with various portions of the texts include Benton C. Price, Kenneth D. Kerri, Lynn Marshall, Wyatt Troxel, Mike Hoover, Bruce Grant and Jack Hoffbuhr. Their comments and suggestions are appreciated and have improved the current edition.

Many thanks also to the staff of the Fallbrook Sanitary District, Fallbrook, California, especially Virginia Grossman, Nancy Hector, Joyce Shand, Mike Page, and Weldon Platt for the numerous times questions were directed their way during the writing of these texts.

The staff of Technomic Publishing Company, Inc., also provided much advice and support during the writing of these texts. First, Melvyn Kohudic, President of Technomic Publishing Company, contacted me several times over the last few years, suggesting that the texts be revised. It was his gentle nudging that finally got the revision underway. Joseph Eckenrode helped work out some of the details in the initial stages and was a constant source of encouragement. Jeff Perini was copy editor for the texts. His keen attention to detail has been of great benefit to the final product. Leo Motter had the arduous task of final proof reading.

I wish to thank all my friends, but especially those in our Bible study group (Gene and Judy Rau, Floyd and Juanita Miller, Dick and Althea Birchall, and Mark and Penny Gray) and our neighbors, Herb and Laurie Pilz, who have all had to live with this project as it progressed slowly chapter by chapter, but who remained a source of strength and support when the project sometimes seemed overwhelming.

Lastly, the many students who have been in my classes or seminars over the years have had no small part in the final form these books have taken. The format and content of these texts is in response to their questions, problems, and successes over the years.

To all of these I extend my heartfelt thanks.

How To Use These Books

The *Mathematics for Water and Wastewater Treatment Plant Operators* series includes three texts and two workbooks:

- Basic Math Concepts for Water and Wastewater Plant Operators

- Applied Math for Water Plant Operators

- Workbook—Applied Math for Water Plant Operators

- Applied Math for Wastewater Plant Operators

- Workbook—Applied Math for Wastewater Plant Operators

Basic Math Concepts

All the basic math you will need to become adept in water and wastewater calculations has been included in the Basic Math Concepts text. This section has been expanded considerably from the basic math included in the first edition. For this reason, students are provided with more methods by which they may solve the problems.

Many people have weak areas in their math skills. It is therefore advisable to take the skills test at the beginning of each chapter in the basic math book to pinpoint areas that require review or study. If possible, it is best to resolve these weak areas <u>before</u> beginning either of the applied math texts. However, when this is not possible, the Basic Math Concepts text can be used as a reference resource for the applied math texts. For example, when making a calculation that includes tank volume, you may wish to refer to the basic math section on volumes.

Applied Math Texts and Workbooks

The applied math texts and workbooks are companion volumes. There is one set for water treatment plant operators and another for wastewater treatment plant operators. Each applied math text has two sections:

- Chapters 1 through 6 present various calculations **grouped by type of math problem**. Perhaps 70 percent of all water and wastewater calculations are represented by these six types. Chapter 7 groups various types of pumping problems into a single chapter. The calculations presented in these seven chapters are common to the water and wastewater fields and have therefore been included in both applied math texts.

  Since the calculations described in Chapters 1 through 6 represent the heart of water and wastewater treatment math, if possible, it is advisable that you master these general types of calculations before continuing with other calculations. Once completed, a review of these calculations in subsequent chapters will further strengthen your math skills.

- The remaining chapters in each applied math text include calculations **grouped by unit processes**. The calculations are presented in the order of the flow through a plant. Some of the calculations included in these chapters are not incorporated in Chapters 1 through 7, since they do not fall into any general problem-type grouping. These chapters are particularly suited for use in a classroom or seminar setting, where the math instruction must parallel unit process instruction.

The workbooks support the applied math texts section by section. They have also been vastly expanded in this edition so that the student can build strength in each type of calculation. A detailed answer key has been provided for all problems. The workbook pages have been perforated so that they may be used in a classroom setting as hand-in assignments. The pages have also been hole-punched so that the student may retain the pages in a notebook when they are returned.

The workbooks may be useful in preparing for a certification exam. However, because these texts include both fundamental and advanced calculations, and because the requirements for each certification level vary somewhat from state to state, it is advisable that you <u>first determine the types of problems to be covered in your exam</u>, then focus on those types of calculations in these texts.

1 *Applied Volume Calculations*

SUMMARY

The **general equation** for most volume calculations is:

$$\text{Volume} = \left[\begin{array}{l}\text{Representative} \\ \text{Surface Area}\end{array}\right]\left[\begin{array}{l}\text{Depth or} \\ \text{Height}\end{array}\right]$$

1. Tank volume calculations.
Most tank volume calculations are for tanks that are either rectangular or cylindrical in shape.

Rectangular Tank

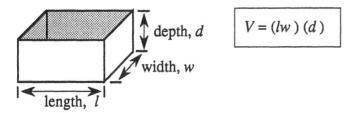

$$V = (lw)(d)$$

Cylindrical Tank

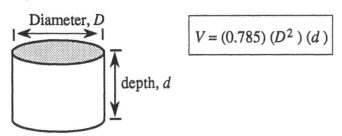

$$V = (0.785)(D^2)(d)$$

2. Channel or pipeline volume calculations are very similar to tank volume calculations. The principal shapes are shown below.

Portion of a Rectangular Channel

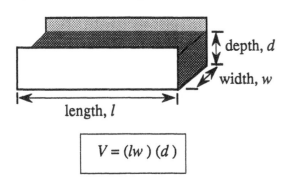

$$V = (lw)(d)$$

Three general types of water and wastewater volume calculations are:

- Tank Volume

- Channel or Pipeline Volume

- Pit, Trench, or Pond Volume

Each of these calculations is simply a specific application of volume calculations. For a more detailed discussion of volume calculations, refer to Chapter 11 in *Basic Math Concepts*.

For many calculations, the volumes must be expressed in terms of **gallons**. To convert from cubic feet to gallons volume, a factor of 7.48 gal/cu ft is used. Refer to Chapter 8 of *Basic Math Concepts* for a detailed discussion of cubic feet to gallons conversions.

SUMMARY—Cont'd

Portion of a Trapezoidal Channel

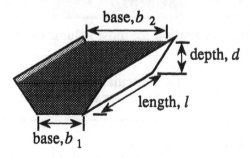

$$V = \frac{(b_1 + b_2)(d)(l)}{2}$$

Portion of a Pipeline

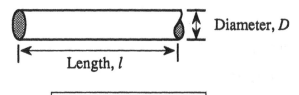

$$V = (0.785)(D^2)(l)$$

3. **Other volume calculations** involving ditches or ponds depend on the shape of the ditch or pond. A pit or trench is often rectangular in shape. A pond or oxidation ditch may have a trapezoidal cross section.

NOTES:

1.1 TANK VOLUME CALCULATIONS

The two common tank shapes in water and wastewater treatment are rectangular and cylindrical tanks.

Rectangular Tank:

$$V = (lw)(d)$$

Where:

V = Volume, cu ft
l = Length, ft
w = Width, ft
d = Depth, ft

Cylindrical Tank:

$$V = (0.785)(D^2)(d)$$

Where:

V = Volume, cu ft
D = Diameter, ft
d = Depth, ft

The volume of these tanks can be expressed in cubic feet or gallons. The equations shown above are for cubic feet volume. Since each cubic foot of water contains 7.48 gallons, to convert cubic feet volume to gallons volume, multiply by 7.48 gal/cu ft, as illustrated in Example 2. As an alternative, you may wish to include the 7.48 gal/cu ft factor in the volume equation, as shown in Example 3.

Example 1: (Tank Volume)
❑ The dimensions of a tank are given below. Calculate the volume of the tank in cubic feet.

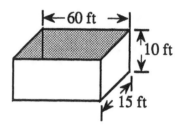

$$\text{Vol., cu ft} = (lw)(d)$$
$$= (60 \text{ ft})(15 \text{ ft})(10 \text{ ft})$$
$$= \boxed{9000 \text{ cu ft}}$$

Example 2: (Tank Volume)
❑ A tank is 25 ft wide, 75 ft long, and can hold water to a depth of 10 ft. What is the volume of the tank, in gallons?

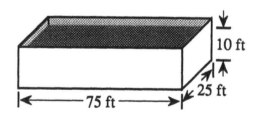

$$\text{Vol., cu ft} = (lw)(d)$$
$$= (75 \text{ ft})(25 \text{ ft})(10 \text{ ft})$$
$$= 18{,}750 \text{ cu ft}$$

Now convert cu ft volume to gal:

$$(18{,}750 \text{ cu ft})(7.48 \text{ gal/cu ft}) = \boxed{140{,}250 \text{ gal}}$$

Example 3: (Tank Volume)
❏ The diameter of a tank is 60 ft. When the water depth is 25 ft, what is the volume of water in the tank, in gallons?

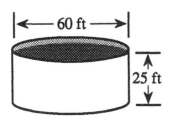

$$\text{Vol., gal} = (0.785)\,(D^2)\,(d)\,(7.48\ \text{gal/cu ft})$$

$$= (0.785)\,(60\ \text{ft})\,(60\ \text{ft})\,(25\ \text{ft})\,(7.48\ \text{gal/cu ft})$$

$$= \boxed{528{,}462\ \text{gal}}$$

Example 4: (Tank Volume)
❏ A tank is 12 ft wide and 20 ft long. If the depth of water is 11 ft, what is the volume of water in the tank?

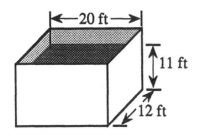

$$\text{Vol., gal} = (lw)\,(d)\,(7.48\ \text{gal/cu ft})$$

$$= (20\ \text{ft})\,(12\ \text{ft})\,(11\ \text{ft})\,(7.48\ \text{gal/cu ft})$$

$$= \boxed{19{,}747\ \text{gal}}$$

1.2 CHANNEL OR PIPELINE VOLUME CALCULATIONS

Channel or pipeline volume calculations are similar to tank volume calculations.* The equations to be used in calculating these volumes are given below.

Channel or pipeline volumes may be expressed as cubic feet, as shown in Example 1, or as gallons, shown in Examples 2-4.

Channel with Rectangular Cross Section:

$$\text{Volume,} \atop \text{cu ft} = (lw)\,(d)$$

Channel with Trapezoidal Cross Section:

$$\text{Volume,} \atop \text{cu ft} = \frac{(b_1 + b_2)\,(d)\,(l)}{2}$$

Pipeline with Circular Cross Section:

$$\text{Volume,} \atop \text{cu ft} = (0.785)\,(D^2)\,(l)$$

Example 1: (Channel or Pipe Volume)
❑ Calculate the volume of water in the section of rectangular channel shown below when the water is 4 ft deep.

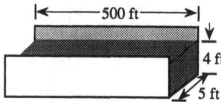

$$
\begin{aligned}
\text{Vol., cu ft} &= (lw)\,(d) \\
&= (5\text{ ft})\,(4\text{ ft})\,(500\text{ ft}) \\
&= \boxed{10{,}000\text{ cu ft}}
\end{aligned}
$$

Example 2: (Channel or Pipe Volume)
❑ Calculate the volume of water in the section of trapezoidal channel shown below when the water depth is 4 ft.

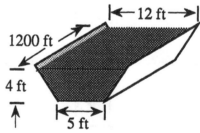

$$
\begin{aligned}
\text{Vol., gal} &= \frac{(b_1 + b_2)\,(d)\,(l)}{2}\,(7.48\text{ gal/cu ft}) \\
&= \frac{(5\text{ ft} + 12\text{ ft})\,(4\text{ ft})\,(1200\text{ ft})}{2}\,(7.48\text{ gal/cu ft}) \\
&= (8.5)\,(4)\,(1200)\,(7.48) \\
&= \boxed{305{,}184\text{ gal}}
\end{aligned}
$$

---

* For a detailed review of volume calculations, refer to Chapter 11 in *Basic Math Concepts.*

Example 3: (Channel or Pipe Volume)

❑ A new section of 12-inch diameter pipe is to be disinfected before it is put into service. If the length of pipeline is 2000 ft, how many gallons of water will be needed to fill the pipeline?

Vol., gal $= (0.785) (D^2) (l) (7.48 \text{ gal/cu ft})$

$= (0.785) (1 \text{ ft}) (1 \text{ ft}) (2000 \text{ ft}) (7.48 \text{ gal/cu ft})$

$= \boxed{11{,}744 \text{ gal}}$

Example 4: (Channel and Pipe Volume)

❑ A section of 6-inch diameter pipeline is to be filled with chlorinated water for disinfection. If 1320 ft of pipeline is be disinfected, how many gallons of water will be required?

Vol., gal $= (0.785) (D^2) (l) (7.48 \text{ gal/cu ft})$

$= (0.785) (0.5 \text{ ft}) (0.5 \text{ ft}) (1320 \text{ ft}) (7.48 \text{ gal/cu ft})$

$= \boxed{1{,}938 \text{ gal}}$

1.3 OTHER VOLUME CALCULATIONS

PIT OR TRENCH VOLUMES

These volume calculations are similar to tank and channel volume calculations, with one exception—the volume is often expressed as cubic yards rather than cubic feet or gallons.

There are two approaches to calculating cubic yard volume:

- Calculate the cubic feet volume, then convert to cubic yards volume. (See Example 1.)

$$\frac{(cu\ ft)}{27\ cu\ ft/cu\ yd} = cu\ yds$$

- Express all dimensions in yards so that the resulting volume calculated will be cubic yards. (See Example 2.)

$$(yds)\ (yds)\ (yds) = cu\ yds$$

Example 1: (Other Volume Calculations)
❑ A trench is to be excavated 2.5 ft wide, 4 ft deep and 900 ft long. What is the cubic yards volume of the trench?

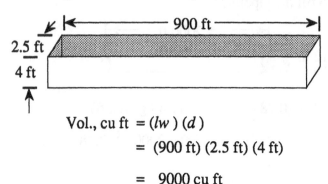

$$\begin{aligned} Vol.,\ cu\ ft &= (lw)\ (d) \\ &= (900\ ft)\ (2.5\ ft)\ (4\ ft) \\ &= 9000\ cu\ ft \end{aligned}$$

Now convert cu ft volume to cu yds:

$$\frac{9000\ cu\ ft}{27\ cu\ ft/cu\ yds} = \boxed{333\ cu\ yds}$$

Example 2: (Other Volume Calculations)
❑ What is the cubic yard volume of a trench 500 ft long, 2.25 ft wide and 4 ft deep?

Convert dimensions in ft to yds before beginning the volume calculation:

$$Length: \quad \frac{500\ ft}{3\ ft/yd} = 166.7\ yds$$

$$Width: \quad \frac{2.25\ ft}{3\ ft/yd} = 0.75\ yds$$

$$Depth: \quad \frac{4\ ft}{3\ ft/yd} = 1.33\ yds$$

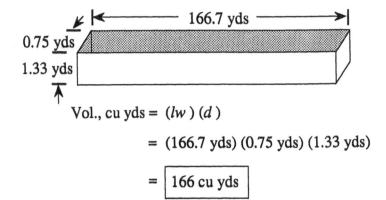

$$\begin{aligned} Vol.,\ cu\ yds &= (lw)\ (d) \\ &= (166.7\ yds)\ (0.75\ yds)\ (1.33\ yds) \\ &= \boxed{166\ cu\ yds} \end{aligned}$$

Example 3: (Other Volume Calculations)
❑ Calculate the volume of the oxidation ditch shown below, in cu ft. The cross section of the ditch is trapezoidal.

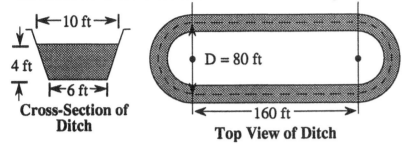

Cross-Section of Ditch

Top View of Ditch

$$\begin{aligned}
\text{Total Volume} &= \text{(Trapezoidal Area)(Total Length)} \\[6pt]
&= \left[\frac{(b_1+b_2)\,(h)}{2}\right]\left[\begin{array}{l}\text{(Length of)}+\text{(Length Around)}\\ \text{2 Sides}\quad\ \text{2 Half Circles}\end{array}\right] \\[6pt]
&= \left[\frac{(6\text{ ft}+10\text{ ft})\,(4\text{ ft})}{2}\right]\Big[320\text{ ft}+(3.14)\,(80\text{ ft})\Big] \\[6pt]
&= \Big[(8\text{ ft})\,(4\text{ ft})\Big]\Big[571.2\text{ ft}\Big] \\[6pt]
&= \boxed{18,278\text{ cu ft}}
\end{aligned}$$

Example 4: (Other Volume Calculations)
❑ A pond is 5 ft deep with side slopes of 2:1 (2 ft horizontal: 1 ft vertical). Given the following data, calculate the volume of the pond in cubic feet.

(Top View of Pond)

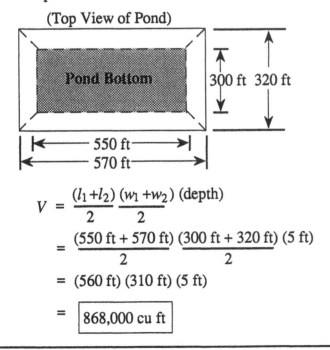

Pond Bottom

300 ft   320 ft

550 ft

570 ft

$$\begin{aligned}
V &= \frac{(l_1+l_2)}{2}\frac{(w_1+w_2)}{2}\,(\text{depth}) \\[6pt]
&= \frac{(550\text{ ft}+570\text{ ft})}{2}\frac{(300\text{ ft}+320\text{ ft})}{2}\,(5\text{ ft}) \\[6pt]
&= (560\text{ ft})\,(310\text{ ft})\,(5\text{ ft}) \\[6pt]
&= \boxed{868,000\text{ cu ft}}
\end{aligned}$$

OXIDATION DITCH OR POND VOLUMES

Many times oxidation ditches and ponds are trapezoidal in configuration. Examples 3 and 4 illustrate these calculations.

In Example 3, the oxidation ditch has sloping sides (trapezoidal cross section). The total volume of the oxidation ditch is the trapezoidal area times the total length:

$$\text{Total Vol.} = \frac{(b_1+b_2)}{2}\,(d)\,(\text{Total Length})$$

(The total length is measured at the center of the ditch; it is equal to the length of the two straight lengths plus two half-circle lengths. Note that the length around the two half circles is equal to the circumference of <u>one</u> full circle.)

WHEN ALL SIDES SLOPE

In many calculations of trapezoidal volume, such as for a trapezoidal channel, only two of the sides slope and the ends are vertical. To calculate the volume for such a shape, the following equation is normally used:

$$V = \frac{(b_1+b_2)\,(d)\,(l)}{2}$$

Another way of thinking of this calculation is average width times the depth of water times the length :

$$V = (\text{aver. width})\,(d)\,(\text{length})$$

In Example 4, however, since both length and width sides are trapezoidal, **the equation must include average length and average width dimensions:**

$$\boxed{V = \frac{(l_1+l_2)}{2}\frac{(w_1+w_2)}{2}\,(\text{depth})}$$

NOTES:

2 *Flow and Velocity Calculations*

SUMMARY

1. Instantaneous Flow Rates

The $Q=AV$ equation can be used to estimate flow rates through channels, tanks and pipelines. In all $Q=AV$ calculations, the units on the left side of the equation (Q) must match the combined units on the right side of the equation (A and V) with respect to <u>volume</u> (cubic feet or gallons) and <u>time</u> (sec, min, hrs, or days). The $Q=AV$ equations in this summary will be expressed in terms of cubic feet per minute.

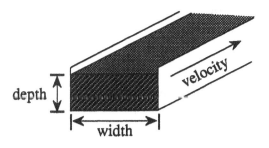

$$Q_{\text{cfm}} = AV_{\text{fpm}}$$

$$\underset{\text{cfm}}{Q} = \underset{\text{ft}}{(\text{width})} \; \underset{\text{ft}}{(\text{depth})} \; \underset{\text{fpm}}{(\text{velocity})}$$

Flow Through A Trapezoidal Channel

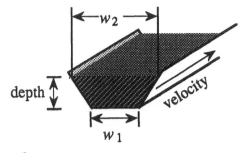

$$Q_{\text{cfm}} = AV_{\text{fpm}}$$

$$\underset{\text{cfm}}{Q} = \frac{(w_1 + w_2)}{2} \; \underset{\text{ft}}{(\text{depth})} \; \underset{\text{fpm}}{(\text{velocity})}$$

There are several ways to determine flow and velocity. Various flow metering devices may be used to measure water or wastewater flows at a particular moment (instantaneous flow) or over a specified time period (total flow). Instantaneous flow can also be determined using the $Q=AV$ equation.

This chapter includes discussions of $Q=AV$, velocity, average flow rates, and flow conversions.

SUMMARY—Cont'd

Flow Into Or Out Of A Tank

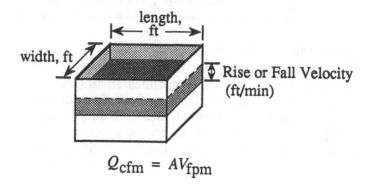

$$Q_{cfm} = AV_{fpm}$$

$$\underset{cfm}{Q} = \underset{ft}{(length)} \; \underset{ft}{(width)} \; \underset{fpm}{(rise\ or\ fall\ velocity)}$$

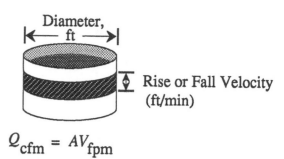

$$Q_{cfm} = AV_{fpm}$$

$$\underset{cfm}{Q} = (0.785) \; (D^2) \; \underset{fpm}{(rise\ or\ fall\ velocity)}$$

Flow Through A Pipeline—When Flowing Full

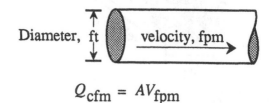

$$Q_{cfm} = AV_{fpm}$$

$$\underset{cfm}{Q} = (0.785) \; (D^2) \; \underset{fpm}{(velocity)}$$

SUMMARY—Cont'd

Flow Through A Pipeline—When Flowing When Less Than Full

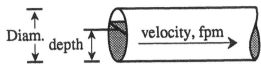

$$Q = \underset{\text{factor}}{\text{(new)}} (D^2) \underset{\text{fpm}}{\text{(velocity)}}$$
$$\underset{\text{cfm}}{} $$

*Based on d/D Table

2. Velocity Calculations

The $Q=AV$ equation can be used to determine the velocity of water at a particular moment. Use the same $Q=AV$ equation, fill in the known information, then solve for velocity.

$$Q = AV$$

Another method for determining velocity is to time the movement of a float or dye through the water.

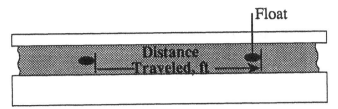

$$\text{Velocity} = \frac{\text{Distance Traveled, ft}}{\text{Duration of Test, min}}$$

$$\text{Velocity} = \frac{\text{ft}}{\text{min}}$$

The $Q=AV$ equation can also be used to determine velocity changes due to differences in pipe diameters. The AV in one pipe is equal to the AV in the other pipe.

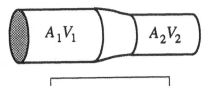

$$A_1V_1 = A_2V_2$$

SUMMARY—Cont'd

3. Average Flow Rates

The average flow rate may be calculated using two methods—one utilizing several different flows which are then averaged; and one utilizing a total flow and the time over which the flow is measured.

$$\text{Average Flow} = \frac{\text{Total of all Sample Flows}}{\text{Number of Samples}}$$

Or

$$\text{Average Flow} = \frac{\text{Total Flow}}{\text{Time Flow Measured}}$$

4. Flow Conversions

The box method may be used to convert from one flow expression to another. When using the box method of conversions, **multiply** when moving from a smaller box to a larger box, and **divide** when moving from a larger box to a smaller box. The common conversions are shown below:*

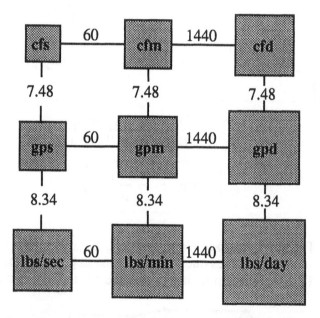

Dimensional analysis may also be used in making flow conversions. For a review of dimensional analysis, refer to Chapter 15 in *Basic Math Concepts*.

---

* The factors shown in the diagram have the following units associated with them: 60 min/sec, 1440 min/day, 7.48 gal/cu ft, and 8.34 lbs/gal.

NOTES:

2.1 INSTANTANEOUS FLOW RATES CALCULATIONS

The flow rate through channels and pipelines is normally measured by some type of flow metering device. However the flow rate for any particular moment can also be determined by using the $Q=AV$ equation.

The flow rate (Q) is equal to the cross-sectional area (A) of the channel or pipeline multiplied by the velocity through the channel or pipeline. There are two important considerations in these calculations:

1. Remember that volume is calculated by multiplying the representative area by a third dimension, often depth or height.* The $Q=AV$ calculation is essentially a volume calculation. The length dimension is a <u>velocity length</u> (length/time):

Vol = (Cross Sectional) (3rd Dim.)
         Area

Q =   A   V

2. The units used for volume and time must be the same on both sides of the equation, as shown in the diagram to the right.

FLOW THROUGH A RECTANGULAR CHANNEL

The principal difference among various $Q=AV$ calculations is the shape of the cross-sectional area. Channels normally have rectangular or trapezoidal cross sections, whereas pipelines have circular cross sections. Examples 1-3 illustrate the $Q=AV$ calculation when the channel is rectangular.

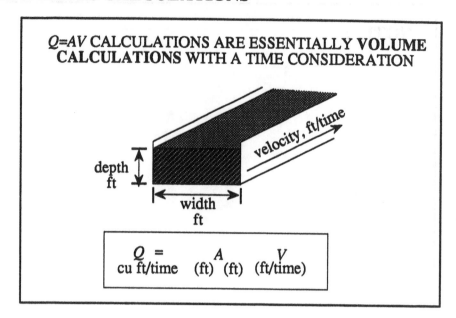

$Q=AV$ CALCULATIONS ARE ESSENTIALLY **VOLUME** CALCULATIONS WITH A TIME CONSIDERATION

$$\begin{array}{cccc} Q & = & A & V \\ \text{cu ft/time} & & \text{(ft)} \ \text{(ft)} & \text{(ft/time)} \end{array}$$

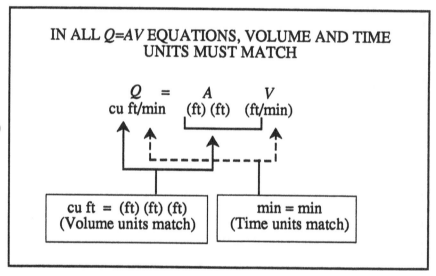

IN ALL $Q=AV$ EQUATIONS, VOLUME AND TIME UNITS MUST MATCH

$$\begin{array}{cccc} Q & = & A & V \\ \text{cu ft/min} & & \text{(ft) (ft)} & \text{(ft/min)} \end{array}$$

cu ft = (ft) (ft) (ft)	min = min
(Volume units match)	(Time units match)

Example 1: (Instantaneous Flow)
❑ A channel 3 ft wide has water flowing to a depth of 2.5 ft. If the velocity through the channel is 2 fps, what is the cfs flow rate through the channel?

$$Q_{cfs} = AV_{fps}$$

$$= (3 \text{ ft}) (2.5 \text{ ft}) (2 \text{ fps})$$

$$= \boxed{15 \text{ cfs}}$$

---

* For a review of volume calculations, refer to Chapter 11 in *Basic Math Concepts*.

Example 2: (Instantaneous Flow)
❑ A channel 40 inches wide has water flowing to a depth of 1.5 ft. If the velocity of the water is 2.3 fps, what is the cfs flow in the channel?

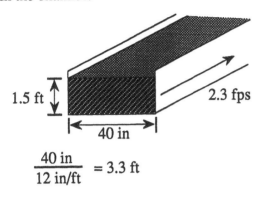

$$\frac{40 \text{ in}}{12 \text{ in/ft}} = 3.3 \text{ ft}$$

$$Q_{\text{cfs}} = AV_{\text{fps}}$$

$$= (3.3 \text{ ft}) (1.5 \text{ ft}) (2.3 \text{ fps})$$

$$= \boxed{11.4 \text{ cfs}}$$

DIMENSIONS SHOULD BE EXPRESSED AS FEET

The dimensions in a $Q=AV$ calculation should always be expressed in feet because (ft) (ft) (ft) = cu ft. Therefore, when dimensions are given as inches, first convert all dimensions to feet before beginning the $Q=AV$ calculation.

Note that velocity may be written in either of two forms:

- fps or fpm
- ft/sec or ft/min

The first form is shorter. The second form is useful when dimensional analysis* is desired.

Example 3: (Instantaneous Flow)
❑ A channel 3 ft wide has water flowing at a velocity of 1.5 fps. If the flow through the channel is 8.1 cfs, what is the depth of the water?

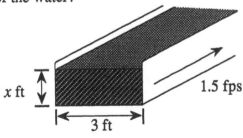

$$Q_{\text{cfs}} = AV_{\text{fps}}$$

$$8.1 \text{ cfs} = (3 \text{ ft}) (x \text{ ft}) 1.5 \text{ fps})$$

$$\frac{8.1}{(3)(1.5)} = x \text{ ft}$$

$$\boxed{1.8 \text{ ft}} = x$$

CALCULATING OTHER UNKNOWN VARIABLES

There are four variables in $Q=AV$ calculations for rectangular channels: flow rate, width, depth, and velocity. In Examples 1 and 2, the unknown variable was flow rate, Q. However, any of the other variables can also be unknown.**Example 3 illustrates a calculation when depth is the unknown factor. Section 2.2 of this chapter illustrates calculations when velocity is the unknown factor.

---

* The concept of dimensional analysis is discussed in Chapter 15 in *Basic Math Concepts*.

** For a review of solving for the unknown variable, refer to Chapter 2 in *Basic Math Concepts*. These type problems are primarily theoretical, since channel size is normally a given and water depth is measured.

FLOW THROUGH A TRAPEZOIDAL CHANNEL

Calculating the flow rate for a trapezoidal channel is similar to calculating the flow rate for a rectangular channel **except that the cross-sectional area, *A*, is a trapezoid.***

$$A = \text{(average) (water)}$$
$$\text{width} \quad \text{depth}$$

$$A = \frac{(w_1 + w_2)}{2} \text{(depth)}$$

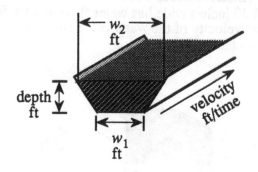

Q=AV FOR A TRAPEZOIDAL CHANNEL

$$Q = \underbrace{\qquad A \qquad}_{} \quad \underbrace{V}_{}$$

$$\underset{\text{cfm}}{Q} = \frac{(w_1 + w_2)}{2} \underset{\text{ft}}{\text{(depth)}} \underset{\text{fpm}}{\text{(velocity)}}$$

Example 4: (Instantaneous Flow)

❑ A trapezoidal channel has water flowing to a depth of 2 ft. The width of the channel at the water surface is 6 ft and the width of the channel at the bottom is 4 ft. What is the cfm flow rate in the channel if the velocity is 132 fpm?

$$\underset{\text{cfm}}{Q} = \frac{(w_1 + w_2)}{2} \underset{\text{ft}}{\text{(depth)}} \underset{\text{fpm}}{\text{(velocity)}}$$

$$= \frac{(4 \text{ ft} + 6 \text{ ft})}{2} (2 \text{ ft}) (132 \text{ fpm})$$

$$= (5 \text{ ft}) (2 \text{ ft}) (132 \text{ fpm})$$

$$= \boxed{1320 \text{ cfm}}$$

---

* For a review of trapezoid area calculations, refer to Chapter 10 in *Basic Math Concepts*.

Example 5: (Instantaneous Flow)
❏ A trapezoidal channel is 3 ft wide at the bottom and 5.5 ft wide at the water surface. The water depth is 30 inches. If the flow velocity through the channel is 168 ft/min, what is the cfm flow rate through the channel?

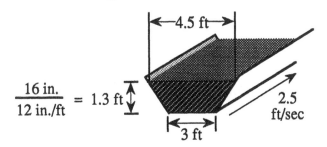

$$\frac{30 \text{ in.}}{12 \text{ in./ft}} = 2.5 \text{ ft}$$

$$\underset{\text{cfm}}{Q} = \frac{(w_1 + w_2)}{2} \underset{\text{ft}}{(\text{depth})} \underset{\text{fpm}}{(\text{velocity})}$$

$$= \frac{(3 \text{ ft} + 5.5 \text{ ft})(2.5 \text{ ft})(168 \text{ fpm})}{2}$$

$$= \boxed{1785 \text{ cfm}}$$

WHEN DATA IS NOT GIVEN IN DESIRED TERMS

Many times the data to be used in a *Q=AV* equation is not in the form desired. For example, dimensions might be given in inches rather than in feet, as desired. Or perhaps the velocity is expressed as fps yet the flow rate is desired in cfm. (The time element does not match— seconds vs. minutes.) These type calculations are illustrated in Examples 5 and 6.

Example 6: (Instantaneous Flow)
❏ A trapezoidal channel has water flowing to a depth of 16 inches. The width of the channel at the bottom is 3 ft and the width of the channel at the water surface is 4.5 ft. If the velocity of flow through the channel is 2.5 ft/sec, what is the cfm flow through the channel?

$$\frac{16 \text{ in.}}{12 \text{ in./ft}} = 1.3 \text{ ft}$$

First calculate the flow rate that matches the velocity time frame:

$$\underset{\text{cfs}}{Q} = \frac{(3 \text{ ft} + 4.5 \text{ ft})(1.3 \text{ ft})(2.5 \text{ fps})}{2}$$

$$= 12.2 \text{ cfs}$$

Now convert cfs flow rate to cfm flow rate:

$$(12.2 \text{ cfs}) \frac{(60 \text{ sec})}{\text{min}} = \boxed{732 \text{ cfm}}$$

When the velocity and flow rate time frames do not match, you must convert one of the terms to match the other.

Since flow rate conversions are quite common, you may find it easiest to leave the velocity expression as is and then convert the flow rate to match the velocity time frame. Example 6 illustrates such a process.

FLOW INTO A TANK

Flow through a tank can be considered a type of $Q=AV$ calculation.* If the discharge valve to a tank were closed, the water level would begin to rise. Timing how fast the water rises would give you an indication of the **velocity of flow into the tank**. The $Q=AV$ equation could then be used to determine the flow rate into the tank, as illustrated in Example 7.

If the influent valve to the tank were closed, rather than the discharge valve, and a pump continued discharging water from the tank, the water level in the tank would begin to drop. The rate of this drop in water level could be timed so that **the velocity of flow from the tank** could be calculated. Then the $Q=AV$ equation could be used to determine the flow rate out of the tank, as illustrated in Example 8.

Example 7: (Instantaneous Flow)

❑ A tank is 12 ft by 12 ft. With the discharge valve closed, the influent to the tank causes the water level to rise 1.25 feet in one minute. What is the gpm flow into the tank?

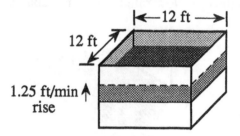

First, calculate the cfm flow rate:

$$Q_{cfm} = AV_{fpm}$$

$$= (12 \text{ ft}) (12 \text{ ft}) (1.25 \text{ fpm})$$

$$= 180 \text{ cfm}$$

Then convert cfm flow rate to gpm flow rate:

$$(180 \text{ cfm}) (7.48 \text{ gal/cu ft}) = \boxed{1346 \text{ gpm}}$$

Example 8: (Instantaneous Flow)

❑ A tank is 8 ft wide and 10 ft long. The influent valve to the tank is closed and the water level drops 2.8 ft in 2 minutes. What is the gpm flow from the tank?

$$\underline{\text{Drop:}} = \frac{2.8 \text{ ft}}{2 \text{ min}}$$

$$= 1.4 \text{ ft/min}$$

First, calculate the cfm flow rate:

$$Q_{cfm} = AV_{fpm}$$

$$= (10 \text{ ft}) (8 \text{ ft}) (1.4 \text{ fpm})$$

$$= 112 \text{ cfm}$$

Then convert cfm flow rate to gpm flow rate:

$$(112 \text{ cfm}) (7.48 \text{ gal/cu ft}) = \boxed{838 \text{ gpm}}$$

* This is the same type of calculation described in Chapter 7 as pump capacity calculations.

Example 9: (Instantaneous Flow)

❑ The discharge valve to a 30-ft diameter tank is closed. If the water rises at a rate of 10 inches in 5 minutes. What is the gpm flow into the tank?

The same basic method is used to determine the flow rate when the tank is cylindrical in shape. Examples 9 and 10 illustrate the calculation for cylindrical tanks.

Rise:*
(10 in. = 0.83 ft)

$$= \frac{0.83 \text{ ft}}{5 \text{ min}}$$

$$= 0.17 \text{ ft/min}$$

First calculate the cfm flow into the tank:

$$Q_{cfm} = AV_{fpm}$$

$$= (0.785)(30 \text{ ft})(30 \text{ ft})(0.17 \text{ ft/min})$$

$$= 120 \text{ cfm}$$

Then convert cfm flow rate to gpm flow rate:

(120 cfm) (7.48 gal/cu ft) = $\boxed{898 \text{ gpm}}$

Example 10: (Instantaneous Flow)

❑ A pump discharges into a 2-ft diameter barrel. If the water level in the barrel rises 2 ft in 30 seconds, what is the gpm flow into the barrel?

Rise: $= \dfrac{2 \text{ ft}}{30 \text{ sec}}$

$$= 4 \text{ ft/min}$$

First calculate the cfm flow into the tank:

$$Q_{cfm} = AV_{fpm}$$

$$= (0.785)(2 \text{ ft})(2 \text{ ft})(4 \text{ fpm})$$

$$= 12.6 \text{ cfm}$$

Then convert cfm flow rate to gpm flow rate:

(12.6 cfm) (7.48 gal/cu ft) = $\boxed{94 \text{ gpm}}$

* Refer to Chapter 8, "Linear Measurement Conversions", in *Basic Math Concepts*.

FLOW THROUGH A PIPELINE—WHEN FLOWING FULL

The flow rate through a pipeline can be calculated using the $Q=AV$ equation. The cross-sectional area is a circle, so the area, A, is represented by (0.785) (D^2). **Pipe diameters should generally be expressed as feet** to avoid errors in terms.

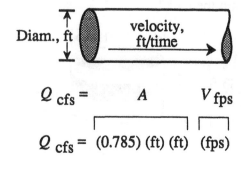

$Q=AV$ CALCULATIONS FOR A PIPELINE FLOWING FULL

Diam., ft

velocity, ft/time

$Q_{cfs} =$ A V_{fps}

$Q_{cfs} =$ (0.785) (ft) (ft) (fps)

Example 11: (Instantaneous Flow)
❏ The flow through a 6-inch diameter pipeline is moving at a velocity of 3 ft/sec. What is the cfs flow rate through the pipeline? (Assume the pipe is flowing full.)

6 in = 0.5 ft

3 fps

$$Q_{cfs} = AV_{fps}$$
$$= (0.785)\ (0.5\ ft)\ (0.5\ ft)\ (3\ fps)$$
$$= \boxed{0.59\ cfs}$$

Example 12: (Instantaneous Flow)
❑ An 8-inch diameter pipeline has water flowing at a velocity of 3.4 fps. What is the gpm flow rate through the pipeline?

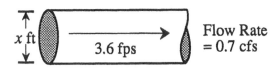

8 in = 0.67 ft

3.4 fps

First, calculate the cfs flow rate:

$$Q_{cfs} = AV_{fps}$$

$$= (0.785)(0.67 \text{ ft})(0.67 \text{ ft})(3.4 \text{ fps})$$

$$= 1.2 \text{ cfs}$$

Then convert cfs flow rate to gpm flow rate:

$$(1.2 \, \frac{\text{cu ft}}{\text{sec}})(60 \, \frac{\text{sec}}{\text{min}})(7.48 \, \frac{\text{gal}}{\text{cu ft}}) = \boxed{539 \text{ gpm}}$$

Example 13: (Instantaneous Flow)
❑ The flow through a pipeline is 0.7 cfs. If the velocity of flow is 3.6 ft/sec and the pipe is flowing full, what is the diameter (inches) of the pipeline?

x ft

3.6 fps

Flow Rate
= 0.7 cfs

First calculate the diameter in feet:

$$Q_{cfs} = AV_{fps}$$

$$0.7 \text{ cfs} = (0.785)(x^2 \text{ sq ft})(3.6 \text{ fps})$$

$$\frac{0.7}{(0.785)(3.6)} = x^2$$

$$0.25 \text{ ft}^2 = x^2$$

$$0.5 \text{ ft} = x$$

Now convert feet to inches:

$$(0.5 \text{ ft})(12 \, \frac{\text{in.}}{\text{ft}}) = \boxed{6 \text{ inches}}$$

SOLVING FOR OTHER UNKNOWN VARIABLES

There are three variables in $Q=AV$ calculations for pipelines: flow rate (Q), diameter (D), and velocity (V). In Examples 11 and 12, the unknown factor is flow rate. In Example 13, the unknown factor is pipeline diameter. (Section 2.2 illustrates calculations when velocity is the unknown variable.)

When the diameter is the unknown variable, first solve for x^2. Then, by taking the square root* of both sides of the equation, x may be determined.

---

* For a review of square roots, refer to Chapter 13 in *Basic Math Concepts*.

FLOW THROUGH A PIPELINE—WHEN FLOWING LESS THAN FULL

Calculating the flow rate through a pipeline flowing less than full is similar to calculating the flow rate for a pipeline flowing full with one exception—**instead of using 0.785 as a factor in the area calculation, a different factor is used.** This factor is based on the ratio of water depth (d) to the pipe diameter (D). Calculate the d/D value, then use the table to determine the factor to be used instead of 0.785.

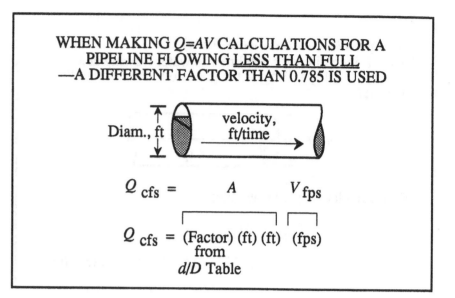

WHEN MAKING $Q=AV$ CALCULATIONS FOR A PIPELINE FLOWING <u>LESS THAN FULL</u> —A DIFFERENT FACTOR THAN 0.785 IS USED

$$Q_{cfs} = A \quad V_{fps}$$

$$Q_{cfs} = \text{(Factor) (ft) (ft)} \quad \text{(fps)}$$
from
d/D Table

depth/Diameter Table							
d/D	Factor	*d/D*	Factor	*d/D*	Factor	*d/D*	Factor
0.01	0.0013	0.26	0.1623	0.51	0.4027	0.76	0.6404
0.02	0.0037	0.27	0.1711	0.52	0.4127	0.77	0.6489
0.03	0.0069	0.28	0.1800	0.53	0.4227	0.78	0.6573
0.04	0.0105	0.29	0.1890	0.54	0.4327	0.79	0.6655
0.05	0.0147	0.30	0.1982	0.55	0.4426	0.80	0.6736
0.06	0.0192	0.31	0.2074	0.56	0.4526	0.81	0.6815
0.07	0.0242	0.32	0.2167	0.57	0.4625	0.82	0.6893
0.08	0.0294	0.33	0.2260	0.58	0.4724	0.83	0.6969
0.09	0.0350	0.34	0.2355	0.59	0.4822	0.84	0.7043
0.10	0.0409	0.35	0.2450	0.60	0.4920	0.85	0.7115
0.11	0.0470	0.36	0.2546	0.61	0.5018	0.86	0.7186
0.12	0.0534	0.37	0.2642	0.62	0.5115	0.87	0.7254
0.13	0.0600	0.38	0.2739	0.63	0.5212	0.88	0.7320
0.14	0.0668	0.39	0.2836	0.64	0.5308	0.89	0.7384
0.15	0.0739	0.40	0.2934	0.65	0.5404	0.90	0.7445
0.16	0.0811	0.41	0.3032	0.66	0.5499	0.91	0.7504
0.17	0.0885	0.42	0.3130	0.67	0.5594	0.92	0.7560
0.18	0.0961	0.43	0.3229	0.68	0.5687	0.93	0.7612
0.19	0.1039	0.44	0.3328	0.69	0.5780	0.94	0.7662
0.20	0.1118	0.45	0.3428	0.70	0.5872	0.95	0.7707
0.21	0.1199	0.46	0.3527	0.71	0.5964	0.96	0.7749
0.22	0.1281	0.47	0.3627	0.72	0.6054	0.97	0.7785
0.23	0.1365	0.48	0.3727	0.73	0.6143	0.98	0.7816
0.24	0.1449	0.49	0.3827	0.74	0.6231	0.99	0.7841
0.25	0.1535	0.50	0.3927	0.75	0.6318	1.00	0.7854

Example 14: **(Instantaneous Flow)**
❏ The flow through a 6-inch diameter pipeline is moving at a velocity of 3 ft/sec. If the water is flowing at a depth of 4 inches, what is the cfs flow rate through the pipeline?

First use the *d/D* ratio to determine the factor to be used instead of 0.785 in the *Q=AV* calculation:

$$\frac{d}{D} = \frac{4}{6} = 0.67$$

The factor shown in the table corresponding to a *d/D* of 0.67 is 0.5594. Now calculate the flow rate using *Q=AV*:

$$Q_{cfs} = AV_{fps}$$

$$= (0.5594)(0.5\ ft)(0.5\ ft)(3\ fps)$$

$$= \boxed{0.42\ cfs}$$

Examples 14 and 15 illustrate use of the *Q = AV* equation when the pipeline is flowing less than full.

Example 15: **(Instantaneous Flow)**
❏ An 8-inch diameter pipeline has water flowing at a velocity of 3.4 fps. What is the gpm flow rate through the pipeline if the water is flowing at a depth of 5 inches?

First, determine the factor to be used instead of 0.785. Since *d/D*=0.63, the factor listed in the table is 0.5212. Now calculate the flow rate using *Q=AV*. Although gpm flow rate is desired, first calculate cfs flow rate, then convert cfs to gpm flow rate:

$$Q_{cfs} = AV_{fps}$$

$$= (0.5212)(0.67\ ft)(0.67\ ft)(3.4\ fps)$$

$$= 0.8\ cfs$$

Then convert cfs flow rate to gpm flow rate:

$$\frac{(0.8\ cu\ ft)}{sec}\frac{(60\ sec)}{min}\frac{(7.48\ gal)}{cu\ ft} = \boxed{359\ gpm}$$

2.2 VELOCITY CALCULATIONS

VELOCITY USING *Q=AV*

The *Q=AV* equation may be used to estimate the velocity of flow in a channel or pipeline. Write the equation as usual, filling in the known data, then solve for the unknown factor (velocity in this case).

Be sure that the volume and time expressions match on both sides of the equation. For instance, if the velocity is desired in ft/sec, then the flow rate should be converted to cfs before beginning the *Q=AV* calculation.
Examples 1 and 2 illustrate a velocity estimate for a pipeline.

Example 1: (Velocity Calculations)
❑ A channel has a rectangular cross section. The channel is 4 ft wide with water flowing to a depth of 1.8 ft. If the flow rate through the channel is 9050 gpm, what is the velocity of the water in the channel (ft/sec)?

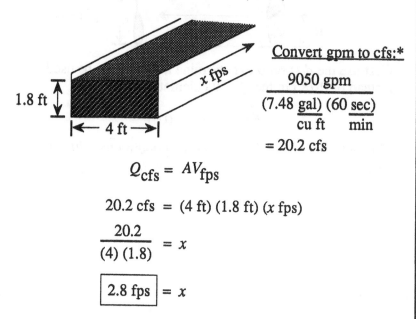

Convert gpm to cfs:*

$$\frac{9050 \text{ gpm}}{\frac{(7.48 \text{ gal})}{\text{cu ft}} \frac{(60 \text{ sec})}{\text{min}}}$$

= 20.2 cfs

$$Q_{cfs} = AV_{fps}$$

20.2 cfs = (4 ft) (1.8 ft) (x fps)

$$\frac{20.2}{(4)(1.8)} = x$$

$$\boxed{2.8 \text{ fps}} = x$$

Example 2: (Velocity Calculations)
❑ A 6-inch diameter pipe flowing full delivers 280 gpm. What is the velocity of flow in the pipeline (ft/sec)?

6 in = 0.5 ft

x fps

Convert gpm to cfs flow:

$$\frac{280 \text{ gpm}}{\frac{(7.48 \text{ gal})}{\text{cu ft}} \frac{(60 \text{ sec})}{\text{min}}}$$

= 0.62 cfs

$$Q_{cfs} = AV_{fps}$$

0.62 cfs = (0.785) (0.5 ft) (0.5 ft) (x fps)

$$\frac{0.62}{(0.785)(0.5)(0.5)} = x$$

$$\boxed{3.2 \text{ fps}} = x$$

---

* For a review of flow conversions, refer to Chapter 8 in *Basic Math Concepts*.

Example 3: (Velocity Calculations)

❑ A float travels 300 ft in a channel in 2 min 14 sec. What is the estimated velocity in the channel (ft/sec)?

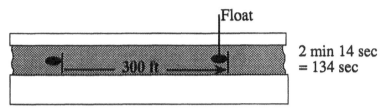

Float

300 ft

2 min 14 sec = 134 sec

$$\text{Velocity ft/sec} = \frac{\text{Distance, ft}}{\text{Time, sec}}$$

$$= \frac{300 \text{ ft}}{134 \text{ sec}}$$

$$= \boxed{2.24 \text{ ft/sec}}$$

Example 4: (Velocity Calculations)

❑ A fluorescent dye is used to estimate the velocity of flow in a sewer. The dye is injected in the water at one manhole and the travel time to the next manhole 400 ft away is noted. The dye first appears at the downstream manhole in 128 seconds. The dye continues to be visible until a total elapsed time of 148 seconds. What is the ft/sec velocity of flow through the pipeline?

Dye injected (Manhole 1)   Manhole 2

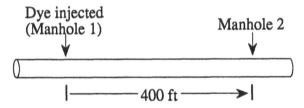

400 ft

First calculate the average travel time of the dye:

$$\frac{128 \text{ sec} + 148 \text{ sec}}{2} = 138 \text{ sec}$$

Then calculate the ft/sec velocity:

$$\text{Velocity ft/sec} = \frac{400 \text{ ft}}{138 \text{ sec}}$$

$$= \boxed{2.9 \text{ ft/sec}}$$

VELOCITY USING THE FLOAT OR DYE METHOD

The $Q = AV$ calculation estimates the theoretical velocity of flow in a channel or pipeline. Actual velocities in the pipeline can be measured by metering devices. Velocities can also be estimated by the use of a float or dye placed in the water. Then, by timing the distance traveled using a float or dye, the velocity of flow can be determined:

$$\text{Velocity ft/sec} = \frac{\text{Distance, ft}}{\text{Time, sec}}$$

A float is perhaps less accurate in estimating velocities in a pipeline than use of fluorescent tracer dyes. In channels, floats move along with the faster surface waters and can be as much as 10 or 15 percent faster than the actual average flow rate. Some floats designed for use in channels include segments that extend into the water, thus responding to a more average velocity through the channel.

In pipelines, floats can become entangled or slowed down by obstructions.

Tracer dyes tend to give a better estimate of velocity. Since some of the dye will travel faster and some slower, you will need to determine the **average time** required to travel from one point to the next. To calculate the average time for travel:

Average Time, sec =	Total elapsed time 'til dye first appears	+	Total elapsed time 'til dye no longer seen
		2	

USING $Q=AV$ TO ESTIMATE CHANGES IN VELOCITY

In addition to estimating flow in a channel or pipeline, the $Q=AV$ equation can be used to estimate the change in velocity as the water flows from one diameter pipeline to another.

When water flows from a larger diameter pipe to a smaller diameter pipe, the velocity increases. (The water must move faster since the same amount of water is flowing through a smaller space.) Example 5 illustrates this calculation.

FLOW RATE (Q) IN PIPES REMAIN CONSTANT

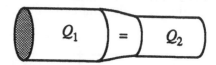

Since the total flow in the pipeline must remain constant:

$$Q_1 = Q_2$$

$$A_1 V_1 = A_2 V_2$$

$$A_1 V_1 = A_2 V_2$$

Example 5: (Velocity Calculations)

❑ The velocity in a 12-inch diameter pipeline is 3.8 ft/sec. If the 12-inch pipeline flows into a 10-inch diameter pipeline, what is the velocity in the 10-inch pipeline?

12 in = 1 ft $A_1 V_1 = A_2 V_2$ $\dfrac{10\ in}{12\ in/ft} = 0.83\ ft$

$$A_1 V_1 = A_2 V_2$$

$$(0.785)\,(1\ ft)\,(1\ ft)\,(3.8\ fps) = (0.785)\,(0.83\ ft)\,(0.83\ ft)\,(x\ fps)$$

$$\frac{(0.785)\,(1)\,(1)\,(3.8)}{(0.785)\,(0.83)\,(0.83)} = x\ fps$$

$$\boxed{5.5\ fps} = x$$

Example 6: (Velocity Calculations)
❏ The velocity in a 6-inch diameter pipe is 4.8 ft/sec. If the flow travels from a 6-inch pipeline to an 8-inch pipeline, what is the velocity in the 8-inch pipeline?

$$\frac{6 \text{ in}}{12 \text{ in/ft}} = 0.5 \text{ ft} \qquad A_1V_1 \quad = \quad A_2V_2 \qquad \frac{8 \text{ in}}{12 \text{ in/ft}} = 0.67 \text{ ft}$$

$$A_1V_1 = A_2V_2$$

$$(0.785)\,(0.5 \text{ ft})\,(0.5 \text{ ft})\,(4.8 \text{ fps}) = (0.785)\,(0.67 \text{ ft})\,(0.67 \text{ ft})\,(x \text{ fps})$$

$$\frac{(0.785)\,(0.5)\,(0.5)\,(4.8)}{(0.785)\,(0.67)\,(0.67)} = x \text{ fps}$$

$$\boxed{2.7 \text{ fps}} = x$$

When water flows from a smaller diameter pipe to a larger diameter pipe, the velocity decreases. Example 6 illustrates this principle.

WHEN *Q* DATA IS GIVEN FOR ONE PIPE AND *AV* DATA GIVEN FOR THE OTHER

In Examples 5 and 6, the *AV* of one pipeline was set equal to the *AV* of the second pipeline:

$$\boxed{A_1V_1 = A_2V_2}$$

This equation is possible since the same flow travels through both pipes:

$$\boxed{Q_1 = Q_2}$$

Since $Q_1 = A_1V_1$, either of these terms (Q_1 or A_1V_1) can be used on the left side of the equation. Similarly since $Q_2 = A_2V_2$, either of these terms (Q_2 or A_2V_2) can be used interchangeably on the right side of the equation:

$$\boxed{Q_1 = A_2V_2}$$

$$\boxed{A_1V_1 = Q_2}$$

Example 7 illustrates a calculation where the *Q* of one pipeline is set equal to the *AV* of the second pipeline.

Example 7: (Velocity Calculations)
❏ The flow through a 6-inch diameter pipeline is 220 gpm. What is the estimated velocity of flow (fps) through the 4-inch diameter pipeline shown below?

$$A_1V_1 \quad = \quad A_2V_2 \qquad \frac{4 \text{ in}}{12 \text{ in/ft}} = 0.33 \text{ ft}$$

Since flow data is given for the first pipeline and velocity (part of the *AV*) is unknown for the second pipeline, the equation to be used is:

$$Q_1 = A_2V_2$$

Remember that *Q* must be expressed in cfs since the right side of the equation is (sq ft) (ft/sec) or cfs:

$$\frac{220 \text{ gpm}}{(7.48 \text{ gal/cu ft})\,(60 \text{ sec/min})} = 0.49 \text{ cfs}$$

Now complete the *Q=AV* equation:

$$Q_1 = A_2V_2$$

$$0.49 \text{ cfs} = (0.785)\,(0.33 \text{ ft})\,(0.33 \text{ ft})\,(x \text{ fps})$$

$$\frac{0.49}{(0.785)\,(0.33)\,(0.33)} = x$$

$$\boxed{5.7 \text{ fps}} = x$$

2.3 AVERAGE FLOW RATES CALCULATIONS

Flow rates in a treatment system may vary considerably during the course of a day. Calculating an **average flow rate** is a way to determine the **typical flow rate** for a given time frame such as: average daily flow, average weekly flow, average monthly flow, or even average yearly flow.

There are two ways to calculate an average flow rate. In the first method, several flow rate values are used to determine an average value, as illustrated in Examples 1 and 2.

$$\text{Average Flow} = \frac{\text{Tot. of all Sample Flows}}{\text{No. of Samples}}$$

In the second method, a total flow is used (from a totalizer reading) to determine an average flow rate. Examples 3 and 4 illustrate this type of calculation.

$$\text{Average Flow} = \frac{\text{Tot. Flow from Totalizer}}{\text{Time Over Which Flow Measured}}$$

Example 1: (Average Flow Rates)
❑ The following flows were recorded for the week: Monday—8.6 MGD; Tuesday—7.6 MGD; Wednesday—7.2 MGD; Thursday—7.8 MGD; Friday—8.4 MGD; Saturday—8.6 MGD; Sunday—7.5 MGD. What was the average daily flow for the week?

$$\text{Average Daily Flow} = \frac{\text{Total of all Sample Flows}}{\text{Number of Days}}$$

$$= \frac{55.7 \text{ MGD}}{7 \text{ Days}}$$

$$= \boxed{8.0 \text{ MGD}}$$

Example 2: (Average Flow Rates)
❑ The following flows were recorded for the months of September, October and November: September—120.8 MG; October—136.4 MG; November—156.1 MG. What was the average daily flow for this three-month period?

Since average **daily** flow is desired, you must divide by the **number of days** represented by the three-month period, rather than by the number of months represented. (If average monthly flow had been desired, then you would divide by the number of months represented.)

$$\text{Average Daily Flow} = \frac{\text{Total of all Sample Flows}}{\text{Number of Days}}$$

$$= \frac{413.3 \text{ MG}}{91 \text{ days}}$$

$$= \boxed{4.5 \text{ MGD}}$$

Example 3: (Average Flow Rates)
❏ The totalizer reading for the month of November was 142.8 MG. What was the average daily flow (ADF) for the month of November?

Since average daily flow is desired, the denominator must reflect the number of days represented by the totalizer flow:

$$\text{Average Daily Flow} = \frac{\text{Tot. Flow from Totalizer}}{\text{Number of Days}}$$

$$= \frac{142.8 \text{ MG}}{30 \text{ days}}$$

$$= \boxed{4.8 \text{ MGD}}$$

CALCULATING AVERAGE FLOWS USING TOTALIZER INFORMATION

When totalizer data is used to calculate average flows, the number in the denominator depends on the time frame desired. If you desire to know the average **flow per minute**, divide by the **number of minutes** represented by the totalizer flow. If you wish to determine average **daily** flow, divide by the **number of days** represented by the totalizer flow. To calculate the average **weekly** flow, divide by the **number of weeks** represented by the totalizer flow.

Example 4: (Average Flow Rates)
❏ The total flow for one day at a plant was 2,600,000 gallons. What was the average gpm flow for that day?

The average flow per **minute** is desired. Therefore, the denominator must reflect the **number of minutes** represented by the totalizer flow:

$$\text{Average Flow, gpm} = \frac{\text{Tot. Flow from Totalizer}}{\text{Number of Days}}$$

$$= \frac{2,600,000 \text{ gallons}}{1440 \text{ minutes}}$$

$$= \boxed{1805.6 \text{ gpm}}$$

2.4 FLOW CONVERSIONS CALCULATIONS

Flow rates may be expressed in several different ways—cubic feet per second (cfs), cubic feet per minute (cfm), gallons per minute (gpm), gallons per day (gpd), etc. as shown in the diagram to the right.

To excel in water and wastewater math, it is essential that you be able to convert one expression of flow to any other.

The box method was designed to make these conversions easy to visualize. **Moving from a smaller box to a larger box requires multiplication by the factor indicated. Moving from a larger box to a smaller box requires division by the factor indicated.**

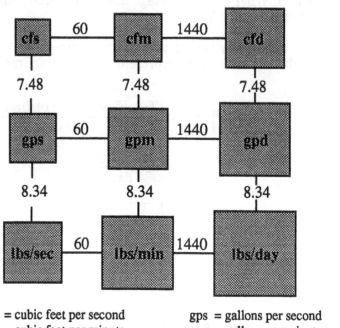

FLOW CONVERSIONS USING THE BOX METHOD*

cfs = cubic feet per second
cfm = cubic feet per minute
cfd = cubic feet per day

gps = gallons per second
gpm = gallons per minute
gpd = gallons per day

Example 1: (Flow Conversions)
❑ Convert a flow of 3 cfs to gpm.

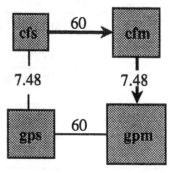

First write the flow rate to be converted (3 cfs). Then any factors to be multiplied must be placed in the numerator with 3 cfs. Any division factors are placed in the denominator. There are two different paths to gpm. **Either path will result in the same answer:**

$$\frac{(3 \text{ cfs}) (60 \text{ } \underline{\text{sec}}) (7.48 \text{ } \underline{\text{gal}})}{\text{min} \quad \text{cu ft}} = \boxed{1346 \text{ gpm}}$$

---

* The factors shown in the diagram have the following units associated with them: 60 min/sec, 1440 min/day, 7.48 gal/cu ft, and 8.34 lbs/gal.

Example 2: (Flow Conversions)
❑ Convert a flow of 45 gps to gpd. Use dimensional analysis to check the set up of the problem.

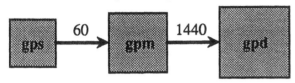

Write the flow rate to be converted, then place multiplication or division factors in the numerator or denominator, as required:

$$(45 \text{ gps}) (60 \frac{\text{sec}}{\text{min}}) (1440 \frac{\text{min}}{\text{cu ft}}) = \boxed{3,888,000 \text{ gpd}}$$

Now use dimensional analysis to check the math set up of this problem:

$$\frac{\text{gal}}{\cancel{\text{sec}}} \cdot \frac{\cancel{\text{sec}}}{\cancel{\text{min}}} \cdot \frac{\cancel{\text{min}}}{\text{day}} = \frac{\text{gal}}{\text{day}}$$

Example 3: (Flow Conversions)
❑ Convert a flow of 3,200,000 gpd to cfm. Use dimensional analysis to check the set up of the problem.

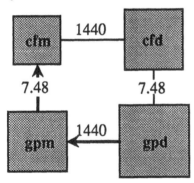

Two different paths may be used from gpd to cfm. Either path will result in the same answer:

$$\frac{3,200,000 \text{ gpd}}{(1440 \frac{\text{min}}{\text{day}}) (7.48 \frac{\text{gal}}{\text{cu ft}})} = \boxed{297 \text{ cfm}}$$

Now use dimensional analysis to check the math set up of this problem:*

$$\frac{\frac{\text{gal}}{\text{day}}}{\frac{\text{min}}{\text{day}} \cdot \frac{\text{gal}}{\text{cu ft}}} = \frac{\cancel{\text{gal}}}{\cancel{\text{day}}} \cdot \frac{\cancel{\text{day}}}{\text{min}} \cdot \frac{\text{cu ft}}{\cancel{\text{gal}}} = \frac{\text{cu ft}}{\text{min}}$$

* For a review of complex fractions, refer to Chapter 3 in *Basic Math Concepts.*

USING DIMENSIONAL ANALYSIS TO CHECK THE MATH SET UP

Dimensional analysis is often used to check the mathematical set up of conversions. Examples 2 and 3 illustrate how to use dimensional analysis in checking the problem set up. Refer to Chapter 15 in *Basic Math Concepts* for a further discussion of dimensional analysis.

QUICK CONVERSIONS

There are two conversion equations used quite frequently in water and wastewater treatment calculations. You would be well advised to memorize these equations for use in quick conversions:

$$\boxed{\begin{array}{l} 1 \text{ MGD} = 1.55 \text{ cfs} \\ 1 \text{ MGD} = 694 \text{ gpm} \end{array}}$$

Should you forget these numbers, you can always derive them yourself using the box method of conversions.

NOTES:

3 *Milligrams per Liter to Pounds per Day Calculations*

SUMMARY

1. The five general types of mg/*L* to lbs/day or lbs calculations are:

   • Chemical Dosage

   • BOD, COD, or SS Loading

   • BOD, COD, or SS Removal

   • Pounds of Solids Under Aeration

   • WAS Pumping Rate

2. All of the calculations listed above use one of two equations:

$$(\text{mg/}L)\ (\text{MGD flow})\ (8.34\ \text{lbs/gal}) = \text{lbs/day}$$

or

$$(\text{mg/}L)\ (\text{MG volume})\ (8.34\ \text{lbs/gal}) = \text{lbs}$$

3. To determine which of the two equations to use, you must first determine whether the mg/*L* concentration pertains to a **flow** or a **tank or pipeline volume**. If the mg/*L* concentration represents a concentration in a flow, then million gallons per day (MGD) flow is used as the second factor. If the concentration pertains to a tank or pipeline volume, then million gallons (MG) volume is used as the second factor.

One of the most frequently used calculations in water and water mathematics is the conversion of milligrams per liter (mg/*L*) concentration to pounds per day (lbs/day) or pounds (lbs) dosage or loading. This calculation is the basis of five general types of calculations, as noted in the summary to the left.

3.1 CHEMICAL DOSAGE CALCULATIONS

CHLORINE DOSAGE

In chemical dosing, a measured amount of chemical is added to the water or wastewater. The amount of chemical required depends on such factors as the type of chemical used, the reason for dosing, and the flow rate being treated.

Two ways to describe the amount of chemical added or required are:

• milligrams per liter (mg/L)

• pounds per day (lbs/day)

To convert from mg/L (or ppm) concentration to lbs/day, use the following equation:

$$\underset{\text{Conc.}}{(mg/L)} \underset{\text{flow}}{(MGD)} \underset{\text{lbs/gal}}{(8.34)} = lbs/day$$

In previous years, parts per million (ppm) was also used as an expression of concentration. In fact, it was used interchangeably with mg/L concentration, since 1 mg/L = 1 ppm.* However, because *Standard Methods* no longer uses ppm, mg/L is the preferred expression of concentration.

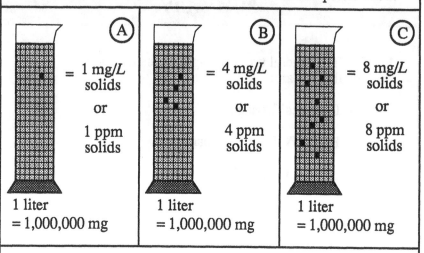

MILLIGRAMS PER LITER IS A MEASURE OF CONCENTRATION

Assume each liter below is divided into 1 million parts. Then:

A = 1 mg/L solids or 1 ppm solids — 1 liter = 1,000,000 mg

B = 4 mg/L solids or 4 ppm solids — 1 liter = 1,000,000 mg

C = 8 mg/L solids or 8 ppm solids — 1 liter = 1,000,000 mg

Assuming the liter in these three examples has been divided into 1 million parts (each part representing 1 milligram, mg), the **concentration of solids** in each liter could be expressed as:

• The number of mg solids per liter (mg/L) or
• The number of mg solids per 1,000,000 mg (ppm).

The concentration of solids shown in diagram A is 1 milligram per liter (1 mg/L). The solids concentration shown in diagrams B and C are 4 mg/L and 8 mg/L, respectively.

Example 1: (Chemical Dosage)
❑ Determine the chlorinator setting (lbs/day) needed to treat a flow of 3 MGD with a chlorine dose of 4 mg/L.

First write the equation. Then fill in the information given:

$$\underset{\text{Conc.}}{(mg/L)} \underset{\text{flow}}{(MGD)} \underset{\text{lbs/gal}}{(8.34)} = lbs/day$$

$$(4 \text{ mg/L}) (3 \text{ MGD}) (8.34 \text{ lbs/gal}) = lbs/day$$

$$= \boxed{100 \text{ lbs/day}}$$

$$* \quad \frac{1mg}{L} = \frac{1 \text{ mg}}{1,000,000 \text{ mg}} = \frac{1 \text{ lb}}{1,000,000 \text{ lbs}} = \frac{1 \text{ part}}{1,000,000 \text{ parts}} = 1 \text{ ppm}$$

Example 2: (Chemical Dosage)

❑ Determine the chlorinator setting (lbs/day) if a flow of 3.8 MGD is to be treated with a chlorine dose of 2.7 mg/L.

Write the equation then fill in the information given:

$$\underset{\text{Conc.}}{(mg/L)} \; \underset{\text{flow}}{(MGD)} \; \underset{\text{lbs/gal}}{(8.34)} \; = \; lbs/day$$

$$(2.7 \; mg/L) \; (3.8 \; mg/L) \; (8.34 \; lbs/gal) \; = \; lbs/day$$

$$= \boxed{85.6 \; lbs/day}$$

Example 3: (Chemical Dosage)

❑ What should the chlorinator setting be (lbs/day) to treat a flow of 2 MGD if the chlorine demand is 10 mg/L and a chlorine residual of 2 mg/L is desired?

First, write the mg/L to lbs/day equation:

$$\underset{\text{Conc.}}{(mg/L)} \; \underset{\text{flow}}{(MGD)} \; \underset{\text{lbs/gal}}{(8.34)} \; = \; lbs/day$$

In this problem the unknown value is lbs/day. Information is given for each of the other two variables: mg/L and flow. Notice that information for the mg/L dose is given only indirectly, as chlorine demand and residual and can be found using the equation:

$$\underset{mg/L}{Cl_2 \; Dose} = \underset{mg/L}{Cl_2 \; Demand} + \underset{mg/L}{Cl_2 \; Residual}$$

$$= 10 \; mg/L + 2 \; mg/L$$

$$= 12 \; mg/L$$

The mg/L to lbs/day calculation may now be completed:

$$(12 \; mg/L) \; (2 \; MGD) \; (8.34 \; lbs/gal) \; = \boxed{200 \; lbs/day}$$

CHLORINE DOSAGE, DEMAND, AND RESIDUAL

In some chlorination calculations, the mg/L chlorine dose is not given directly but indirectly as chlorine demand and residual information.

Chlorine dose depends on two considerations—the chlorine demand and the desired chlorine residual such that:

$$\underset{mg/L}{Dose} = \underset{mg/L}{Demand} + \underset{mg/L}{Resid.}$$

The **chlorine demand** is the amount of chlorine used in reacting with various components of the water such as harmful organisms and other organic and inorganic substances. When the chlorine demand has been satisfied, these reactions stop.

In some cases, such as perhaps during pretreatment, chlorinating just enough to meet some or all of the chlorine demand is sufficient. However, in other cases, it is desirable to have an additional amount of chlorine available for disinfection.

Using the equation shown above, if you are given information about any two of the variables, you can determine the value of the third variable. For example, if you know that the chlorine dose is 3 mg/L and the chlorine residual is 0.5 mg/L, the chlorine demand must therefore be 2.5 mg/L:

$$3 \; mg/L = 2.5 \; mg/L + 0.5 \; mg/L$$

If chlorine demand and residual are known, then chlorine dose (mg/L) can be determined, as illustrated in Example 3.

CHEMICAL DOSAGE FOR OTHER CHEMICALS

Examples 1-3 illustrated chemical dosage calculations for chlorine. The same method is used in calculating dosages for other chemicals, as shown in Examples 4 and 5.

Example 4: (Chemical Dosage)
❑ A jar test indicates that the best dry alum dose is 12 mg/L. If the flow is 3.5 MGD, what is the desired alum feed rate? (lbs/day)

$$\underset{\text{Conc.}}{(mg/L)} \; \underset{\text{flow}}{(MGD)} \; \underset{\text{lbs/gal}}{(8.34)} = \text{lbs/day}$$

$$(12 \text{ mg/L}) \; (3.5 \text{ MGD}) \; (8.34 \text{ lbs/gal}) = \text{lbs/day}$$

$$= \boxed{350 \text{ lbs/day}}$$

Example 5: (Chemical Dosage)
❑ To dechlorinate a wastewater, sulfur dioxide is to be applied at a level 3 mg/L more than the chlorine residual. What should the sulfonator feed rate be (lbs/day) for a flow of 4 MGD with a chlorine residual of 4.2 mg/L?

Since the chlorine residual is 4.2 mg/L, the sulfur dioxide dosage should be 4.2 + 3 = 7.2 mg/L:

$$\underset{\text{Conc.}}{(mg/L)} \; \underset{\text{flow}}{(MGD)} \; \underset{\text{lbs/gal}}{(8.34)} = \text{lbs/day}$$

$$(7.2 \text{ mg/L}) \; (4 \text{ MGD}) \; (8.34 \text{ lbs/gal}) = \text{lbs/day}$$

$$= \boxed{240 \text{ lbs/day}}$$

CALCULATING mg/L GIVEN lbs /day

In some chemical dosage calculations, you will know the dosage in lbs/day and the flow rate, but the mg/L dosage will be unknown. Approach these problems as any other mg/L to lbs/day problem:

• Write the equation,

• Fill in the known information,

• Solve for the unknown value.

Example 6: (Chemical Dosage)
❑ The chlorine feed rate at a plant is 175 lbs/day. If the flow is 2,450,000 gpd, what is this dosage in mg/L?

$$\underset{\text{Conc.}}{(mg/L)} \; \underset{\text{flow}}{(MGD)} \; \underset{\text{lbs/gal}}{(8.34)} = \text{lbs/day}$$

$$(x \text{ mg/L}) \; (2.45 \text{ MGD}) \; (8.34 \text{ lbs/gal}) = 175 \text{ lbs/day}$$

$$x = \frac{175 \text{ lbs/day}}{(2.45 \text{ MGD}) \; (8.34 \text{ lbs/gal})}$$

$$x = \boxed{8.6 \text{ mg/L}}$$

Example 7: (Chemical Dosage)
❑ A storage tank is to be disinfected with a 50 mg/*L* chlorine solution. If the tank holds 70,000 gallons, how many pounds of chlorine (gas) will be needed?

$$\frac{(mg/L)\ (MG)\ (8.34)}{Conc.\ \ Vol\ \ lbs/gal} = lbs$$

(50 mg/*L*) (0.07 MG) (8.34 lbs/gal) = lbs

= 29.2 lbs

Example 8: (Chemical Dosage)
❑ To neutralize a sour digester, one pound of lime is to be added for every pound of volatile acids in the digester liquor. If the digester contains 250,000 gal of sludge with a volatile acid (VA) level of 2,300 mg/*L*, how many pounds of lime should be added?

Since the VA concentration is 2300 mg/*L*, the lime concentration should also be 2300 mg/*L*:

$$\frac{(mg/L)\ (MG)\ (8.34)}{Conc.\ \ Vol\ \ lbs/gal} = lbs$$

(2300 mg/*L*) (0.25 MG) (8.34 lbs/gal) = lbs

= 4,796 lbs

CHEMICAL DOSAGE IN WELLS, TANKS, RESERVOIRS, OR PIPELINES

Wells are disinfected (chlorinated) during and after construction and also after any well or pump repairs. Tanks and reservoirs are chlorinated after initial inspection and after any time they have been drained for cleaning, repair or maintenance. Similarly, a pipeline is chlorinated after initial installation and after any repair.

Digesters may also require chemical dosing, although the chemical used is not chlorine but lime or some other chemical.

For calculations such as these, use the mg/*L* to lbs equation:

$$\frac{(mg/L)\ (MG)\ (8.34)}{Conc.\ \ Vol\ \ lbs/gal} = lbs$$

Notice that this equation is very similar to that used in Examples 1-6. The only difference is that MG volume is used rather than MGD flow; therefore, the result is lbs rather than lbs/day. (When dosing a volume, there is no time factor consideration.) Examples 7-8 illustrate these calculations.

HYPOCHLORITE COMPOUNDS

When chlorinating water or wastewater with chlorine gas, you are chlorinating with 100% available chlorine. Therefore, if the chlorine demand and residual requires 50 lbs/day chlorine, the chlorinator setting would be just that—50 lbs/24 hrs.

Many times, however, a chlorine compound called hypochlorite is used to chlorinate water or wastewater. Hypochlorite compounds contain chlorine and are similar to a strong bleach. They are available in liquid form or as powder or granules. Calcium hypochlorite, sometimes referred to as HTH is the most commonly used dry hypochlorite. It contains about 65% available chlorine. Sodium hypochlorite, or liquid bleach, contains about 12-15% available chlorine as commercial bleach or 3-5.25% as household bleach.

Because hypochlorite is not 100% pure chlorine, **more lbs/day must be fed into the system to obtain the same amount of chlorine for disinfection.**

To calculate the lbs/day hypochlorite required:

1. First calculate the lbs/day chlorine required.

$$\frac{(mg/L)\ (MGD)\ (8.34)}{Conc.\quad flow\quad lbs/gal} = lbs/day$$

2. Then calculate the lbs/day hypochlorite needed by dividing the lbs/day chlorine by the percent available chlorine.

$$\frac{Chlorine,\ lbs/day}{\dfrac{\%\ Available}{100}} = \frac{Hypochlorite}{lbs/day}$$

Example 9: (Chemical Dosage)

❑ A total chlorine dosage of 12 mg/L is required to treat a particular water. If the flow is 1.2 MGD and the hypochlorite has 65% available chlorine how many lbs/day of hypochlorite will be required?

First, calculate the lbs/day chlorine required using the mg/L to lbs/day equation:

$$\frac{(mg/L)\ (MGD)\ (8.34)}{Conc.\quad flow\quad lbs/gal} = lbs/day$$

$$(12\ mg/L)\ (1.2\ MGD)\ (8.34\ lbs/gal) = lbs/day$$

$$= \boxed{120\ lbs/day}$$

Now calculate the lbs/day hypochlorite required. Since only 65% of the hypochlorite is chlorine, more than 120 lbs/day will be required:

$$\frac{120\ lbs/day\ Cl_2}{\dfrac{65\ Avail.\ Cl_2}{100}} = \boxed{\begin{array}{c}185\ lbs/day\\ Hypochlorite\end{array}}$$

Example 10: (Chemical Dosage)

❑ A wastewater flow of 850,000 gpd requires a chlorine dose of 25 mg/L. If sodium hypochlorite (15% available chlorine) is to be used, how many lbs/day of sodium hypochlorite are required? How many gal/day of sodium hypochlorite is this?

First, calculate the lbs/day chlorine required:

$$\frac{(mg/L)\ (MGD)\ (8.34)}{Conc.\quad flow\quad lbs/gal} = lbs/day$$

$$(25\ mg/L)\ (0.85\ MGD)\ (8.34\ lbs/gal) = \boxed{\begin{array}{c}177\ lbs/day\\ Chlorine\end{array}}$$

Then calculate the lbs/day sodium hypochlorite:

$$\frac{177\ lbs/day\ Cl_2}{\dfrac{15\ Avail.\ Cl_2}{100}} = \boxed{\begin{array}{c}1180\ lbs/day\\ Hypochlorite\end{array}}$$

Then calculate the gal/day sodium hypochlorite:

$$\frac{1180\ lbs/day}{8.34\ lbs/gal} = \boxed{\begin{array}{c}141\ gal/day\\ Sodium\ Hypochlorite\end{array}}$$

Example 11: (Chemical Dosage)
❏ A flow of 800,000 gpd requires a chlorine dose of 9 mg/L. If chlorinated lime (34% available chlorine) is to be used, how many lbs/day of chlorinated lime will be required?

$$\frac{(mg/L)\ (MGD)\ (8.34)}{Conc.\quad flow\quad lbs/gal} = lbs/day$$

$$(9\ mg/L)\ (0.8\ MGD)\ (8.34\ lbs/gal) = lbs/day$$

$$= \boxed{\begin{array}{c} 60\ lbs/day \\ Chlorine \end{array}}$$

Then calculate the lbs/day chlorinated lime needed:

$$\frac{60\ lbs/day\ Cl_2}{\dfrac{34}{100}\ Avail.\ Cl_2} = \boxed{\begin{array}{c} 176\ lbs/day \\ Chlorinated\ Lime \end{array}}$$

Example 12: (Chemical Dosage)
❏ A small reservoir holds 70 acre-feet of water. To treat the reservoir for algae control, 0.5 mg/L of copper is required. How many pounds of copper sulfate will be required if the copper sulfate to be used contains 25% copper?

Before the mg/L to lbs equation can be used, the reservoir volume must be converted from ac-ft to cu ft to gal:

$$(70\ ac\text{-}ft)\ \left(43,560\ \frac{cu\ ft}{ac\text{-}ft}\right) = 3,049,200\ cu\ ft$$

$$(3,049,200\ cu\ ft)\ \left(7.48\ \frac{gal}{cu\ ft}\right) = 22,808,016\ gal$$

Now calculate the lbs/day copper required:

$$(0.5\ mg/L)\ (22.8\ MG)\ (8.34\ lbs/gal) = \boxed{\begin{array}{c} 95\ lbs \\ Copper \end{array}}$$

And then the lbs/day copper sulfate required:

$$\frac{95\ lbs\ Copper}{\dfrac{25}{100}\ Avail.\ Copper} = \boxed{\begin{array}{c} 380\ lbs \\ Copper\ Sulfate \end{array}}$$

OTHER CHEMICAL COMPOUNDS

Other chemical compounds used in water and wastewater treatment are like hypochlorite compounds. For example, chlorinated lime contains only about 34% available chlorine. And copper sulfate pentahydrate contains about 25% copper (the chemical of interest for algae control).

Calculating the lbs or lbs/day of chlorinated lime, copper sulfate, or other similar compound, you follow the same procedure as with the hypochlorite problems:

1. First calculate the lbs/day of chemical desired (such as chlorine or copper). Using the usual mg/L to lbs/day or lbs equations:

$$\boxed{\frac{(mg/L)\ (MGD)\ (8.34)}{Conc\quad flow\quad lbs/gal} = lbs/day}$$

or

$$\boxed{\frac{(mg/L)\ (MG)\ (8.34)}{Conc.\quad Vol\quad lbs/gal} = lbs}$$

2. Then calculate the lbs/day or lbs compound required:

$$\boxed{\frac{Chemical,\ lbs/day}{\dfrac{\%\ Available}{100}} = \begin{array}{c} Compound \\ lbs/day \end{array}}$$

Examples 11 and 12 illustrate these calculations. Note that in Example 11 a flow is being dosed, and in Example 12 a reservoir is being dosed.

3.2 LOADING CALCULATIONS—BOD, COD AND SS

When calculating BOD (Biochemical Oxygen Demand), COD (Chemical Oxygen Demand), or SS (Suspended Solids) loading on a treatment system, the following equation is used:

(mg/L) (MGD) (8.34) = lbs/day
Conc.   flow   lbs/gal

Loading on a system is usually calculated as lbs/day. Given the BOD, COD, or SS concentration and flow information, the lbs/day loading may be calculated as demonstrated in Examples 1-3.

Example 1: (Loading Calculations)
❑ Calculate the BOD loading (lbs/day) on a stream if the secondary effluent flow is 2.5 MGD and the BOD of the secondary effluent is 20 mg/L.

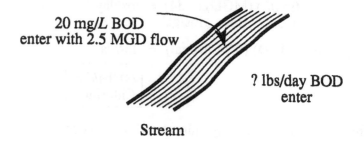

20 mg/L BOD
enter with 2.5 MGD flow

? lbs/day BOD
enter

Stream

First, select the appropriate equation:

(mg/L) (MGD flow) (8.34 lbs/gal) = lbs/day

Then fill in the information given in the problem:

(20 mg/L) (2.5 MGD) (8.34 lbs/gal) = 417 lbs/day
BOD

Example 2: (Loading Calculations)
❑ The suspended solids concentration of the wastewater entering the primary system is 480 mg/L. If the plant flow is 3,600,000 gpd, how many lbs/day suspended solids enter the primary system?

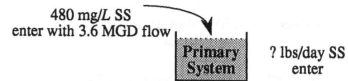

480 mg/L SS
enter with 3.6 MGD flow

Primary
System

? lbs/day SS
enter

First write the equation:

(mg/L) (MGD flow) (8.34 lbs/gal) = lbs/day

Then fill in the data given in the problem:

(480 mg/L) (3.6 MGD) (8.34 lbs/gal) = 14,412 lbs/day
SS

Example 3: (Loading Calculations)
❑ The flow to an aeration tank is 7 MGD. If the COD concentration of the water is 110 mg/*L*, how many pounds of COD are applied to the aeration tank daily?

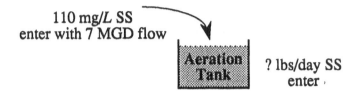

110 mg/*L* SS
enter with 7 MGD flow

? lbs/day SS
enter

Use the mg/*L* to lbs/day equation to solve the problem:

(mg/*L*) (MGD flow) (8.34 lbs/gal) = lbs/day

(110 mg/*L*) (7 MGD) (8.34 lbs/gal) = | 6422 lbs/day
       COD   COD |

Example 4: (Loading Calculations)
❑ The daily flow to a trickling filter is 4,500,000 gpd. If the BOD concentration of the trickling filter influent is 213 mg/*L*, how many lbs BOD enter the trickling filter daily?

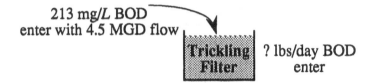

213 mg/*L* BOD
enter with 4.5 MGD flow

? lbs/day BOD
enter

Write the equation, fill in the given information, then solve for the unknown value:

(mg/*L*) (MGD flow) (8.34 lbs/gal) = lbs/day

(213 mg/*L*) (4.5 MGD) (8.34 lbs/gal) = | 7994 lbs/day
                                          BOD |

3.3 BOD AND SS REMOVAL CALCULATIONS, lbs/day

To calculate the pounds of BOD or suspended solids removed each day, you will need to know the mg/L BOD or SS removed and the plant flow. Then you can use the mg/L to lbs/day equation:

(mg/L) (MGD) (8.34) = lbs/day
Removed flow lbs/gal

For most calculations of BOD or SS removal, you will not be given information stating how many mg/L BOD or SS have been removed. This is something you will calculate based on the mg/L concentrations entering (influent) and leaving (effluent) the system.

The influent BOD or SS concentration indicates how much BOD or SS is entering the system. The effluent concentration indicates how much is still in the wastewater (the part not removed). The mg/L SS or BOD removed would therefore be:

Influent   _ Effluent = Removed
SS mg/L   SS mg/L   SS mg/L

Once you have determined the mg/L BOD or SS removed, you can then continue with the usual mg/L to lbs/day equation to calculate lbs/day BOD or SS removed. Examples 2-4 illustrate this calculation.

Example 1: (BOD and SS Removal)
❑ If 130 mg/L suspended solids are removed by a primary clarifier, how many lbs/day suspended solids are removed when the flow is 7.4 MGD?

130 mg/L
SS Removed

(mg/L) (MGD flow) (8.34 lbs/gal) = lbs/day

(130 mg/L) (7.4 MGD) (8.34 lbs/gal) = | 8023 lbs/day SS Removed |

Example 2: (BOD and SS Removal)
❑ The flow to a trickling filter is 3.7 MGD. If the primary effluent has a BOD concentration of 180 mg/L and the trickling filter effluent has a BOD concentration of 28 mg/L, how many pounds of BOD are removed daily?

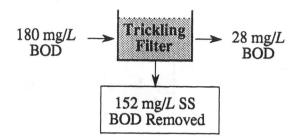

After calculating mg/L BOD removed, you can now calculate lbs/day BOD removed:

(mg/L) (MGD flow) (8.34 lbs/gal) = lbs/day
Removed   Removed

(152 mg/L) (3.7 MGD) (8.34 lbs/gal) = | 4690 lbs/day BOD Removed |

Example 3: (BOD and SS Removal)
❑ The flow to a primary clarifier is 2.7 MGD. If the influent to the clarifier has a suspended solids concentration of 230 mg/*L* and the primary effluent has 110 mg/*L* SS, how many lbs/day suspended solids are removed by the clarifier?

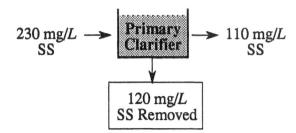

Now calculate lbs/day SS removed:

(mg/*L*) (MGD flow) (8.34 lbs/gal) = lbs/day

(120 mg/*L*) (2.7 MGD) (8.34 lbs/gal) = ⎥2702 lbs/day SS Removed

Example 4: (BOD and SS Removal)
❑ The flow to a trickling filter is 4,600,000 gpd, with a BOD concentration of 195 mg/*L*. If the BOD of the trickling filter effluent is 98 mg/*L*, how many lbs/day BOD are removed by the trickling filter ?

195 mg/*L* BOD → Trickling Filter → 98 mg/*L* BOD
↓
97 mg/*L* SS BOD Removed

Now calculate lbs/day BOD removed:

(mg/*L*) (MGD flow) (8.34 lbs/gal) = lbs/day

(97 mg/*L*) (4.6 MGD) (8.34 lbs/gal) = ⎥3721 lbs/day BOD Removed

3.4 POUNDS OF SOLIDS UNDER AERATION CALCULATIONS

In any activated sludge system it is important to control the amount of solids under aeration (solids inventory). The suspended solids in an aerator are called Mixed Liquor Suspended Solids (MLSS). To calculate the pounds of suspended solids in the aeration tank, you will need to know the mg/L concentration of the MLSS. Then mg/L MLSS can be expressed as lbs MLSS, using the mg/L to lbs equation:

$$
\underset{\text{MLSS}}{(mg/L)} \; \underset{\text{Vol}}{(MG)} \; \underset{\text{lbs/gal}}{(8.34)} = lbs
$$

Notice that the mixed liquor suspended solids concentration is **concentration within a tank**. Therefore, the equation using **MG volume** is used.

Another important measure of solids in the aeration tank is the amount of volatile suspended solids.* The volatile solids content of the aeration tank is used as an estimate of the microorganism population of the aeration tank. The Mixed Liquor Volatile Suspended Solids (MLVSS) usually comprises about 70% of the MLSS. The other 30% of the MLSS are fixed (inorganic) solids. To calculate the lbs MLVSS, use the mg/L to lbs equation:

$$
\underset{\text{MLVSS}}{(mg/L)} \; \underset{\text{Vol}}{(MG)} \; \underset{\text{lbs/gal}}{(8.34)} = lbs
$$

Example 1: (lbs Solids Under Aeration)
❏ An aeration tank has a volume of 450,000 gallons. If the mixed liquor suspended solids are 1820 mg/L, how many pounds of suspended solids are in the aerator?

Aeration Tank
820 mg/L MLSS

Vol = 0.45 MG

$(mg/L) \; (MG \; vol) \; (8.34 \; lbs/gal) = lbs$

$(1820 \; mg/L) \; (0.45 \; MG) \; (8.34 \; lbs/gal) = \boxed{6830 \; lbs \\ MLSS}$

Example 2: (lbs Solids Under Aeration)
❏ The volume of an oxidation ditch is 23,040 cubic feet. If the MLVSS concentration is 3800 mg/L, how many pounds of volatile suspended solids are under aeration?

Before the mg/L to lbs equation can be used cubic feet aeration tank volume must be converted to million gallons:

Aeration Tank
3800 mg/L

Vol = $(23,040 \; cu \; ft) \; (7.48 \; \dfrac{gal}{cu \; ft})$

   = 172,339 gal

or = 0.17 MG

The lbs volatile solids calculation can now be completed:

$(3800 \; mg/L) \; (0.17 \; MG \; vol) \; (8.34 \; lbs/gal) = \boxed{5388 \; lbs \\ MLVSS}$

---

* For a discussion of volatile suspended solids calculations, refer to Chapter 6, Efficiency and Other Percent Calculations.

Example 3: (lbs Solids Under Aeration)
❏ The aeration tank of a conventional activated sludge plant has a mixed liquor volatile suspended solids concentration of 2300 mg/L. If the aeration basin is 110 ft long, 35 ft wide and has wastewater to a depth of 13 ft, how many pounds of MLVSS are under aeration?

**Aeration Tank
2300 mg/***L*

$$\text{Vol} = (110 \text{ ft}) (35 \text{ ft}) (13 \text{ ft}) (7.48 \frac{\text{gal}}{\text{cu ft}})$$

$$= 374,374 \text{ gal}$$

$$\text{or} = 0.37 \text{ MG}$$

Now calculate lbs MLVSS using the usual equation and fill in the given information:

$$(2300 \text{ mg/}L) (0.37 \text{ MG}) (8.34 \text{ lbs/gal}) = \boxed{7097 \text{ lbs} \\ \text{MLVSS}}$$

Example 4: (lbs Solids Under Aeration)
❏ An aeration tank is 90 ft long and 40 ft wide. The depth of wastewater in the tank is 16 ft. If the concentration of MLSS is 1980 mg*L*, how many pounds of MLSS are under aeration?

**Aeration Tank
x mg/***L*

$$\text{Vol} = (90 \text{ ft}) (40 \text{ ft}) (16 \text{ ft}) (7.48 \frac{\text{gal}}{\text{cu ft}})$$

$$= 430,848 \text{ gal}$$

$$\text{or} = 0.43 \text{ MG}$$

Now fill in the mg/*L* to lbs equation with known information:

$$(1980 \text{ mg/}L) (0.43 \text{ MG}) (8.34 \text{ lbs/gal}) = \boxed{7101 \text{ lbs MLSS}}$$

3.5 WAS PUMPING RATE CALCULATIONS

Waste Activated Sludge (WAS) pumping rate calculations are calculations that involve mg/*L* and flow. Therefore the equation used in these calculations is:

$$\frac{(mg/L)\ (MGD)\ (8.34)}{flow\quad lbs/gal} = lbs/day$$

In WAS pumping rate calculations, the "WAS SS" refers to the suspended solids content of the Waste Activated Sludge being pumped away, and "MGD flow" refers to the WAS pumping rate of the sludge being wasted.

Sometimes waste activated sludge SS is not known but return activated sludge SS is known. Remember that **RAS SS and WAS SS are the same measurement**. It is a measurement taken of secondary clarifier sludge. This sludge is either pumped back to the aerator (RAS) or wasted (WAS).

WAS PUMPING RATE CALCULATIONS ARE mg/*L* TO lbs/day PROBLEMS:

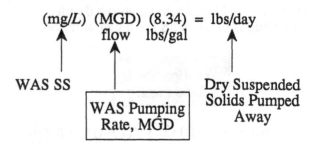

Example 1: (WAS Pumping Rate Calculations)
❑ The WAS suspended solids concentration is 5860 mg/*L*. If 3800 lbs/day dry solids are to be wasted, (a) What must the WAS pumping rate be, in MGD? (b) What is this rate expressed in gpm?

(a) First calculate the MGD pumping rate required, using the mg/*L* to lbs/day equation:

(mg/*L*) (MGD flow) (8.34 lbs/gal) = lbs/day

$$\underset{mg/L\quad flow\quad lbs/day}{(5860)\ (x\ MGD)\ (8.34)} = 3800\ lbs/day$$

$$x = \frac{3800\ lbs/day}{\underset{mg/L\quad lbs/gal}{(5600)\ (8.34)}}$$

$$x = 0.0814\ MGD$$

(b) Then convert the MGD flow to gpm flow: *

$$0.0814\ MGD = 81,400\ gpd$$

$$= \frac{81,400\ gpd}{1440\ min/day}$$

$$= \boxed{57\ gpm}$$

---

* Refer to Chapter 8 in *Basic Math Concepts* for a review of flow conversions.

Example 2: (WAS Pumping Rate Calculations)
❑ It has been determined that 4770 lbs/day of dry solids must be removed from the secondary system. If the WAS SS concentration is 7340 mg/*L*, what must be the WAS pumping rate, in gpm?

First calculate the MGD pumping rate required:

$$\underset{\substack{\text{mg/}L \quad \text{flow} \quad \text{lbs/day}}}{(7340)\ (x\ \text{MGD})\ (8.34)} = 4700\ \text{lbs/day}$$

$$x = \frac{4700\ \text{lbs/day}}{\underset{\text{mg/}L \quad \text{lbs/gal}}{(7340)\ (8.34)}}$$

$$x = 0.0768\ \text{MGD}$$

Then convert MGD pumping rate to gpm pumping rate:

$$0.0768\ \text{MGD} = 76,800\ \text{gpd}$$

$$= \frac{76,800\ \text{gpd}}{1440\ \text{min/day}}$$

$$= \boxed{53\ \text{gpm}}$$

Example 3: (WAS Pumping Rate Calculations)
❑ The WAS suspended solids concentration is 6980 mg/*L*. If 5300 lbs/day dry sludge solids are to be wasted, what must be the WAS pumping rate, in gpm?

First calculate the MGD pumping rate required:

$$(\text{mg/}L)\ (\text{MGD flow})\ (8.34\ \text{lbs/gal}) = \text{lbs/day}$$

$$\underset{\substack{\text{mg/}L \quad \text{flow} \quad \text{lbs/day}}}{(6980)\ (x\ \text{MGD})\ (8.34)} = 5300\ \text{lbs/day}$$

$$x = \frac{5300\ \text{lbs/day}}{\underset{\text{mg/}L \quad \text{lbs/gal}}{(6980)\ (8.34)}}$$

$$x = 0.091\ \text{MGD}$$

Then convert the MGD flow to gpm flow:

$$0.091\ \text{MGD} = 91,000\ \text{gpd}$$

$$= \frac{91,000\ \text{gpd}}{1440\ \text{min/day}}$$

$$= \boxed{63\ \text{gpm}}$$

NOTES:

4 *Loading Rate Calculations*

SUMMARY

1. The **hydraulic loading rate** of a treatment system is a measure of the flow treated per square foot of surface area. The most common expression of hydraulic loading rate is gpd/sq ft. Recirculated flows are included as part of the gpd flow to the system.

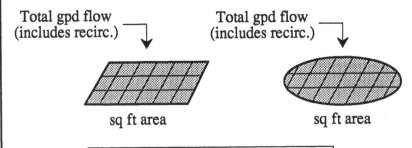

Total gpd flow (includes recirc.)

sq ft area

Total gpd flow (includes recirc.)

sq ft area

$$\text{Hydraulic Loading Rate} = \frac{\text{Flow, gpd}}{\text{Area, sq ft}}$$

Hydraulic loading rate for ponds is generally expressed as inches/day:

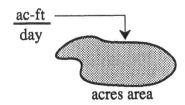

$$\frac{\text{ac-ft}}{\text{day}}$$

acres area

$$\text{Hydraulic Loading Rate} = \frac{\text{Flow, ac-ft/day}}{\text{Area, ac}}$$

$$= \text{ft/day}$$

Then, multiplying by 12 in./ft, the hydraulic loading rate can be expressed in in./day.

$$\text{Hydraulic Loading Rate} = \frac{\text{in.}}{\text{day}}$$

There are several calculations that measure the water and solids loading on the treatment system. When the water and solids loading consistently exceed design values, the efficiency of the treatment system begins to deteriorate.

Calculations that reflect various types of **water loading** on the system include:

- Hydraulic Loading Rate (gpd/sq ft)

- Surface Loading Rate (gpd/sq ft, gpm/sq ft or in./day)

- Filtration Rate (gpm/sq ft or in./min)

- Backwash Rate (gpm/sq ft or in./min)

- Unit Filter Run Volume, UFRV, (gal/sq ft)

- Weir Overflow Rate (gpd/ft)

Calculations that reflect various types of **solids loading** are:

- Organic Loading Rate (lbs BOD/day/1000 cu ft)

- Food/Microorganism Ratio (lbs BOD/day/lb MLVSS)

- Solids Loading Rate (lbs SS/day/sq ft)

- Digester Loading Rate (lbs VS/day/cu ft)

- Digester Volatile Solids Loading Ratio (lbs VS/day Added/lb VS in Digester)

- Population Loading and Population Equivalent

SUMMARY—Cont'd

2. **Surface overflow rate** is a calculation similar to hydraulic loading rate—flow rate per unit area. The difference between these calculations pertains to recirculation rates. Recirculation is not included in the surface overflow rate calculation.

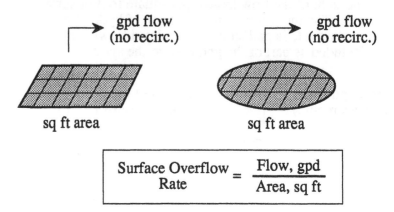

$$\text{Surface Overflow Rate} = \frac{\text{Flow, gpd}}{\text{Area, sq ft}}$$

3. **Filtration rate** is a measure of the gallons per minute flow filtered by each square foot of filter.

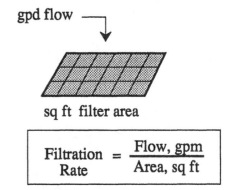

$$\text{Filtration Rate} = \frac{\text{Flow, gpm}}{\text{Area, sq ft}}$$

4. **Backwash rate** is a very similar calculation to filtration rate. It is the gallons per minute of backwash water flowing through each square foot of filter area.

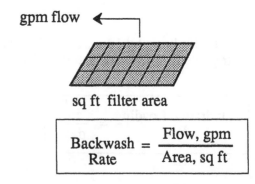

$$\text{Backwash Rate} = \frac{\text{Flow, gpm}}{\text{Area, sq ft}}$$

SUMMARY—Cont'd

5. **Unit filter run volume** (UFRV) is a measure of the total gallons of water filtered by each square foot of filter surface area during a filter run.

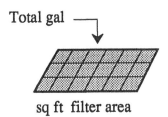

Total gal

sq ft filter area

$$\text{UFRV} = \frac{\text{Total gal}}{\text{Area, sq ft}}$$

6. The **weir overflow rate** is a measure of the gallons per day flowing over each foot of weir.

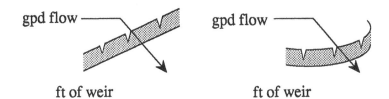

gpd flow   gpd flow

ft of weir   ft of weir

$$\text{Weir Overflow Rate} = \frac{\text{Flow, gpd}}{\text{ft of weir}}$$

7. The **organic loading rate** on a system is the pounds per day of BOD applied to each 1000 cu ft volume.

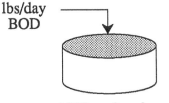

lbs/day
BOD

1000 cu ft volume

$$\text{Organic Loading} = \frac{\text{BOD, lbs/day}}{\text{Volume, 1000 cu ft}}$$

SUMMARY—Cont'd

8. The **food/microorganism ratio** indicates the relative balance between the food entering the secondary system and the microorganisms present.

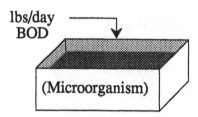

lbs/day
BOD

(Microorganism)

lbs MLVSS in Aeration Tank

$$\text{F/M} = \frac{\text{BOD entering, lbs/day}}{\text{MLVSS, lbs}}$$

9. **Solids loading rate** indicates the pounds of solids that are removed daily per square foot of secondary clarifier surface area.

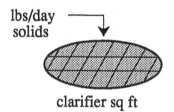

lbs/day
solids

clarifier sq ft

$$\text{Solids Loading Rate} = \frac{\text{Solids, lbs/day}}{\text{Area, sq ft}}$$

10. The **digester loading rate** measures the lbs/day volatile solids entering the digester per cubic foot of digester volume.

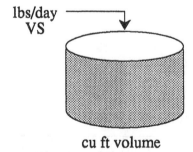

lbs/day
VS

cu ft volume

$$\text{Digester Loading Rate} = \frac{\text{VS, lbs/day}}{\text{Dig. Volume, cu ft}}$$

11. The **digester volatile solids loading ratio** is used to determine the seed sludge required in the digester. It can also be used to determine the balance between volatile solids added to the digester and the volatile solid in the digester.

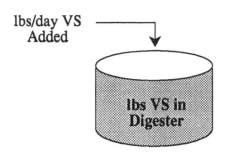

lbs/day VS Added

lbs VS in Digester

$$\text{Volatile Solids Loading Ratio} = \frac{\text{VS Added, lbs/day}}{\text{VS in Digester, lbs}}$$

12. **Population loading** is a calculation most often associated with wastewater ponds. It is the number of people served per acre of pond area.

$$\text{Population Loading} = \frac{\text{People Served by the System}}{\text{Pond Area, acres}}$$

Population equivalent is a calculation that expresses the organic content of a wastewater (BOD) in terms of an equivalent number of people using the system. This calculation assumes that for each person using the system, about 0.2 lbs/day BOD enter the system.

$$\text{Population Equivalent} = \frac{(\text{mg}/L \text{ BOD})(\text{MGD flow})(8.34 \text{ lbs/gal})}{0.2 \text{ lbs/day BOD/person}}$$

4.1 HYDRAULIC LOADING RATE CALCULATIONS

Hydraulic loading rate is a term used to indicate the total flow, in gpd, loaded or entering each square foot of water surface area. It is the total gpd flow to the process divided by the water surface area of the tank or pond.

As shown in the diagram to the right **recirculated flows must be included** as part of the total flow (total Q) to the process.

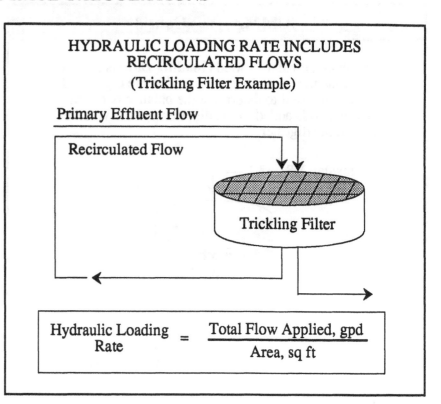

HYDRAULIC LOADING RATE INCLUDES RECIRCULATED FLOWS

(Trickling Filter Example)

Primary Effluent Flow

Recirculated Flow

Trickling Filter

$$\text{Hydraulic Loading Rate} = \frac{\text{Total Flow Applied, gpd}}{\text{Area, sq ft}}$$

Example 1: (Hydraulic Loading)
❏ A trickling filter 80 ft in diameter treats a primary effluent flow of 1.8 MGD. If the recirculated flow to the clarifier is 0.3 MGD, what is the hydraulic loading on the trickling filter?

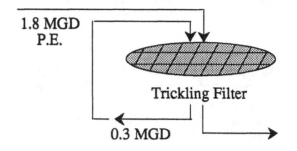

1.8 MGD
P.E.

Trickling Filter

0.3 MGD

$$\text{Hydraulic Loading Rate} = \frac{\text{Flow, gpd}}{\text{Area, sq ft}}$$

$$= \frac{2,100,000 \text{ gpd}}{(0.785)(80 \text{ ft})(80 \text{ ft})}$$

$$= \boxed{418 \text{ gpd/sq ft}}$$

Example 2: (Hydraulic Loading)
❏ If 50,000 gpd are pumped to a 30 ft diameter gravity thickener, what is the hydraulic loading rate on the thickener?

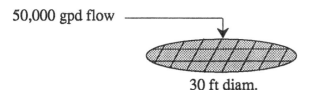

50,000 gpd flow

30 ft diam.

$$\text{Hydraulic Loading Rate} = \frac{\text{Flow, gpd}}{\text{Area, sq ft}}$$

$$= \frac{50,000 \text{ gpd}}{(0.785)(30\text{ ft})(30\text{ ft})}$$

$$= \boxed{71 \text{ gpd/sq ft}}$$

WHEN THERE IS NO RECIRCULATION

When there is no recirculated flow, the total flow applied is simply the flow to the unit process.

Example 3: (Hydraulic Loading)
❏ A rotating biological contactor treats a flow of 2.8 MGD. The manufacturer's data indicates a media surface area of 800,000 sq ft. What is the hydraulic loading rate on the RBC?

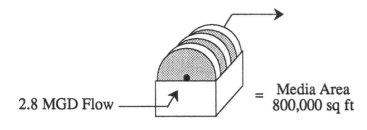

2.8 MGD Flow

= Media Area 800,000 sq ft

$$\text{Hydraulic Loading Rate} = \frac{\text{Flow, gpd}}{\text{Area, sq ft}}$$

$$= \frac{2,800,000 \text{ gpd}}{800,000 \text{ sq ft}}$$

$$= \boxed{3.5 \text{ gpd/sq ft}}$$

HYDRAULIC LOADING FOR ROTATING BIOLOGICAL CONTACTORS (RBC)

When calculating the hydraulic loading rate on a rotating biological contactor, use the **sq ft area of the media** rather than the sq ft area of the water surface, as in other hydraulic loading calculations. The RBC manufacturer provides media area information.

HYDRAULIC LOADING FOR PONDS

When calculating hydraulic loading for wastewater ponds, the answer is generally expressed as in./day rather than gpd/sq ft.

There are two ways to calculate in./day hydraulic loading, depending on how the flow to the pond is expressed.

HYDRAULIC LOADING RATE FOR PONDS, in./day

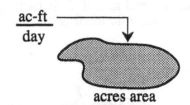

$$\frac{\text{ac-ft}}{\text{day}}$$

acres area

$$\text{Hydraulic Loading Rate} = \frac{\text{Flow, ac-ft/day}}{\text{Area, ac}}$$

$$= \text{ft/day}$$

Then, by multiplying by 12 in./ft, the hydraulic loading rate can be expressed in in./day.

$$\boxed{\text{Hydraulic Loading Rate} = \frac{\text{in.}}{\text{day}}}$$

If the flow to the pond is expressed in gpd:

1. Set up the hydraulic loading equation as usual.

2. Convert gpd flow to cubic feet per day flow (cfd or ft^3/day). This is done by dividing gpd by 7.48 gal/cu ft.*

3. Cancel terms to obtain ft/day hydraulic loading.**

$$\frac{\text{cfd}}{\text{sq ft}} = \frac{\text{ft}^3/\text{day}}{\text{ft}^2} = \text{ft/day}$$

4. Then convert ft/day to in./day by multiplying by 12 in./ft.

Example 4: (Hydraulic Loading)
❑ A pond receives a flow of 1,980,000 gpd. If the surface area of the pond is 700,000 sq ft, what is the hydraulic loading in in./day?

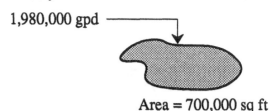

1,980,000 gpd

Area = 700,000 sq ft

$$\text{Hydraulic Loading Rate} = \frac{1,980,000 \text{ gpd}}{700,000 \text{ sq ft}}$$

Convert gpd flow to ft^3/day flow (1,980,000 gpd ÷ 7.48 gal/cu ft):

$$= \frac{264,706 \text{ ft}^3/\text{day}}{700,000 \text{ ft}^2}$$

$$= 0.4 \text{ ft/day}$$

Then convert to in/day:

$$(0.4 \text{ ft/day}) (12 \text{ in./ft}) \quad \boxed{5 \text{ in./day}}$$

---

* For a review of flow conversions, refer to Chapter 8 in *Basic Math Concepts*.

** To review cancellation of terms, refer to Chapter 15, Dimensional Analysis, in *Basic Math Concepts*.

Example 5: (Hydraulic Loading)
❏ A pond receives a flow of 2,400,000 gpd. If the surface area of the pond is 15 acres, what is the hydraulic loading in in/day?

 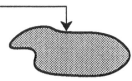

2,400,000 gpd

$$\text{Area} = 15 \text{ ac}$$
$$\text{or} \quad (15 \text{ ac}) (43{,}560 \text{ sq ft/ac})$$
$$= 653{,}400 \text{ sq ft}$$

$$\frac{\text{Hydraulic Loading}}{\text{Rate}} = \frac{2{,}400{,}000 \text{ gpd}}{653{,}400 \text{ sq ft}}$$

Convert gpd flow to ft^3/day flow (2,400,000 gpd ÷ 7.48 gal/cu ft):

$$= \frac{320{,}856 \text{ ft}^3/\text{day}}{653{,}400 \text{ ft}^2}$$

$$= 0.5 \text{ ft/day}$$

Then convert to in/day:

$$(0.5 \text{ ft/day}) (12 \text{ in./ft}) = \boxed{6 \text{ in./day}}$$

Example 6: (Hydraulic Loading)
❏ A 25-acre pond receives a flow of 6.2 acre-feet/day. What is the hydraulic loading on the pond in in/day?

Use the equation for hydraulic loading which includes acre-ft/day flow:

$$\frac{\text{Hydraulic Loading}}{\text{Rate}} = \frac{6.2 \text{ ac-ft/day}}{25 \text{ ac}}$$

$$= 0.25 \text{ ft/day}$$

Then convert ft/day to in./day:

$$(0.25 \text{ ft/day}) (12 \text{ in./ft}) = \boxed{3 \text{ in./day}}$$

If the flow to the pond is expressed in acre-feet/day:

1. Set up the hydraulic loading equation in a slightly different form. Instead of gpd/sq ft, use the form of acre-ft/day flow per acres area.

2. Canceling terms results in ft/day hydraulic loading.

$$\frac{\text{Flow, ac-ft/day}}{\text{Area, ac}} = \text{ft/day}$$

3. Then convert ft/day to in./day.

This calculation is illustrated in Example 6.

4.2 SURFACE OVERFLOW RATE CALCULATIONS

Surface overflow rate is used to determine loading on clarifiers. It is similar to hydraulic loading rate—flow per unit area. However, hydraulic loading rate measures the total water entering the process (plant flow plus recirculation) whereas **surface overflow rate measures only the water overflowing the process (plant flow only).**

As indicated in the diagram to the right, **surface overflow rate calculations do not include recirculated flows.** This is because recirculated flows are taken from the bottom of the clarifier and hence do not flow up and out of the clarifier (overflow).

Since surface overflow rate is a measure of flow (Q) divided by area (A), surface overflow is an indirect measure of the **upward velocity** of water as it overflows the clarifier:*

$$V = \frac{Q}{A}$$

This calculation is important in maintaining proper clarifier operation since settling solids will be drawn upward and out of the clarifier if surface overflow rates are too high.

Other terms used synonymously with surface overflow rate are:

• Surface Loading Rate, and

• Surface Settling Rate

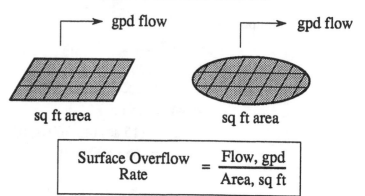

SURFACE OVERFLOW RATE DOES NOT INCLUDE RECIRCULATED FLOWS

$$\text{Surface Overflow Rate} = \frac{\text{Flow, gpd}}{\text{Area, sq ft}}$$

Surface overflow rate for wastewater calculations is normally expressed as gpd/sq ft, as shown above. However this calculation for water systems is often expressed as gpm/sq ft.

$$\text{Surface Overflow Rate} = \frac{\text{Flow, gpm}}{\text{Area, sq ft}}$$

Example 1: (Surface Overflow Rate)
❑ A circular clarifier has a diameter of 60 ft. If the primary effluent flow is 2.3 MGD, what is the surface overflow rate in gpd/sq ft?

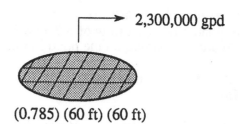

2,300,000 gpd

(0.785) (60 ft) (60 ft)

$$\text{Surface Overflow Rate} = \frac{\text{Flow, gpd}}{\text{Area, sq ft}}$$

$$= \frac{2,300,000 \text{ gpd}}{(0.785)\,(60 \text{ ft})\,(60 \text{ ft})}$$

$$= \boxed{814 \text{ gpd/sq ft}}$$

---

* Refer to Chapter 2 for a review of $Q = AV$ problems.

Example 2: (Surface Overflow Rate)

❏ A sedimentation basin 75 ft by 20 ft receives a flow of 1.3 MGD. What is the surface overflow rate in gpd/sq ft?

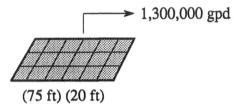

1,300,000 gpd

(75 ft) (20 ft)

$$\text{Surface Overflow Rate} = \frac{\text{Flow, gpd}}{\text{Area, sq ft}}$$

$$= \frac{1,300,000 \text{ gpd}}{(75 \text{ ft}) (20 \text{ ft})}$$

$$= \boxed{867 \text{ gpd/sq ft}}$$

Example 3: (Surface Overflow Rate)

❏ The flow to a sedimentation tank is 3.2 MGD. If the length of the basin is 90 ft and the width is 45 ft, what is the surface overflow rate in gpd/sq ft?

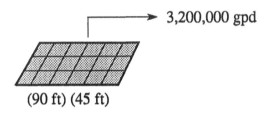

3,200,000 gpd

(90 ft) (45 ft)

$$\text{Surface Overflow Rate} = \frac{\text{Flow, gpd}}{\text{Area, sq ft}}$$

$$= \frac{3,200,000 \text{ gpd}}{(90 \text{ ft}) (45 \text{ ft})}$$

$$= \boxed{790 \text{ gpd/sq ft}}$$

4.3 FILTRATION RATE CALCULATIONS

The calculation of filtration rate or filter loading rate is similar to that of hydraulic loading rate. It is gpm filtered by each square foot of filter area.

$$\text{Filtration Rate} = \frac{\text{Flow, gpm}}{\text{Area, sq ft}}$$

Example 1: (Filtration Rate)
❏ A filter 20 ft by 25 ft receives a flow of 1940 gpm. What is the filtration rate in gpm/sq ft?

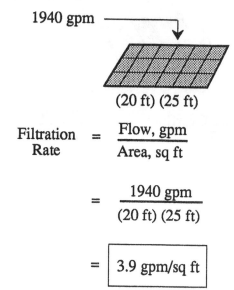

1940 gpm

(20 ft) (25 ft)

$$\text{Filtration Rate} = \frac{\text{Flow, gpm}}{\text{Area, sq ft}}$$

$$= \frac{1940 \text{ gpm}}{(20 \text{ ft}) (25 \text{ ft})}$$

$$= \boxed{3.9 \text{ gpm/sq ft}}$$

Example 2: (Filtration Rate)
❏ A filter 20 ft by 35 ft treats a flow of 1530 gpm. What is the filtration rate in gpm/sq ft?

1530 gpm

(20 ft) (35 ft)

$$\text{Filtration Rate} = \frac{\text{Flow, gpm}}{\text{Area, sq ft}}$$

$$= \frac{1530 \text{ gpm}}{(20 \text{ ft}) (35 \text{ ft})}$$

$$= \boxed{2.2 \text{ gpm/sq ft}}$$

Example 3: (Filtration Rate)

❑ A filter 25 ft by 30 ft treats a flow of 3.3 MGD. What is the filtration rate in gpm/sq ft?

$$\frac{3{,}300{,}000 \text{ gpd}}{1440 \text{ min/day}}$$

= 2292 gpm

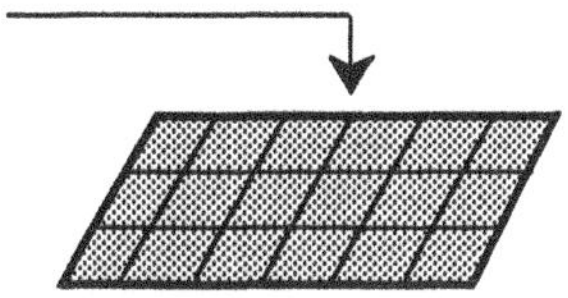

(20 ft) (35 ft)

$$\begin{array}{rl} \text{Filtration} \\ \text{Rate} \end{array} = \frac{\text{Flow, gpm}}{\text{Area, sq ft}}$$

$$= \frac{2292 \text{ gpm}}{(25 \text{ ft}) (30 \text{ ft})}$$

$$= \boxed{3.1 \text{ gpm/sq ft}}$$

Example 4: (Filtration Rate)

❑ A filter has a surface area of 35 ft by 25 ft. If the filter receives a flow of 2,912,000 gpd, what is the filtration rate in gpm/sq ft?

$$\frac{2{,}912{,}000 \text{ gpd}}{1440 \text{ min/day}}$$

= 2022 gpm

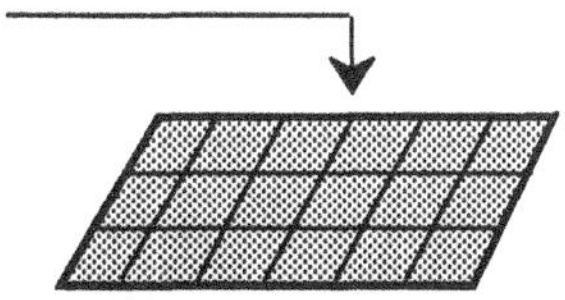

(35 ft) (25 ft)

$$\begin{array}{rl} \text{Filtration} \\ \text{Rate} \end{array} = \frac{\text{Flow, gpm}}{\text{Area, sq ft}}$$

$$= \frac{2022 \text{ gpm}}{(35 \text{ ft}) (25 \text{ ft})}$$

$$= \boxed{2.3 \text{ gpm/sq ft}}$$

4.4 BACKWASH RATE CALCULATIONS

A filter backwash rate is a measure of the gpm flowing upward through each sq ft of filter surface area. The calculation of backwash rate is similar to filtration rate.

$$\text{Backwash Rate} = \frac{\text{Flow, gpm}}{\text{Area, sq ft}}$$

Example 1: (Backwash Rate)
❑ A filter with a surface area of 150 sq ft has a backwash flow rate of 2900 gpm. What is the filter backwash rate in gpm/sq ft?

2900 gpm ←

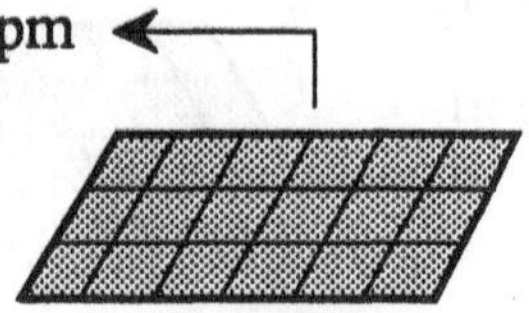

(10 ft) (15 ft)

$$\text{Backwash Rate} = \frac{\text{Flow, gpm}}{\text{Area, sq ft}}$$

$$= \frac{2900 \text{ gpm}}{150 \text{ sq ft}}$$

$$= \boxed{19.3 \text{ gpm/sq ft}}$$

Example 2: (Backwash Rate)
❑ A filter 25 ft by 10 ft has a backwash rate of 3400 gpm. What is the backwash rate in gpm/sq ft?

3400 gpm ←

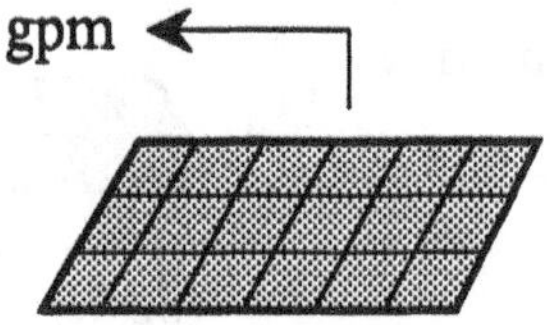

(25 ft) (10 ft)

$$\text{Backwash Rate} = \frac{\text{Flow, gpm}}{\text{Area, sq ft}}$$

$$= \frac{3400 \text{ gpm}}{(25 \text{ ft}) (10 \text{ ft})}$$

$$= \boxed{13.6 \text{ gpm/sq ft}}$$

Example 3: (Backwash Rate)

❑ A filter 15 ft by 15 ft has a backwash flow rate of 3150 gpm. What is the filter backwash rate in gpm/sq ft?

3150 gpm ←

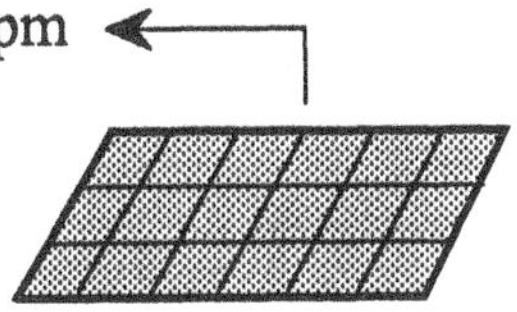

(15 ft) (15 ft)

$$\text{Backwash Rate} = \frac{\text{Flow, gpm}}{\text{Area, sq ft}}$$

$$= \frac{3150 \text{ gpm}}{(15 \text{ ft}) (15 \text{ ft})}$$

$$= \boxed{14 \text{ gpm/sq ft}}$$

Example 4: (Backwash Rate)

❑ A filter 20 ft long and 15 ft wide has a backwash flow rate of 4.64 MGD. What is the filter backwash rate in gpm/sq ft?

$$\frac{4,640,000 \text{ gpd}}{1440 \text{ min/day}}$$

$$= 3222 \text{ gpm}$$

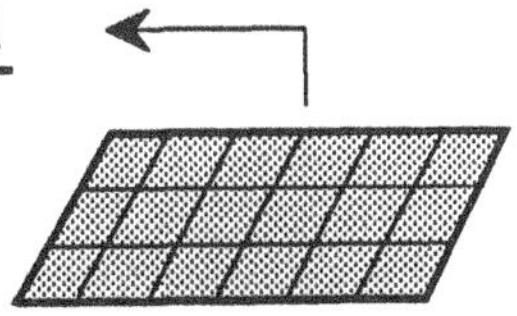

(20 ft) (15 ft)

$$\text{Backwash Rate} = \frac{\text{Flow, gpm}}{\text{Area, sq ft}}$$

$$= \frac{3222 \text{ gpm}}{(20 \text{ ft}) (15 \text{ ft})}$$

$$= \boxed{10.7 \text{ gpm/sq ft}}$$

WHEN THE FLOW RATE IS EXPRESSED AS GPD

Normally the backwash flow rate is expressed as gpm. If it is expressed in any other flow rate terms, simply convert the given flow rate to gpm.* For example, if gpd flow rate is given, convert the gpd flow rate as follows:

$$\boxed{\frac{\text{Flow Rate, gpd}}{1440 \text{ min/day}} = \text{Flow Rate, gpm}}$$

* For a review of flow rate conversions, refer to Chapter 8 in *Basic Math Concepts*.

4.5 UNIT FILTER RUN VOLUME CALCULATIONS

The unit filter run volume (UFRV) calculation indicates the total gallons passing through each square foot of filter surface area during an entire filter run. The equation to be used in these calculations is shown to the right.

As the performance of the filter begins to deteriorate, the UFRV value will begin to decline as well.

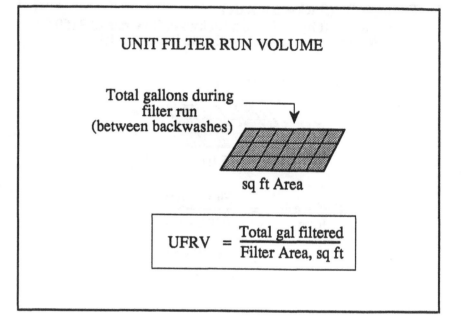

UNIT FILTER RUN VOLUME

Total gallons during filter run (between backwashes)

sq ft Area

$$UFRV = \frac{\text{Total gal filtered}}{\text{Filter Area, sq ft}}$$

Example 1: (UFRV)
❑ The total water filtered during a filter run (between backwashes) is 2,950,000 gal. If the filter is 15 ft by 20 ft, what is the unit filter run volume (UFRV)?

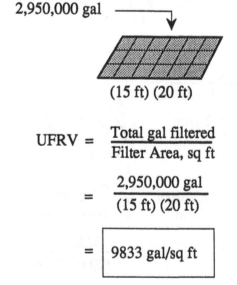

2,950,000 gal

(15 ft) (20 ft)

$$UFRV = \frac{\text{Total gal filtered}}{\text{Filter Area, sq ft}}$$

$$= \frac{2,950,000 \text{ gal}}{(15 \text{ ft}) (20 \text{ ft})}$$

$$= \boxed{9833 \text{ gal/sq ft}}$$

Example 2: (UFRV)

❑ The total water filtered during a filter run is 3,220,000. If the filter is 20 ft by 20 ft what is the UFRV?

3,220,000 gal

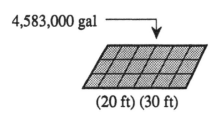

(20 ft) (20 ft)

$$\text{UFRV} = \frac{\text{Total gal filtered}}{\text{Filter Area, sq ft}}$$

$$= \frac{3,220,000 \text{ gal}}{(20 \text{ ft}) (20 \text{ ft})}$$

$$= \boxed{8050 \text{ gal/sq ft}}$$

Example 3: (UFRV)

❑ The total water filtered during a filter run is 4,583,000? If the filter is 20 ft by 30 ft, what is the unit filter run volume?

4,583,000 gal

(20 ft) (30 ft)

$$\text{UFRV} = \frac{\text{Total gal filtered}}{\text{Filter Area, sq ft}}$$

$$= \frac{4,583,000 \text{ gal}}{(20 \text{ ft}) (30 \text{ ft})}$$

$$= \boxed{7638 \text{ gal/sq ft}}$$

4.6 WEIR OVERFLOW RATE CALCULATIONS

Weir overflow rate is a measure of the gallons per day flowing over each foot of weir.

$$\begin{array}{c}\text{Weir Overflow} \\ \text{Rate}\end{array} = \frac{\text{Flow, gpd}}{\text{Weir Length, ft}}$$

CALCULATING WEIR CIRCUMFERENCE

In some calculations of weir overflow rate, you will have to calculate the total weir length given the weir diameter. To calculate the length of weir around the clarifier, you need to know the relationship between the diameter and circumference of a circle. **The distance around any circle (circumference) is about three times the distance across the circle (diameter).** In fact, the circumference is (3.14) (Diameter).* Therefore, given a diameter, the total ft of weir can be calculated as:

$$\begin{array}{c}\text{Weir} \\ \text{Length,} \\ \text{ft}\end{array} = (3.14)\,(\text{Weir Diam.,}) \atop \text{ft}$$

Example 1: (Weir Overflow Rate)
❏ A rectangular clarifier has a total of 100 ft of weir. What is the weir overflow rate in gpd/ft when the flow is 1.2 MGD?

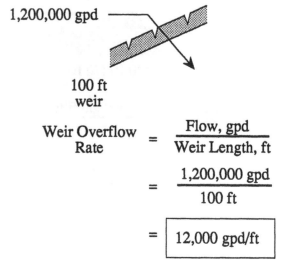

1,200,000 gpd

100 ft
weir

$$\begin{array}{c}\text{Weir Overflow} \\ \text{Rate}\end{array} = \frac{\text{Flow, gpd}}{\text{Weir Length, ft}}$$

$$= \frac{1{,}200{,}000 \text{ gpd}}{100 \text{ ft}}$$

$$= \boxed{12{,}000 \text{ gpd/ft}}$$

Example 2: (Weir Overflow Rate)
❏ A circular clarifier receives a flow of 3.38 MGD. If the diameter of the weir is 80 ft, what is the weir overflow rate in gpd/ft?

The total ft of weir is not given directly in this problem. However, weir diameter is given (80 ft) and from that information we can determine the length of the weir.

3,380,000 gpd

ft weir:
= (3.14) (80 ft)
= 251 ft

$$\begin{array}{c}\text{Weir Overflow} \\ \text{Rate}\end{array} = \frac{\text{Flow, gpd}}{\text{Weir Length, ft}}$$

$$= \frac{3{,}380{,}000 \text{ gpd}}{251 \text{ ft}}$$

$$= \boxed{13{,}466 \text{ gpd/ft}}$$

---

* For a review of circumference calculations, refer to Chapter 9, Linear Measurement, in *Basic Math Concepts.*.

Example 3: (Weir Overflow Rate)
❑ The flow to a circular clarifier is 2.12 MGD. If the diameter of the weir is 60 ft, what is the weir overflow rate in gpd/ft?

2,120,000 gpd

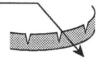

ft weir:
= (3.14) (60 ft)
= 188 ft

$$\text{Weir Overflow Rate} = \frac{\text{Flow, gpd}}{\text{Weir Length, ft}}$$

$$= \frac{2,120,000 \text{ gpd}}{188 \text{ ft}}$$

$$= \boxed{11,277 \text{ gpd/ft}}$$

Example 4: (Weir Overflow Rate)
❑ A rectangular sedimentation basin has a total weir length of 80 ft. If the flow to the basin is 1.3 MGD, what is the weir loading rate in gpm/ft?

$$\frac{1,300,000 \text{ gpd}}{1440 \text{ min/day}}$$

= 903 gpm

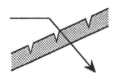

$$\text{Weir Loading Rate} = \frac{\text{Flow, gpm}}{\text{Weir Length, ft}}$$

$$= \frac{903 \text{ gpm}}{80 \text{ ft weir}}$$

$$= \boxed{11.3 \text{ gpm/ft}}$$

WEIR LOADING RATE

Weir overflow rate is a term most often associated with wastewater clarifier calculations. A similar calculation often used for water system clarifiers is weir loading rate, expressed as gpm/ft.

4.7 ORGANIC LOADING RATE CALCULATIONS

When calculating the pounds of BOD entering a wastewater process daily, you are calculating the organic load on the process—the food entering that process. Organic loading for trickling filters is calculated as lbs BOD/day per 1000 cu ft media:

$$\text{Organic Loading Rate} = \frac{\text{BOD, lbs/day}}{\text{Vol., 1000 cu ft}}$$

In most instances BOD will be expressed as a mg/L concentration and must be converted to lbs BOD/day.* Therefore the equation given above can be expanded as:

$$\text{O.L. Rate} = \frac{\text{(mg/L BOD) (MGD) (8.34)}}{\text{flow \quad lbs/gal}}{1000 \text{ cu ft}}$$

Note that the "1000" in the denominator of both equations is a unit of measure ("thousand cu ft") and is **not part of the numerical calculation.**

To determine the number of 1000 cu ft, **find the thousands comma and place a decimal** at that position. In Example 1, 19,233 cu ft is 19.233 units of 1000 cu ft. So 19.233 is placed in the denominator.

Example 1: (Organic Loading Rate)
❑ A trickling filter 70 ft in diameter with a media depth of 5 feet receives a flow of 1,150,000 gpd. If the BOD concentration of the primary effluent is 230 mg/L, what is the organic loading on the trickling filter?

BOD, lbs/day

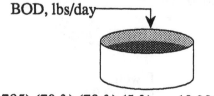

(0.785) (70 ft) (70 ft) (5 ft) = 19,233 cu ft

$$\text{Organic Loading Rate} = \frac{\text{BOD, lbs/day}}{\text{Volume, 1000 cu ft}}$$

$$= \frac{\text{(230 mg/L) (1.15 MGD) (8.34 lbs/gal)}}{19.233 \quad 1000\text{-cu ft}}$$

$$= \boxed{\frac{115 \text{ lbs BOD/day}}{1000 \text{ cu ft}}}$$

Example 2: (Organic Loading Rate)
❑ A100-ft diameter trickling filter with a media depth of 4 ft receives a primary effluent flow of 1.65 MGD with a BOD of 105 mg/L. What is the organic loading on the trickling filter?

BOD, lbs/day

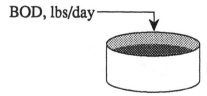

(0.785) (100 ft) (100 ft) (4 ft) = 31,400 cu ft

$$\text{Organic Loading Rate} = \frac{\text{BOD, lbs/day}}{1000 \text{ cu ft}}$$

$$= \frac{\text{(105 mg/L) (1.65 MGD) (8.34 lbs/gal)}}{31.4 \quad 1000\text{-cu ft}}$$

$$= \boxed{\frac{46 \text{ lbs BOD/day}}{1000 \text{ cu ft}}}$$

* For a review of milligrams per liter (mg/L) to pounds per day (lbs/day) BOD, refer to Chapter 3.

Example 3: (Organic Loading Rate)
❑ The flow to a 3-acre wastewater pond is 90,000 gpd. The influent BOD concentration is 125 mg/L. What is the organic loading to the pond?

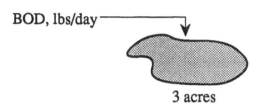

BOD, lbs/day

3 acres

$$\text{Organic Loading Rate} = \frac{\text{BOD, lbs/day}}{\text{Area, ac}}$$

$$= \frac{(125 \text{ mg/L}) (0.09 \text{ MGD}) (8.34 \text{ lbs/gal})}{3 \text{ ac}}$$

$$= \boxed{\frac{31.3 \text{ BOD/day}}{\text{ac}}}$$

ORGANIC LOADING FOR PONDS

Organic loading to ponds is generally calculated as lbs BOD/day per acre of pond surface area.

$$\boxed{\text{Organic Loading Rate} = \frac{\text{BOD, lbs/day}}{\text{Area, ac}}}$$

Example 4: (Organic Loading Rate)
❑ A rotating biological contactor (RBC) receives a flow of 3.6 MGD. If the soluble BOD of the influent wastewater to the RBC is 122 mg/L, and the surface area of the media is 600,000 sq ft, what is the organic loading rate?

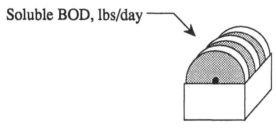

Soluble BOD, lbs/day

Media = 600,000 sq ft

$$\text{Organic Loading Rate} = \frac{\text{Sol. BOD, lbs /day}}{\text{Area, 1000 sq ft}}$$

$$= \frac{(122 \text{ mg/L}) (3.6 \text{ MGD}) (8.34 \text{ lbs/gal})}{600 \text{ 1000-sq ft}}$$

$$= \boxed{\frac{6.1 \text{ lbs/day Sol. BOD}}{1000 \text{ sq ft}}}$$

ORGANIC LOADING FOR ROTATING BIOLOGICAL CONTACTORS

There are two different aspects to calculating organic loading on rotating biological contactors:

1. **Soluble BOD** is used to measure organic content rather than total BOD, and

2. The calculation of organic loading is **per 1000 sq ft media** rather than 1000 cu ft media as with trickling filters.

4.8 FOOD/MICROORGANISM RATIO CALCULATIONS

In order for the activated sludge process to operate properly, there must be a balance between food entering the system (as measured by BOD or COD) and micro-organisms in the aeration tank. The best F/M ratio for a particular system depends on the type of activated sludge process and the characteristics of the wastewater entering the system.

COD is sometimes used as the measure of food entering the system.* Since the COD test can be completed in only a few hours, compared with 5 days for a BOD test, the COD more accurately reflects the current food loading on the system.

Note that the F/M equation is given in two forms: the simplified equation and the expanded equation. If BOD and MLVSS data is given as lbs/day and lbs, then the simplified equation should be used. In most instances, however, BOD and MLVSS data will be given as mg/L, and a calculation of mg/L to lbs/day or lbs will be required as shown in the expanded equation.

FOOD SUPPLY (BOD OR COD) AND
MICROORGANISMS (MLVSS)
MUST BE IN BALANCE

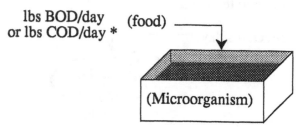

lbs BOD/day
or lbs COD/day *   (food)

(Microorganism)

lbs MLVSS in Aerator

Simplified Equation:

$$F/M = \frac{BOD,\ lbs/day}{MLVSS,\ lbs}$$

Expanded Equation:

$$F/M = \frac{(mg/L\ BOD)\ (MGD\ Flow)\ (8.34\ lbs/gal)}{(mg/L\ MLVSS)\ (Aer\ Vol,\ MG)\ (8.34\ lbs/gal)}$$

Example 1: (F/M Ratio)
❑ An activated sludge aeration tank receives a primary effluent flow of 2,100,000 gpd with a BOD concentration of 158 mg/L. The mixed liquor volatile suspended solids is 1840 mg/L and the aeration tank volume is 300,000 gallons. What is the current F/M ratio?

Since BOD and MLVSS data is given in mg/L, the expanded equation will be needed.**

$$F/M = \frac{(mg/L\ BOD)\ (MGD)\ (8.34\ lbs/gal)}{(mg/L\ MLVSS)\ (Aer\ Vol,\ MG)\ (8.34\ lbs/gal)}$$

$$= \frac{(158\ mg/L\ )\ (2.1\ MGD)\ (8.34\ lbs/gal)}{(1840\ mg/L\ )\ (0.3\ MG)\ (8.34\ lbs/gal)}$$

$$= \boxed{0.6}$$

---

* COD may be used as the measure of food if there is generally a good correlation in BOD and COD characteristics of the wastewater. If not, the COD will not accurately reflect the microbiological content of the aeration tank.
** It is sometimes desirable to calculate the value in the numerator and denominator <u>before</u> calculating the final answer (see Example 2) , since lbs BOD/day or lbs COD/day and lbs MLVSS may be required for other calculations.

Example 2: (F/M Ratio)
❏ The volume of an aeration tank is 200,000 gallons. The aeration tank receives a primary effluent flow of 2,320,000 gpd, with a COD concentration of 100 mg/L. If the mixed liquor volatile suspended solids is 1900 mg/L, what is the current F/M ratio?

Since COD and MLVSS information is given as mg/L, the expanded form of the equation is used in the calculation:

$$F/M = \frac{(mg/L\ COD)\ (MGD\ flow)\ (8.34\ lbs/gal)}{(lbs\ MLVSS)\ (Aer\ Vol,\ MG)\ (8.34\ lbs/gal)}$$

$$= \frac{(100\ mg/L)\ (2.32\ MGD)\ (8.34\ lbs/gal)}{(1900\ mg/L)\ (0.2\ MGD)\ (8.34\ lbs/gal)}$$

$$= \frac{(1935\ lbs\ COD/day)}{(3169\ lbs\ MLVSS)}$$

$$= \boxed{0.6}$$

CALCULATING MLVSS USING THE F/M RATIO

The F/M ratio calculation can be used to calculate the desired pounds of MLVSS to be maintained in the aerator. Use the same F/M equation, fill in the given information, then solve for the unknown value (MLVSS).*

Example 3: (F/M Ratio)
❏ The desired F/M ratio at a particular activated sludge plant is 0.5 lbs BOD/1 lb mixed liquor volatile suspended solids. If the 3 MGD primary effluent flow has a BOD of 165 mg/L how many lbs of MLVSS should be maintained in the aeration tank?

$$F/M = \frac{BOD,\ lbs/day}{MLVSS,\ lbs}$$

Fill in the equation with the information known. Since BOD is given as mg/L, the lbs/day BOD must be written in expanded form.

$$0.5 = \frac{(165\ mg/L)\ (3\ MGD)\ (8.34\ lbs/gal)}{x\ lbs\ MLVSS}$$

Then solve for the unknown value:*

$$x = \frac{(165)\ (3)\ (8.34)}{0.5}$$

$$= \boxed{8257\ lbs\ MLVSS}$$

---

* To review solving for the unknown value, refer to Chapter 2 in *Basic Math Concepts*.

4.9 SOLIDS LOADING RATE CALCULATIONS

The solids loading rate indicates the lbs/day solids loaded to each square foot of clarifier surface area. This calculation is used to determine solids loading on activated sludge secondary clarifiers and gravity sludge thickeners. The general solids loading rate equation is:

$$\text{Solids Loading Rate} = \frac{\text{Solids Applied, lbs/day}}{\text{Surface Area, sq ft}}$$

In expanded form, the equation includes the mg/L to lbs/day calculation in the numerator* and surface area calculation in the denominator:**

$$\text{S.L.R.} = \frac{(\underset{mg/L}{\text{MLSS}})(\underset{\text{flow}}{\text{MGD}})(8.34)}{(0.785)(D^2)}$$

The vast majority of solids coming into the secondary clarifier comes in as mixed liquor suspended solids (MLSS) from the aeration tank. A negligible amount of suspended solids enter the clarifier by the primary effluent flow. (Remember, up to 70% of the suspended solids are removed by the primary system.)

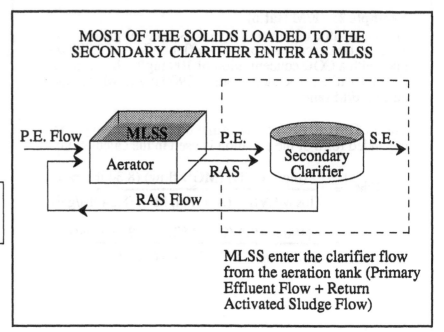

MOST OF THE SOLIDS LOADED TO THE SECONDARY CLARIFIER ENTER AS MLSS

MLSS enter the clarifier flow from the aeration tank (Primary Effluent Flow + Return Activated Sludge Flow)

Example 1: (Solids Loading Rate)
❑ A secondary clarifier is 90 ft in diameter and receives a combined primary effluent (P.E.) and return activated sludge (RAS) flow of 4.9 MGD. If the MLSS concentration in the aeration tank is 3100 mg/L, what is the solids loading rate on the secondary clarifier in lbs/day/sq ft?

Solids, lbs/day

sq ft Area

$$\text{Solids Loading Rate, lbs/day/sq ft} = \frac{(\text{MLSS mg/}L)(\text{MGD Flow})(8.34 \text{ lbs/gal})}{(0.785)(D^2)}$$

$$= \frac{(3100 \text{ mg/}L)(4.9 \text{ MGD})(8.34 \text{ lbs/gal})}{(0.785)(90 \text{ ft})(90 \text{ ft})}$$

$$= \boxed{\frac{19.9 \text{ lbs solids/day}}{\text{sq ft}}}$$

---

* For a review of mg/L to lbs/day calculations refer to Chapter 3.

** Secondary clarifiers are typically circular. For a rectangular clarifier, the surface area would be (*l*) (*w*). Area calculations are discussed in Chapter 10 of *Basic Math Concepts*.

Example 2: (Solids Loading Rate)
❏ A secondary clarifier 80 ft in diameter receives a primary effluent flow of 3.15 MGD and a return sludge flow of 0.8 MGD. If the MLSS concentration is 3650 mg/L, what is the solids loading rate on the clarifier in lbs/day/sq ft?

Solids, lbs/day

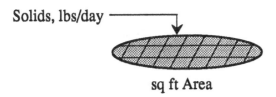

sq ft Area

$$\text{Solids Loading Rate, lbs/day/sq ft} = \frac{\text{Solids, lbs/day}}{\text{Area, sq ft}}$$

$$= \frac{(3650 \text{ mg/}L)\ (3.95 \text{ MGD})\ (8.34 \text{ lbs/gal})}{(0.785)\ (80 \text{ ft})\ (80 \text{ ft})}$$

$$= \boxed{\frac{23.9 \text{ lbs solids/day}}{\text{sq ft}}}$$

Example 3: (Solids Loading Rate)
❏ The total flow to a 70-ft diameter clarifier is 4,600,000 gpd (P.E. + RAS flows). If the MLSS concentration is 2500 mg/L, what is the solids loading rate on the clarifier in lbs/day/sq ft?

Solids, lbs/day

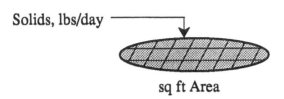

sq ft Area

$$\text{Solids Loading Rate, lbs/day/sq ft} = \frac{\text{Solids, lbs/day}}{\text{Area, sq ft}}$$

$$= \frac{(2500 \text{ mg/}L)\ (4.6 \text{ MGD})\ (8.34 \text{ lbs/gal})}{(0.785)\ (70 \text{ ft})\ (70 \text{ ft})}$$

$$= \boxed{\frac{24.9 \text{ lbs solids/day}}{\text{sq ft}}}$$

4.10 DIGESTER LOADING RATE CALCULATIONS

Sludge is sent to a digester in order to break down or stabilize the organic portion of the sludge. Therefore, it is the organic part of the sludge (the volatile solids portion) that is of interest when calculating solids loading on a digester.

Digester loading rate is a measure of the pounds of volatile solids* entering each cubic foot of digester volume daily, as illustrated in the diagram to the right.

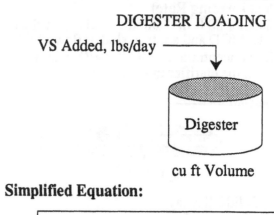

DIGESTER LOADING

VS Added, lbs/day

Digester

cu ft Volume

Simplified Equation:

$$\text{Digester Loading, lbs VS/day/cu ft} = \frac{\text{VS Added, lbs/day}}{\text{Volume, cu ft}}$$

Expanded Equation:

$$\text{Digester Loading, lbs VS/day/cu ft} = \frac{(\text{Sludge, lbs/day})\frac{(\%\ \text{Solids})}{100}\frac{(\%\ \text{VS})}{100}}{(0.785)(D^2)(\text{Water Depth, ft})}$$

Example 1: (Digester Loading)

❑ A digester 40 ft in diameter with an operating depth of 20 ft receives 92,700 lbs/day raw sludge. If the sludge contains 6% solids with 69% volatile matter, what is the digester loading in lbs VS added/day/cu ft volume?

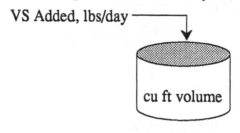

VS Added, lbs/day

cu ft volume

$$\text{Digester Loading, lbs VS/day/cu ft} = \frac{(\text{Sludge, lbs/day})\frac{(\%\ \text{Solids})}{100}\frac{(\%\ \text{VS})}{100}}{(0.785)(D^2)(\text{Water Depth, ft})}$$

$$= \frac{(92,700\ \text{lbs/day})(0.06)(0.69)}{(0.785)(40\ \text{ft})(40\ \text{ft})(20\ \text{ft})}$$

$$= \boxed{0.15\ \frac{\text{lbs/day VS}}{\text{cu ft}}}$$

* For a review of calculating percent solids and percent volatile solids, refer to Chapter 6.

Example 2: (Digester Loading)
❏ A digester 40 ft in diameter operating at a depth of 18 ft receives 180,000 lbs/day sludge with 5% total solids and 72% volatile solids. What is the digester loading in lbs VS/day/cu ft?

$$\text{Digester Loading, lbs VS/day/cu ft} = \frac{(\text{Sludge, lbs/day}) \dfrac{(\% \text{ Solids})}{100} \dfrac{(\% \text{ VS})}{100}}{(0.785) (D^2) (\text{Water Depth, ft})}$$

$$= \frac{(180,000 \text{ lbs/day}) (0.05) (0.72)}{(0.785) (40 \text{ ft}) (40 \text{ ft}) (18 \text{ ft})}$$

$$= \boxed{0.29 \ \frac{\text{lbs VS/day}}{\text{cu ft}}}$$

Example 3: (Digester Loading)
❏ A digester 50 ft in diameter operating at a depth of 20 ft receives 34,300 gpd sludge with 5.5% solids and 70% volatile solids. What is the digester loading in lbs VS/day/cu ft? (Assume the sludge weighs 8.34 lbs/gal.)

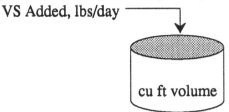

VS Added, lbs/day

cu ft volume

$$\text{Digester Loading, lbs VS/day/cu ft} = \frac{(\text{Sludge, gpd}) (8.34 \text{ lbs/gal}) \dfrac{(\% \text{ Sol.})}{100} \dfrac{(\% \text{ VS})}{100}}{(0.785) (D^2) (\text{Water Depth, ft})}$$

$$= \frac{(34,300 \text{ gpd}) (8.34 \text{ lbs/gal}) (0.055) (0.70)}{(0.785) (50 \text{ ft}) (50 \text{ ft}) (20 \text{ ft})}$$

$$= \boxed{0.28 \ \frac{\text{lbs VS/day}}{\text{cu ft}}}$$

GIVEN GPD OR GPM SLUDGE PUMPED TO DIGESTER

In Examples 1 and 2, the sludge pumped to the digester was expressed as lbs/day. Many times, however, sludge pumped to the digester is expressed as gpd or gpm. When this is the case, convert the gpd or gpm pumping rate to lbs/day* and continue as in Examples 1 and 2. You can make the gpd to lbs/day conversion a separate calculation, or you can incorporate it into the numerator of the equation as shown in Example 3.

---

* To review flow conversions, refer to Chapter 8 in *Basic Math Concepts.*.

4.11 DIGESTER VOLATILE SOLIDS LOADING RATIO CALCULATIONS

One way of expressing digester loading was described in the previous section—lbs/day volatile solids added per cu ft digester volume.

Another way to express digester loading is lbs/day volatile solids* added per lb of volatile solids under digestion (in the digester).

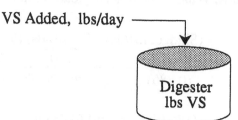

VOLATILE SOLIDS LOADING RATIO COMPARES VS ADDED WITH VS IN THE DIGESTER

VS Added, lbs/day ⟶

Digester
lbs VS

Simplified Equation:

$$\text{VS Loading Ratio} = \frac{\text{VS Added, lbs/day}}{\text{VS in Digester, lbs}}$$

Expanded Equation:**

$$\text{VS Loading Ratio} = \frac{(\text{Sludge Added, lbs/day})\left(\frac{\% \text{ Sol}}{100}\right)\left(\frac{\% \text{ VS}}{100}\right)}{(\text{Sludge in Dig., lbs})\left(\frac{\% \text{ Sol}}{100}\right)\left(\frac{\% \text{ VS}}{100}\right)}$$

Example 1: (Volatile Solids Loading Ratio)
❏ A total of 52,500 lbs/day sludge is pumped to a 100,000-gallon digester. The sludge being pumped to the digester has total solids content of 5% and volatile solids content of 74%. What is the volatile solids loading on the digester in lbs VS added/day/lb VS in digester?

$$\text{VS Loading Ratio} = \frac{\text{VS Added, lbs/day}}{\text{VS in Digester, lbs}}$$

$$= \frac{(52,500 \text{ lbs/day})\left(\frac{5}{100}\right)\left(\frac{74}{100}\right)}{\underbrace{(100,000 \text{ gal})(8.34 \text{ lbs/gal})}\left(\frac{6}{100}\right)\left(\frac{58}{100}\right)}$$

This is lbs digester sludge

$$= \boxed{\frac{0.067 \text{ lbs/day VS Added}}{\text{lb VS in Digester}}}$$

---

* For a review of calculating percent solids and percent volatile solids, refer to Chapter 6.
** The two hundreds in the numerator and denominator cancel each other out and may be omitted, if desired.

Example 2: (Volatile Solids Loading Ratio)
❑ A total of 22,850 gpd sludge is pumped to an 80,000-gal digester. This sludge has a solids content of 6% and a volatile solids concentration of 72%. The sludge in the digester has a solids content of 5.1% with a 56% volatile solids content. What is the volatile solids loading on the digester in lbs VS added/day/lb VS in digester?

$$\text{VS Loading Ratio} = \frac{\text{VS Added, lbs/day}}{\text{VS in Digester, lbs}}$$

$$= \frac{(22{,}850 \text{ lbs/day})\left(\dfrac{6}{100}\right)\left(\dfrac{72}{100}\right)}{\underbrace{(80{,}000 \text{ gal})(8.34 \text{ lbs/gal})}\left(\dfrac{5.1}{100}\right)\left(\dfrac{56}{100}\right)}$$

This is lbs digester sludge

$$= \boxed{\frac{0.052 \text{ lbs/day VS Added}}{\text{lb VS in Digester}}}$$

Example 3: (Volatile Solids Loading Ratio)
❑ A total of 63,000 gpd sludge is pumped to the digester. The sludge has 4% solids with a volatile solids content of 74%. If the desired VS loading ratio is 0.08 lbs VS added/lb VS in digester, how many lbs VS should be in the digester for this volatile solids load?

$$\text{VS Loading Ratio} = \frac{\text{VS Added, lbs/day}}{\text{VS in Digester, lbs}}$$

This is lbs/day sludge added

$$0.08 = \frac{\overbrace{(63{,}000 \text{ gpd})(8.34 \text{ lbs/day})\left(\dfrac{4}{100}\right)\left(\dfrac{74}{100}\right)}}{x \text{ lbs VS in Digester}}$$

Then solve for x:

$$x = \frac{(63{,}000)(8.34)(0.04)(0.74)}{0.08}$$

$$x = \boxed{\begin{array}{c}194{,}405 \text{ lbs VS} \\ \text{in Digester}\end{array}}$$

CALCULATING OTHER UNKNOWN VALUES

Volatile solids loading ratio calculations have three variables: VS loading ratio, lbs VS added/day, and lbs VS in the digester.

Given a **desired** volatile solids loading ratio, you can calculate the desired lbs of volatile solids in the digester. This type of calculation is used for determining seed sludge requirements for startup of a digester. Example 3 illustrates this calculation.

4.12 POPULATION LOADING AND POPULATION EQUIVALENT

POPULATION LOADING

Population loading is a calculation associated with wastewater treatment by ponds. Population loading is an indirect measure of both water and solids loading to a system. It is calculated as the number of persons served per acre of pond:

$$\text{Population Loading} = \frac{\text{persons}}{\text{acre}}$$

Example 1: (Population Loading)
❑ A 3.5 acre wastewater pond serves a population of 1500. What is the population loading on the pond?

$$\text{Population Loading} = \frac{\text{persons}}{\text{acre}}$$

$$= \frac{1500 \text{ persons}}{3.5 \text{ acres}}$$

$$= \frac{429 \text{ persons}}{\text{acre}}$$

Example 2: (Population Loading)
❑ A wastewater pond serves a population of 4000. If the pond is 16 acres, what is the population loading on the pond?

$$\text{Population Loading} = \frac{\text{persons}}{\text{acre}}$$

$$= \frac{4000 \text{ persons}}{16 \text{ acres}}$$

$$= \frac{250 \text{ persons}}{\text{acre}}$$

Example 3: (Population Equivalent)
❏ A 0.4-MGD wastewater flow has a BOD concentration of 1800 mg/*L* BOD. Using an average of 0.2 lbs/day BOD/person, what is the population equivalent of this wastewater flow?

$$\text{Population Equivalent} = \frac{\text{BOD, lbs/day}}{\text{lbs BOD/day/person}}$$

Convert mg/*L* BOD to lbs/day BOD* then divide by 0.2 lbs BOD/day/person:

$$\text{Population Equivalent} = \frac{(1800 \text{ mg/}L)(0.4 \text{ MGD})(8.34 \text{ lbs/gal})}{0.2 \text{ lbs BOD/day/person}}$$

$$= \boxed{30,024 \text{ people}}$$

POPULATION EQUIVALENT

Industrial or commercial wastewater generally has a higher organic content than domestic wastewater. Population equivalent calculations equate these concentrated flows with the number of people that would produce a domestic wastewater of that strength. For a domestic wastewater system, each person served by the system contributes about 0.17 or 0.2 lbs BOD/day . To determine the population equivalent of a wastewater flow, divide the lbs BOD/day content by the lbs BOD/day contributed per person (e.g. 0.2 lbs BOD/day).

Example 4: (Population Equivalent)
❏ A 100,000 gpd wastewater flow has a BOD content of 2800 mg/*L*. Using an average of 0.2 lbs/day BOD/person, what is the population equivalent of this flow?

$$\text{Population Equivalent} = \frac{\text{BOD, lbs/day}}{\text{lbs BOD/day/person}}$$

$$= \frac{(2800 \text{ mg/}L)(0.1 \text{ MGD})(8.34 \text{ lbs/gal})}{0.2 \text{ lbs BOD/day/person}}$$

$$= \boxed{11,676 \text{ people}}$$

---

* For a review of mg/*L* to lbs/day calculations, refer to Chapter 3.

NOTES:

5 *Detention and Retention Times Calculations*

SUMMARY

1. **Detention time** indicates the amount of time a given flow of water is retained by a unit process. It is calculated as the tank volume divided by the flow rate:

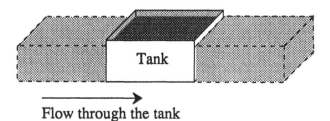

Flow through the tank

$$\text{Detention Time} = \frac{\text{Volume of Tank, gal}}{\text{Flow, gal/time}}$$

2. **Sludge age** is a measure of the average time a suspended solids particle remains under aeration. Sludge age (sometimes called Gould sludge age) is based on the **pounds of solids added daily** to the activated sludge process.

lbs/day SS Added

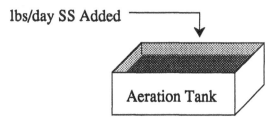

lbs MLSS in Aeration Tank

Simplified Equation:

$$\frac{\text{Sludge Age}}{\text{days}} = \frac{\text{MLSS, lbs}}{\text{SS Added, lbs/day}}$$

Expanded Equation:

$$\frac{\text{Sludge Age}}{\text{days}} = \frac{(\text{MLSS mg}/L)\,(\text{Aer. Vol., MG})\,(8.34\,\text{lbs/gal})}{(\text{P. E. SS, mg}/L)\,(\text{Flow, MGD})\,(8.34\,\text{lbs/gal})}$$

In the previous chapter we focused on calculations that measure the water and solids loading on a system. Now we will examine calculations that measure **how long the water and solids are retained in the system**. Three calculations will be discussed:

- Detention Time or Fill Time

- Sludge Age

- Solids Retention Time (SRT) (also called Mean Cell Residence Time, MCRT)

SUMMARY—Cont'd

3. **Solids Retention Time, SRT** (also called Mean Cell Residence Time, MCRT) is another measure of the length of time a suspended solid particle remains under aeration. However, the SRT calculation is **based on the pounds of solids leaving the activated sludge process** rather than the pounds of solids added, as with sludge age.

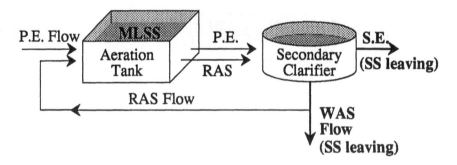

Simplified Equation:

$$\text{Solids Retention Time} = \frac{\text{Suspended Solids in System, lbs**}}{\text{Suspended Solids Leaving System, lbs/day}}$$

Or

$$\text{Solids Retention Time} = \frac{\text{Suspended Solids in System lbs**}}{\text{WAS SS, lbs/day} + \text{S.E.* SS, lbs/day}}$$

Expanded Equation:

$$\text{SRT, days} = \frac{(\text{MLSS mg}/L)\,(\text{Aer. Vol.} + \text{Fin. Clar. Vol., MG})\,(8.34\ \text{lbs/gal})}{\underset{\text{mg}/L \quad\quad \text{MGD} \quad\quad \text{lbs/gal} \quad\quad \text{mg}/L \quad\quad \text{MGD} \quad\quad \text{lbs/gal}}{(\text{WAS SS})\,(\text{WAS Flow})\,(8.34) + (\text{S.E. SS})\,(\text{Plant Flow})\,(8.34)}}$$

---

* S.E. is an abbreviation for Secondary Effluent. P.E. refers to Primary Effluent.

** There are four ways to account for system solids in the SRT calculation (numerator). One commonly used calculation of system solids is given in the SRT equation above. The other three methods are discussed in Chapter 12.

NOTES:

5.1 DETENTION TIME CALCULATIONS

There are two basic ways to consider detention time:

1. Detention time is the length of time required for a given flow rate to pass through a tank.

2. Detention time may also be considered as the length of time required to fill a tank at a given flow rate.

In each case, the calculation of detention time is the same:

$$\text{Detention Time} = \frac{\text{Volume of Tank, gal}}{\text{Flow Rate, gal/time}}$$

THE TWO FACES OF DETENTION TIME

Flow-Through Time:

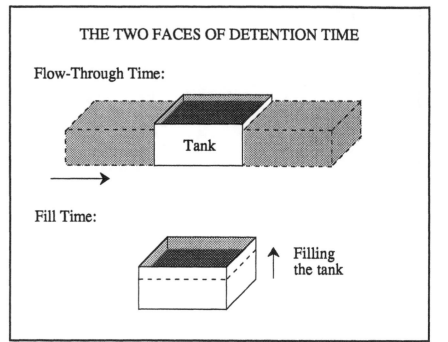

Fill Time:

MATCHING UNITS

There are many possible ways of writing the detention time equation, depending on the time unit desired (seconds, minutes, hours, days) and the expression of volume and flow rate.

When calculating detention time, it is essential that the time and volume units used in the equation are consistent with each other, as illustrated to the right.

BE SURE THE TIME AND VOLUME UNITS MATCH

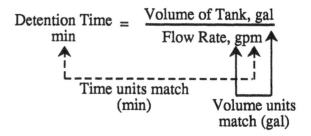

Other examples of detention time equations where time and volume units match include:

$$\text{Detention Time, sec} = \frac{\text{Volume of Tank, cu ft}}{\text{Flow Rate, cfs}}$$

$$\text{Detention Time, hrs} = \frac{\text{Volume of Tank, gal}}{\text{Flow Rate, gph}}$$

$$\text{Detention Time, days} = \frac{\text{Volume of Pond, ac-ft}}{\text{Flow Rate, ac-ft/day}}$$

Example 1: (Detention Time)
❏ The flow to a sedimentation tank 80 ft long, 30 ft wide and 10 ft deep is 3.7 MGD. What is the detention time in the tank in hours?

(80 ft) (30 ft) (10 ft) (7.48 gal/cu ft) = 179,520 gal
Volume

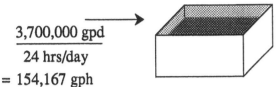

$$\frac{3,700,000 \text{ gpd}}{24 \text{ hrs/day}}$$

= 154,167 gph

First, write the equation so that volume and time units match. Then fill in the equation and solve for the unknown.

$$\frac{\text{Detention Time}}{\text{hrs}} = \frac{\text{Volume of Tank, gal}}{\text{Flow Rate, gph}}$$

$$= \frac{179,520 \text{ gal Volume}}{154,167 \text{ gph}}$$

$$= \boxed{1.2 \text{ hours}}$$

Example 2: (Detention Time)
❏ A flocculation basin is 8 ft deep, 15 ft wide, and 40 ft long. If the flow through the basin is 2.2 MGD, what is the detention time in minutes?

(40 ft) (15 ft) (8 ft) (7.48 gal/cu ft) = 35,904 gal
Volume

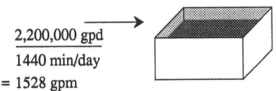

$$\frac{2,200,000 \text{ gpd}}{1440 \text{ min/day}}$$

= 1528 gpm

$$\frac{\text{Detention Time}}{\text{min}} = \frac{\text{Volume of Tank, gal}}{\text{Flow Rate, gpm}}$$

$$= \frac{35,904 \text{ gal Volume}}{1528 \text{ gpm}}$$

$$= \boxed{23 \text{ minutes}}$$

DETENTION TIME AS FLOW THROUGH A TANK

In calculating unit process detention times, you are calculating the length of time it takes the water to flow through that unit process. Detention times are normally calculated for the following basins or tanks:

- Flash mix chambers (sec)

- Flocculation basins (min)

- Sedimentation tanks or clarifiers (hrs),

- Wastewater ponds (days),

- Oxidation ditches (hrs).

There are two key points to remember when calculating detention time:

1. **Tank volume is the numerator (top) of the fraction** and flow rate is the denominator (bottom) of the fraction. Many times students have a difficult time remembering which term belongs in the numerator and which in the denominator. As a memory aid, remember that "V," the <u>victor</u>, <u>is always on top</u>.

2. **Time and volume units must match.** If detention time is desired in minutes, then the flow rate used in the calculation should have the same time frame (cfm or gpm, depending on whether tank volume is expressed as cubic feet or gallons). If detention time is desired in hours, then the flow rate used in the calculation should be cfh or gph.

DETENTION TIME FOR PONDS

Detention time for a pond may be calculated using one of two equations, depending on how the flow rate is expressed:

$$\text{Detention Time, days} = \frac{\text{Pond Volume, gal}}{\text{Flow Rate, gpd}}$$

Or

$$\text{Detention Time, days} = \frac{\text{Pond Volume, ac-ft}}{\text{Flow Rate, ac-ft/day}}$$

For a better understanding of the relative sizes of MGD and ac-ft/day, remember that 1 MGD is equivalent to about 3 ac-ft/day flow.

Examples 3 and 4 illustrate the use of both detention time equations.

Example 3: (Detention Time)
❏ A waste treatment pond is operated at a depth of 5 feet. The average width of the pond is 375 ft and the average length is 610 ft. If the flow to the pond is 570,000 gpd, what is the detention time in days?

(610 ft) (375 ft) (5 ft) (7.48 gal/cu ft) = 8,555,250 gal Volume

570,000 gpd ⟶

$$\frac{\text{Detention Time}}{\text{days}} = \frac{\text{Volume of Pond, gal}}{\text{Flow Rate, gpd}}$$

$$= \frac{8,555,250 \text{ gal Volume}}{570,000 \text{ gpd}}$$

$$= \boxed{15 \text{ days}}$$

Example 4: (Detention Time)
❏ A waste treatment pond is operated at a depth of 6 feet. The volume of the pond is 54 ac-ft. If the flow to the pond is 2.7 ac-ft/day, what is the detention time in days?

2.7 ac-ft/day ⟶

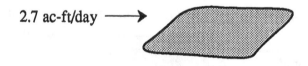

$$\frac{\text{Detention Time}}{\text{days}} = \frac{\text{Volume of Pond, ac-ft}}{\text{Flow Rate, ac-ft/day}}$$

$$= \frac{54 \text{ ac-ft Volume}}{2.7 \text{ ac-ft/day}}$$

$$= \boxed{20 \text{ days}}$$

Example 5: (Detention Time)
❑ A basin 4 ft square is to be filled to the 3 ft level. If the flow to the tank is 3 gpm, how long will it take to fill the tank (in hours)?

(4 ft) (4 ft) (3 ft) (7.48 gal/cu ft) = 359 gal Volume

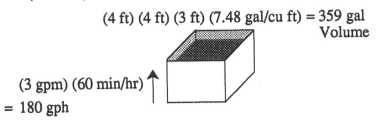

(3 gpm) (60 min/hr) ↑
= 180 gph

$$\text{Fill Time hrs} = \frac{\text{Volume of Tank, gal}}{\text{Flow Rate, gph}}$$

$$= \frac{359 \text{ gal Volume}}{180 \text{ gph}}$$

$$= \boxed{2 \text{ hrs}}$$

DETENTION TIME AS FILL TIME

Another way to think of detention time is the time required to fill a tank or basin at a given flow rate. Regardless of whether you consider detention time flow time through a tank, or fill time, the calculation is precisely the same:

$$\frac{\text{Detention}}{\text{Time}} = \frac{\text{Volume of Tank, gal}}{\text{Flow Rate, gal/time}}$$

In some equations, the word *fill time* is used rather than detention time.

$$\frac{\text{Fill}}{\text{Time}} = \frac{\text{Volume of Tank, gal}}{\text{Flow Rate, gal/time}}$$

In each equation listed above the volume can be given as cubic feet, if desired (cu ft and cu ft/time).

The fill time calculation can also be used to determine the **time remaining before a tank overflows,** as illustrated in Example 6. Such a calculation can be critical during equipment failure conditions.

Example 6: (Detention Time)
❑ A tank has a diameter of 5 ft with an overflow depth at 4 ft. The current water level is 2.8 ft. Water is flowing into the tank at a rate of 4.1 gpm. At this rate, how long will it take before the tank overflows (in min)?

The volume of the tank <u>remaining to be filled</u> is 5 ft in diameter and 1.2 ft deep (4 ft − 2.8 ft = 1.2 ft). Therefore, the fill volume is:

(0.785) (5 ft) (5 ft) (1.2 ft) (7.48 gal/cu ft) = 176 gal Vol.

|← 5 ft →|

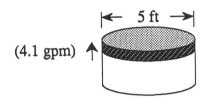

(4.1 gpm) ↑

$$\frac{\text{Time Until Overflow, min}}{} = \frac{\text{Volume of Tank, gal}}{\text{Flow Rate, gpm}}$$

$$= \frac{176 \text{ gal Volume}}{4.1 \text{ gpm}}$$

$$= \boxed{43 \text{ min until overflow}}$$

5.2 SLUDGE AGE CALCULATIONS

Sludge age refers to the average number of days a particle of suspended solids remains under aeration. It is a calculation used to maintain the proper amount of activated sludge in the aeration tank.

When considering sludge age, in effect you are asking, "how many days of suspended solids are in the aeration tank?" If you know how many pounds of suspended solids enter the aeration tank daily and you can determine how many total pounds of suspended solids are in the aeration tank, then you can calculate how many days of solids are in the aeration tank. For example, if 2000 lbs/day SS enter the aeration tank daily and the aeration tank contains 10,000 lbs of suspended solids, then 5 days of solids are in the aeration tank—a sludge age of 5 days.

Notice the similarity of this calculation with that of detention time—sludge age is **solids retained** calculated using units of lbs and lbs/day; detention time is **water retained**, using units of gal and gal/time or cu ft and cu ft/time:

$$\text{Sludge Age} \atop \text{days} = \frac{\text{SS in Tank, lbs}}{\text{SS Added, lbs/day}}$$

$$\text{Detention} \atop \text{Time, min} = \frac{\text{Volume of Tank, gal}}{\text{Flow Rate, gpm}}$$

SLUDGE AGE IS BASED ON SUSPENDED SOLIDS
ENTERING THE AERATION TANK*

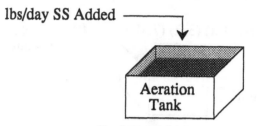

lbs/day SS Added

Aeration Tank

lbs MLSS in Aeration Tank

Simplified Equation:

$$\text{Sludge Age,} \atop \text{days} = \frac{\text{MLSS, lbs}}{\text{SS Added, lbs/day}}$$

Expanded Equation:

$$\text{Sludge Age,} \atop \text{days} = \frac{(\text{MLSS mg}/L)\,(\text{Aer. Vol., MG})\,(8.34 \text{ lbs/gal})}{(\text{P. E. SS, mg}/L)\,(\text{Flow, MGD})\,(8.34 \text{ lbs/gal})}$$

Example 1: (Sludge Age)
❑ An aeration tank has a total of 13,000 lbs of mixed liquor suspended solids. If a total of 2540 lbs/day suspended solids enter the aeration tank in the primary effluent flow, what is the sludge age in the aeration tank?

2540 lbs/day SS

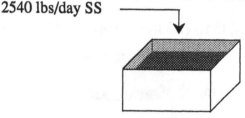

13,000 lbs MLSS

$$\text{Sludge Age} \atop \text{days} = \frac{\text{MLSS, lbs}}{\text{SS Added, lbs/day}}$$

$$= \frac{13,000 \text{ lbs}}{2540 \text{ lbs/day}}$$

$$= \boxed{5.1 \text{ days}}$$

* Sludge age based on solids <u>entering</u> the aeration tank is sometimes referred to as Sludge Age (Gould) to distinguish it from the calculation of Solids Retention Time (or Mean Cell Residence Time), which is based on solids <u>leaving</u> the aeration tank.

Example 2: (Sludge Age)
❏ An aeration tank contains 500,000 gallons of wastewater. The MLSS is 2200 mg/*L*. If the primary effluent flow is 3.7 MGD with a suspended solids concentration of 72 mg/*L*, what is the sludge age?

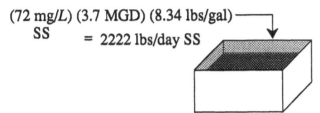

(72 mg/*L*) (3.7 MGD) (8.34 lbs/gal)
SS   = 2222 lbs/day SS

lbs MLSS in Aeration Tank:
(2200 mg/*L*) (0.5 MG) (8.34 lbs/gal) = 9174 lbs
MLSS   MLSS

$$\text{Sludge Age days} = \frac{\text{MLSS, lbs}}{\text{SS Added, lbs/day}}$$

$$= \frac{9174 \text{ lbs MLSS}}{2222 \text{ lbs/day SS}}$$

$$= \boxed{4.1 \text{ days}}$$

Example 3: (Sludge Age)
❏ An aeration tank is 80 ft long, 20 ft wide with wastewater to a depth of 15 ft. The mixed liquor suspended solids concentration is 2800 mg/*L*. If the primary effluent flow is 1.6 MGD with a suspended solids concentration of 65 mg/*L*, what is the sludge age in the aeration tank?

lbs/day SS

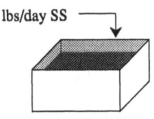

<u>Aeration Tank Volume</u>
(80 ft) (20 ft) (15 ft) (7.48 gal/cu ft) = 179,520 gal

$$\text{Sludge Age days} = \frac{(2800 \text{ mg/}L) (0.18 \text{ MG}) (8.34 \text{ lbs/gal})}{(65 \text{ mg/}L) (1.6 \text{ MGD}) (8.34 \text{ lbs/gal})}$$

$$= \boxed{4.8 \text{ days}}$$

USING SLUDGE AGE TO CALCULATE OTHER UNKNOWNS

The sludge age equation can be used to calculate:

- Desired lbs MLSS, and

- Desired mg/*L* SS

In Examples 1-3, sludge age was the unknown variable. In Examples 4-6, the sludge age equation is used to calculate the desired pounds of MLSS in the aeration tank.

Example 4: (Sludge Age)

❑ A sludge age of 5 days is desired. Assume 1200 lbs/day suspended solids enter the aeration tank in the primary effluent. To maintain the desired sludge age, how many lbs of MLSS must be maintained in the aeration tank?

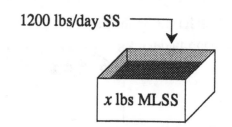

1200 lbs/day SS

x lbs MLSS

$$\frac{\text{Sludge Age}}{\text{days}} = \frac{\text{MLSS, lbs}}{\text{SS Added, lbs/day}}$$

$$5 \text{ days} = \frac{x \text{ lbs MLSS}}{1200 \text{ lbs/day SS Added}}$$

$$(5)(1200) = x$$

$$\boxed{6000 \text{ lbs}} = x$$

Example 5: (Sludge Age)

❑ A sludge age of 5.2 days is desired for an aeration tank 100 ft long, 40 ft wide, with a liquid level of 15 ft. If 1950 lbs/day suspended solids enter the aeration tank in the primary effluent flow, how many lbs of MLSS must be maintained in the aeration tank to maintain the desired sludge age?

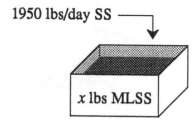

1950 lbs/day SS

x lbs MLSS

$$\frac{\text{Sludge Age}}{\text{days}} = \frac{\text{MLSS, lbs}}{\text{SS Added, lbs/day}}$$

$$5.2 \text{ days} = \frac{x \text{ lbs MLSS}}{1950 \text{ lbs/day SS Added}}$$

$$(5.2)(1950) = x$$

$$\boxed{10{,}140 \text{ lbs}} = x$$

Example 6: (Sludge Age)
❑ The 1.3-MGD primary effluent flow to an aeration tank has a suspended solids concentration of 65 mg/*L*. The aeration tank volume is 180,000 gallons. If a sludge age of 6 days is desired, what is the desired MLSS concentration?

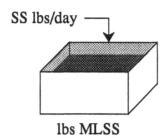

SS lbs/day

lbs MLSS

$$6 \text{ days} = \frac{(x \text{ mg/}L \text{ MLSS})(0.18 \text{ mg})(\cancel{8.34} \text{ lbs/gal})}{(65 \text{ mg/}L)(1.3 \text{ MGD})(\cancel{8.34} \text{ lbs/gal})}$$

After dividing out the 8.34 factor from the numerator and denominator, solve for *x*:

$$x = \frac{(65)(1.3)(6)}{0.18}$$

$$x = \boxed{2817 \text{ mg/}L \text{ MLSS}}$$

Example 7: (Sludge Age)
❑ A total of 5500 lbs of MLSS are desired in the aeration tank. What is the desired MLSS concentration (in mg/*L*) if the aeration tank volume is 250,000 gallons?

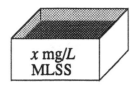

x mg/*L* MLSS

$$(x \text{ mg/}L \text{ MLSS})(0.25 \text{ MG})(8.34) = 5500 \text{ lbs MLSS}$$

$$x = \frac{5500}{(0.25)(8.34)}$$

$$x = \boxed{2638 \text{ mg/}L \text{ MLSS}}$$

CALCULATING MLSS CONCENTRATION GIVEN POUNDS SOLIDS AND AERATION TANK VOLUME

The sludge age equation can be used in some cases, such as in Example 6, to calculate the actual or desired MLSS concentration. However, many times sludge age is not mentioned at all and yet MLSS concentration must be determined.

When converting from mg/*L* to lbs MLSS or vice versa, the following equation is used:*

$$\boxed{(\text{mg/}L)(\text{Vol, MG})(8.34) = \text{lbs} \atop \text{lbs/gal}}$$

Note that this is the same equation used in the sludge age numerator to calculate lbs SS (lbs MLSS) in the aeration tank :

$$\boxed{\underset{\text{MLSS}}{(\text{mg/}L)}\,\underset{\text{MG}}{(\text{Aer.Vol.})}\,\underset{\text{lbs/gal}}{(8.34)} = \underset{\text{MLSS}}{\text{lbs}}}$$

* mg/*L* to lbs/day calculations are discussed in Chapter 3.

5.3 SOLIDS RETENTION TIME CALCULATIONS

Solids Retention Time (SRT), also called Mean Cell Residence Time (MCRT), is a calculation very similar to the sludge age calculation. There are two principal differences in calculating SRT:

1. **The SRT calculation is based on suspended solids leaving the system.** (Sludge age is based on suspended solids entering the system.)

2. **There are four different methods that may be used to calculate lbs MLSS (system solids).**** (In sludge age calculations, only the MLSS concentration and aeration tank volume are used in calculating system solids.)

Examples 1 and 2 illustrate the calculation of SRT, using the combined volume method of estimating system solids.

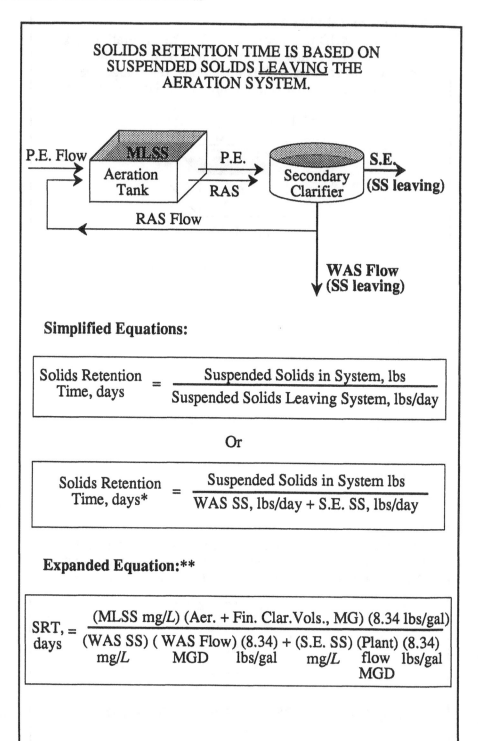

SOLIDS RETENTION TIME IS BASED ON SUSPENDED SOLIDS <u>LEAVING</u> THE AERATION SYSTEM.

Simplified Equations:

$$\text{Solids Retention Time, days} = \frac{\text{Suspended Solids in System, lbs}}{\text{Suspended Solids Leaving System, lbs/day}}$$

Or

$$\text{Solids Retention Time, days*} = \frac{\text{Suspended Solids in System lbs}}{\text{WAS SS, lbs/day} + \text{S.E. SS, lbs/day}}$$

Expanded Equation:**

$$\text{SRT, days} = \frac{(\text{MLSS mg}/L)(\text{Aer.} + \text{Fin. Clar.Vols., MG})(8.34 \text{ lbs/gal})}{\underset{\text{mg}/L \quad\quad \text{MGD} \quad\quad \text{lbs/gal}}{(\text{WAS SS})(\text{WAS Flow})(8.34)} + \underset{\text{mg}/L \quad \text{flow} \quad \text{lbs/gal} \atop \text{MGD}}{(\text{S.E. SS})(\text{Plant})(8.34)}}$$

---

* S.E. is Secondary Effluent.

** There are four ways to account for system solids in the SRT calculation (numerator). The other three methods are discussed in Chapter 12.

Example 1: (Solids Retention Time)
❑ An activated sludge system has a total of 9930 lbs of mixed liquor suspended solids. The suspended solids leaving the final clarifier in the effluent is calculated to be 290 lbs/day. The lbs suspended solids wasted from the final clarifier is 1050 lbs/day. What is the solids retention time, in days?

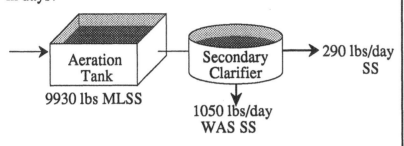

$$\frac{\text{SRT}}{\text{days}} = \frac{\text{MLSS in System, lbs}}{\text{WAS SS, lbs/day} + \text{S.E. SS, lbs/day}}$$

$$= \frac{9930 \text{ lbs MLSS}}{1050 \text{ lbs/day} + 290 \text{ lbs/day}}$$

$$= \boxed{7.4 \text{ days}}$$

Example 2: (Solids Retention Time)
❑ An aeration tank has a volume of 330,000 gal. The final clarifier has a volume of 150,000 gallons. The MLSS concentration in the aeration tank is 2900 mg/L. If a total of 1520 lbs SS/day are wasted and 400 lbs SS/day are in the secondary effluent, what is the solids retention time for the activated sludge system?

$$\frac{\text{SRT}}{\text{days}} = \frac{\text{MLSS in System, lbs}}{\text{WAS SS, lbs/day} + \text{S.E. SS, lbs/day}}$$

$$= \frac{(2900 \text{ mg/}L)(0.48 \text{ MGD})(8.34 \text{ lbs/gal})}{1520 \text{ lbs/day} + 400 \text{ lbs/day}}$$

$$= \frac{11,609 \text{ lbs MLSS}}{1920 \text{ lbs/day SS leaving}}$$

$$= \boxed{6 \text{ days}}$$

Example 3: (Solids Retention Time)
❑ Determine the solids retention time (SRT) given the following data:

Aer. Tank Vol. 1.1 MG MLSS 2500 mg/L
Fin. Clar. Vol. 0.4 MG WAS SS 6150 mg/L
P.E. Flow 3.9 MGD   S.E. SS 15 mg/L
WAS Pumping Rate 80,000 gpd

$$\frac{SRT}{days} = \frac{\text{MLSS in System, lbs}}{\text{SS Leaving System, lbs/day}}$$

$$= \frac{(2500 \text{ mg/}L) (1.5 \text{ MG}) (8.34 \text{ lbs/gal})}{\underset{\text{mg/}L\ \ \ \ \text{MGD}\ \ \text{lbs/gal}}{(6150) (0.08) (8.34)} + \underset{\text{mg/}L\ \ \text{MGD}\ \ \text{lbs/gal}}{(15)\ \ (3.9)\ \ (8.34)}}$$

$$= \frac{31,275 \text{ lbs MLSS}}{4103 \text{ lbs SS} + 488 \text{ lbs SS}}$$

$$= \boxed{6.8 \text{ days}}$$

Example 4: (Solids Retention Time)
❑ Calculate the solids retention time (SRT) given the following data:

Aer. Tank Vol. 300,000 gal MLSS 2400 mg/L
Fin. Clar. Vol. 120,000 gal WAS SS 5900 mg/L
P.E. Flow 2.2 MGD   S.E. SS 20 mg/L
WAS Pumping Rate 19,000 gpd

$$\frac{SRT}{days} = \frac{\text{MLSS in System, lbs}}{\text{SS Leaving System, lbs/day}}$$

$$= \frac{(2400 \text{ mg/}L) (0.42 \text{ MG}) (8.34 \text{ lbs/gal})}{\underset{\text{mg/}L\ \ \ \ \text{MGD}\ \ \text{lbs/gal}}{(5900) (0.019) (8.34)} + \underset{\text{mg/}L\ \ \text{MGD}\ \ \text{lbs/gal}}{(20)\ \ (2.2)\ \ (8.34)}}$$

$$= \frac{8407 \text{ lbs MLSS}}{935 \text{ lbs/day SS} + 367 \text{ lbs/day SS}}$$

$$= \boxed{6.5 \text{ days}}$$

Example 5: (Solids Retention Time)

❏ The volume of an aeration tank is 320,000 gal and the final clarifier is 130,000 gal. The desired SRT for a plant is 7 days. The primary effluent flow is 2 MGD and the WAS pumping rate is 25,000 gpd. If the WAS SS is 5400 mg/L and the secondary effluent SS is 18 mg/L, what is the desired MLSS mg/L?

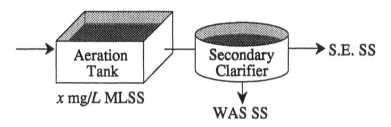

x mg/L MLSS

WAS SS

$$\frac{SRT}{days} = \frac{\text{MLSS in System, lbs}}{\text{SS Leaving System, lbs/day}}$$

$$7 \text{ days} = \frac{(x \text{ mg/L MLSS}) (0.45 \text{ MG}) (8.34 \text{ lbs/gal})}{\underset{\text{mg/L}}{(5400)} \ \underset{\text{MGD}}{(0.025)} \ \underset{\text{lbs/gal}}{(8.34)} + \underset{\text{mg/L}}{(18)} \ \underset{\text{MGD}}{(2)} \ \underset{\text{lbs/gal}}{(8.34)}}$$

$$7 \text{ days} = \frac{(x) (0.45) (8.34)}{1126 + 300}$$

$$7 = \frac{(x) (0.45) (8.34)}{1426}$$

$$\frac{(1426) (7)}{(0.45) (8.34)} = x$$

$$\boxed{\begin{array}{c} 2660 \text{ mg/L} \\ \text{MLSS} \end{array}} = x$$

CALCULATING OTHER UNKNOWN FACTORS

The SRT calculation incorporates many variables:

• SRT

• MLSS, mg/L

• Aeration Tank and Clarifier Volumes

• WAS SS, mg/L

• WAS Pumping Rate, MGD

• S.E. SS, mg/L

• Plant Flow

In Examples 1-4, SRT was the unknown variable. Other variables can also be unknown, as illustrated in Example 5. Regardless of which variable is unknown, use the same equation, fill in the given information, then solve for the unknown value.

NOTES:

6 *Efficiency and Other Percent Calculations*

SUMMARY

1. **Unit process efficiency calculations** refer to the **percent removal** of a water or wastewater constituent such as suspended solids (SS) or biochemical oxygen demand (BOD).

   Simplified Equation:

   $$\% \text{ Removed} = \frac{\text{Part Removed}}{\text{Total}} \times 100$$

   SS Removal Efficiency:

   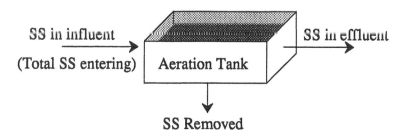

   $$\% \text{ SS Removed} = \frac{\text{SS Removed, mg/}L}{\text{SS Total, mg/}L} \times 100$$

   BOD Removal Efficiency:

   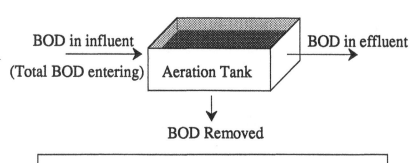

   $$\% \text{ BOD Removed} = \frac{\text{BOD Removed, mg/}L}{\text{BOD Total, mg/}L} \times 100$$

This chapter focuses on various percent calculations in water and wastewater math. The underlying concept in each of these calculations is percent:*

$$\% = \frac{\text{Part}}{\text{Whole}} \times 100$$

Note that **for every percent sign, %, included in an equation, there should be a 100 factor** in the equation as well—either directly under the percent sign or on the opposite side of the equation in the same relative location. (If the percent sign is in the numerator, the 100 factor on the opposite side of the equation will also be in the numerator. And if the percent sign is in the denominator, the 100 factor on the opposite side of the equation will also be in the denominator. This can be verified by moving the 100 factor according to the diagonal rule of movement.)**

The equation above shows the 100 factor on the opposite side of the equation. Written on the same side of the equation it would be:

$$\frac{\%}{100} = \frac{\text{Part}}{\text{Whole}}$$

This concept of the location of the 100 factor is important since you will encounter equations using both means of expression.

* To review percent calculations, refer to Chapter 5 in *Basic Math Concepts*.
** For a review of moving terms from one side of the equation to the other, refer to Chapter 2 in *Basic Math Concepts*.

SUMMARY—Cont'd

Sludge percent calculations include four different calculations (problem types 2-5 below):

2. **Percent Solids and Sludge Pumping Rate**—In making these calculations it is important to distinguish between the terms "solids" and "sludge". Solids refers to dry solids; whereas, sludge refers to water <u>and</u> solids.*

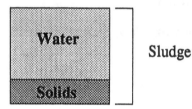

The two most common sludge percent calculations are percent solids and sludge pumping rate.

<u>Using laboratory data:</u>

$$\% \text{ Solids} = \frac{\text{Weight of Solids, grams}}{\text{Total Weight of Sample, grams}} \times 100$$

<u>Using plant data:</u>

$$\% \text{ Solids} = \frac{\text{Solids, lbs/day}}{\text{Sludge, lbs/day}} \times 100$$

The second equation can be used to calculate lbs/day or gpd sludge to be pumped. If desired, the equation can be rearranged as follows:

$$(\text{Sludge, lbs/day}) = \frac{\text{Solids, lbs/day}}{\dfrac{\% \text{ Solids}}{100}}$$

---

* The diagram of sludge and solids is for illustration purposes only and is not to scale. Primary sludge contains about 3-5% solids, and secondary sludges about 1-3% solids. Sludge is actually a <u>mixture</u> of solids and water. The diagram above shows the portions of water and solids <u>if</u> they were separated.

SUMMARY—Cont'd

3. **Mixing Different Percent Solids Sludges**
   When mixing sludges with different percent solids, use the following equation to calculate the percent solids concentration of the resulting sludge:

$$\text{\% Solids of Sludge Mixture} = \frac{\text{Solids in Mixture, lbs/day}}{\text{Sludge Mixture, lbs/day}} \times 100$$

4. **Percent Volatile Solids**—Two equations represent the two most common volatile solids calculations—the first using laboratory data, the second using plant data.*

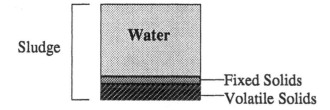

$$\text{\% Volatile Solids} = \frac{\text{Weight Volatile Solids, grams}}{\text{Weight of Total Solids, grams}} \times 100$$

$$\text{\% Volatile Solids} = \frac{\text{Volatile Solids, lbs}}{\text{Total Solids, lbs}} \times 100$$

5. **Seed Sludge (Based on % Digester Volume)**
   In Chapter 4, required digester seed sludge was calculated on the basis of a volatile solids loading ratio. Another way to calculate seed sludge required is based on a percent of the digester volume.

$$\text{\% Seed Sludge} = \frac{\text{Seed Sludge, gal}}{\text{Total Digester Capacity, gal}} \times 100$$

SUMMARY—Cont'd

Chemical dosage percent calculations include solution strength problems and solution mixture problems.

6. **Solution Strength**
The strength of a solution is a measure of the amount of chemical dissolved in the solution.

$$\% \text{ Strength} = \frac{\text{Weight of Chemical}}{\text{Weight of Solution}} \times 100$$

7. **Mixing Different Percent Strength Solutions**

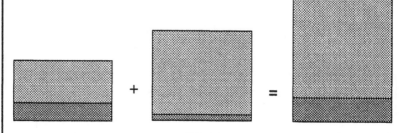

10% Strength Solution + 1% Strength Solution = Solution Mixture (% Strength somewhere between 10% and 1% depending on the quantity contributed by each.)

Simplified Equation:

$$\frac{\% \text{ Strength}}{\text{of Mixture}} = \frac{\text{Chemical in Mixture, lbs}}{\text{Solution Mixture, lbs}} \times 100$$

Expanded Equation:

$$\frac{\% \text{ Strength}}{\text{of Mixture}} = \frac{\text{lbs Chem. from Solution 1} + \text{lbs Chem. from Solution 2}}{\text{lbs Solution 1} + \text{lbs Solution 2}} \times 100$$

SUMMARY—Cont'd

8. **Pump and motor efficiency calculations**
   are based on horsepower input and output.

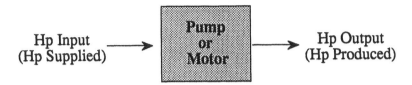

$$\% \text{ Efficiency} = \frac{\text{Hp Output}}{\text{Hp Input}} \times 100$$

6.1 UNIT PROCESS EFFICIENCY CALCULATIONS

The efficiency of a treatment process is its effectiveness in removing various constituents from the water or wastewater. Suspended solids, BOD and COD removal are the most common calculations of unit process efficiency.

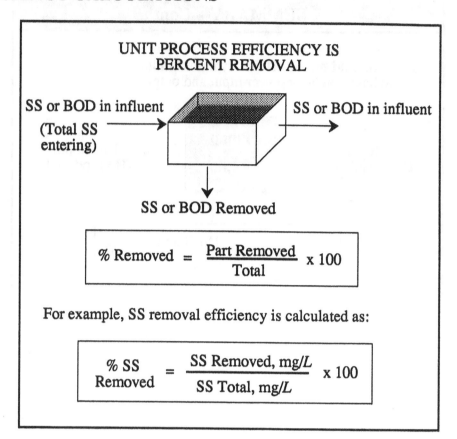

UNIT PROCESS EFFICIENCY IS
PERCENT REMOVAL

SS or BOD in influent
(Total SS entering) → → SS or BOD in influent

SS or BOD Removed

$$\% \text{ Removed} = \frac{\text{Part Removed}}{\text{Total}} \times 100$$

For example, SS removal efficiency is calculated as:

$$\frac{\% \text{ SS}}{\text{Removed}} = \frac{\text{SS Removed, mg/}L}{\text{SS Total, mg/}L} \times 100$$

Example 1: (Unit Process Efficiency)
❑ The suspended solids entering a trickling filter is 190 mg/L. If the suspended solids in the trickling filter effluent is 22 mg/L, what is suspended solids removal efficiency of the trickling filter?

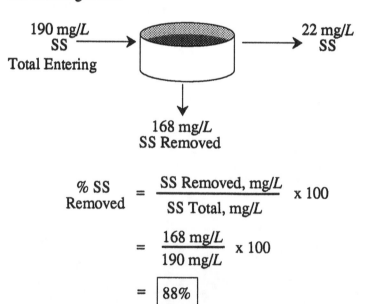

190 mg/L
SS
Total Entering → → 22 mg/L
SS

168 mg/L
SS Removed

$$\frac{\% \text{ SS}}{\text{Removed}} = \frac{\text{SS Removed, mg/}L}{\text{SS Total, mg/}L} \times 100$$

$$= \frac{168 \text{ mg/}L}{190 \text{ mg/}L} \times 100$$

$$= \boxed{88\%}$$

Example 2: (Unit Process Efficiency)
❏ The influent of a primary clarifier has a BOD content of 250 mg/*L*. If the clarifier effluent has a BOD content of 120 mg/*L*, what is the BOD removal efficiency?

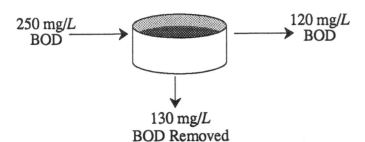

250 mg/*L*
BOD

120 mg/*L*
BOD

130 mg/*L*
BOD Removed

$$\frac{\% \text{ BOD}}{\text{Removed}} = \frac{\text{BOD Removed, mg/}L}{\text{BOD Total, mg/}L} \times 100$$

$$= \frac{130 \text{ mg/}L}{250 \text{ mg/}L} \times 100$$

$$= \boxed{52\%}$$

Example 3: (Unit Process Efficiency)
❏ The suspended solids entering a primary clarifier is 220 mg/*L*. The suspended solids concentration of the primary clarifier effluent is 99 mg/*L*. What is the suspended solids removal efficiency?

$$\frac{\% \text{ SS}}{\text{Removed}} = \frac{\text{SS Removed, mg/}L}{\text{SS Total, mg/}L} \times 100$$

$$= \frac{121 \text{ mg/}L \text{ SS}}{220 \text{ mg/}L \text{ SS}} \times 100$$

$$= \boxed{55\%}$$

6.2 PERCENT SOLIDS AND SLUDGE PUMPING RATE CALCULATIONS

PERCENT SOLIDS

Sludge is composed of water and solids. The vast majority of sludge is water—usually in the range of 93 to 97%.

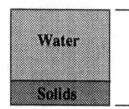

To determine the solids content of a sludge, a sample of sludge is dried overnight in an oven at about 103°- 105° C. **The solids that remain after drying represent the total solids content of the sludge.** This solids content may be expressed as a percent or as a mg/*L* concentration.* Two equations are used to calculate percent solids, depending on whether lab data or plant data is used in the calculation. In both cases, the calculation is **on the basis of solids and sludge weight.**

$$\% \text{ Solids} = \frac{\text{Total Solids, g}}{\text{Sludge Sample, g}} \times 100$$

$$\% \text{ Solids} = \frac{\text{Solids, lbs}}{\text{Sludge, lbs}} \times 100$$

Example 1: (Percent Solids)

❑ The total weight of a sludge sample is 15 grams. (Sludge sample only, not the dish.) If the weight of the solids after drying is 0.62 grams, what is the percent total solids of the sludge?

$$\% \text{ Solids} = \frac{\text{Total Solids, grams}}{\text{Sludge Sample, grams}} \times 100$$

$$= \frac{0.62 \text{ grams}}{15 \text{ grams}} \times 100$$

$$= \boxed{4.1\%}$$

Example 2: (Percent Solids)

❑ A total of 3000 gallons of sludge is pumped to a digester. If the sludge has a 6% solids content, how many lbs/day solids are pumped to the digester?

First, write the % solids equation and fill in the given information:

$$\% \text{ Solids} = \frac{\text{Solids, lbs/day}}{\text{Sludge, lbs/day}} \times 100$$

$$6 = \frac{x \text{ lbs/day Solids}}{(3000 \text{ gal}) (8.34 \text{ lbs/gal})} \times 100$$

$$\frac{(3000) (8.34) (6)}{100} = x$$

$$\boxed{\begin{array}{c} 1501 \text{ lbs/day} \\ \text{Solids} \end{array}} = x$$

---

* 1% solids = 10,000 mg/*L*. For a review of % to mg/*L* conversions, refer to Chapter 8 in *Basic Math Concepts*.

Example 3: (Percent Solids)
❑ A total of 10,000 lbs/day SS are removed from a primary clarifier and pumped to a sludge thickener. If the sludge has a solids content of 3%, how many lbs/day sludge are pumped to the thickener?

$$\% \text{ Solids} = \frac{\text{Solids, lbs/day}}{\text{Sludge, lbs/day}} \times 100$$

$$3 = \frac{10,000 \text{ lbs/day Solids}}{x \text{ lbs/day Sludge}} \times 100$$

$$x = \frac{(10,000)(100)}{3}$$

$$= \boxed{333,333 \text{ lbs/day Sludge}}$$

Example 4: (Percent Solids)
❑ It is anticipated that 200 lbs/day SS will be pumped from the primary clarifier of a new plant. If the primary clarifier sludge has a solids content of 5%, how many gpd sludge will be pumped from the clarifier? (Assume the sludge weighs 8.34 lbs/gal.)

First calculate lbs/day sludge to be pumped using the % solids equation, then convert lbs/day sludge to gpd sludge:

$$\text{Sludge, lbs/day} = \frac{\text{Solids, lbs/day}}{\dfrac{\% \text{ Solids}}{100}}$$

$$x \text{ lbs/day Sludge} = \frac{200 \text{ lbs/day Solids}}{0.05}$$

$$x = 4000 \text{ lbs/day Sludge}$$

Converting lbs/day sludge to gpd sludge:

$$\frac{4000 \text{ lbs/day Sludge}}{8.34 \text{ lbs/gal}} = \boxed{480 \text{ gpd Sludge}}$$

CALCULATING OTHER UNKNOWN VARIABLES

The three variables in percent solids calculations are percent solids, lbs/day solids, and lbs/day sludge. In Example 1, percent solids was the unknown variable. Examples 2-4 illustrate calculations when other variables are unknown. In solving these problems, write the equation as usual, fill in the known information, then solve for the unknown value.*

SLUDGE TO BE PUMPED

As mentioned above, one of the variables in the % solids calculation is lbs/day sludge. Example 3 illustrates a calculation of lbs/day sludge. Example 4 illustrates the calculation of gpd sludge to be pumped.

Because lbs/day sludge is calculated relatively frequently, the % solids equation is often rearranged as follows:

$$\text{Sludge, lbs/day} = \frac{\text{Solids, lbs/day}}{\dfrac{\% \text{ Solids}}{100}}$$

---

* Refer to Chapter 2 in *Basic Math Concepts* for a review of solving for the unknown value.

6.3 MIXING DIFFERENT PERCENT SOLIDS SLUDGES CALCULATIONS

When sludges with different percent solids content are mixed, the resulting sludge has a percent solids content somewhere **between** the solids contents of the original sludges. For example, if a 4% primary sludge is mixed with a 1% secondary sludge, the resulting sludge might have a solids content of about 2 or 3%. The actual percent solids content will depend on how much (lbs) of each sludge is mixed together. If, in the example, most of the sludge is from the secondary sludge (1% solids) and very little from the primary sludge (4% solids), then the resulting sludge would be closer to a 1% sludge (perhaps a 1.5% sludge). If, on the other hand, most of the sludge is primary sludge and very little is secondary sludge, then the resulting sludge mixture might have a solids content closer to 4%—such as 3 or 3.5%.

The actual solids content of a mixture of two or more sludges depends on the pounds of sludge contributed from each source.

As with the sludge thickening equation, remember that if the thickened sludge has a density greater than 8.34 lbs/gal, it must be used instead of 8.34 lbs/gal.*

WHEN SLUDGES ARE MIXED THE MIXTURE HAS A % SOLIDS CONTENT <u>BETWEEN</u> THE TWO ORIGINAL % SOLIDS VALUES

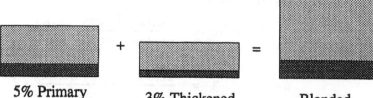

5% Primary Sludge + 3% Thickened Secondary Sludge = Blended Sludge (% Solids Somewhere <u>Between</u> 3% and 5%)

Simplified Equation:

$$\frac{\text{\% Solids of}}{\text{Sludge Mixture}} = \frac{\text{Solids in Mixture, lbs/day}}{\text{Sludge Mixture, lbs/day}} \times 100$$

Expanded Equation:

$$\frac{\text{\% Solids of Sludge Mixture}}{} = \frac{\text{Prim. Sol., lbs/day} + \text{Sec. Sol., lbs/day}}{\text{Prim. Sludge, lbs/day} + \text{Sec. Sludge, lbs/day}} \times 100$$

Example 8: (Mixing Sludges)

❑ A 5% primary sludge flow of 5000 gpd is mixed with a 3% thickened secondary sludge flow of 3500 gpd. What is the percent solids content of the mixed sludge flow?

$$\frac{\text{\% Solids of Sludge Mixture}}{} = \frac{\text{Prim. Sl. Sol., lbs/day} + \text{Sec. Sl. Sol., lbs/day}}{\text{Prim. Sludge, lbs/day} + \text{Sec. Sludge, lbs/day}} \times 100$$

$$= \frac{\underset{\text{Prim. Slud. lbs/gal}}{(5000 \text{ gpd}) (8.34)} \underset{100}{\left(\frac{5}{}\right)} + \underset{\text{Sec. Slud. lbs/gal}}{(3500 \text{ gpd}) (8.34)} \underset{100}{\left(\frac{3}{}\right)}}{\underset{\text{Prim. Sludge}}{(5000 \text{ gpd}) (8.34)} + \underset{\text{Sec. Sludge}}{(3500 \text{ gpd}) (8.34)}} \times 100$$

$$= \frac{2085 \text{ lbs/day Prim Sol.} + 876 \text{ lbs/day Sec. Sol.}}{41,700 \text{ lbs/day Prim Slud.} + 29,190 \text{ lbs/day Sec. Slud.}} \times 100$$

$$= \frac{2961 \text{ lbs/day Solids}}{70,890 \text{ lbs/day Sludge}} \times 100$$

$$= \boxed{4.2\% \text{ Solids}}$$

---

* Refer to Chapter 7 for a review of density and specific gravity.

Example 9: (Mixing Sludges)

❑ Primary and thickened secondary sludges are to be mixed and sent to the digester. The 5700 gpd primary sludge has a solids content of 5.5%; the 4200 gpd thickened secondary sludge has a solids content of 3.8%. What would be the percent solids content of the mixed sludge?

$$\text{\% Solids of Sludge Mixture} = \frac{\text{Prim. Sl. Sol., lbs/day} + \text{Sec. Sl. Sol., lbs/day}}{\text{Prim. Sludge, lbs/day} + \text{Sec. Sludge, lbs/day}} \times 100$$

$$= \frac{(5700 \text{ gpd}) \underset{\text{Prim. Sl. lbs/gal}}{(8.34)} \underset{100}{(\underline{5.5})} + (4200 \text{ gpd}) \underset{\text{Sec. Sl. lbs/gal}}{(8.34)} \underset{100}{(\underline{3.8})}}{(5700 \text{ gpd}) \underset{\text{Prim. Sludge lbs/gal}}{(8.34)} + (4200 \text{ gpd}) \underset{\text{Sec. Sludge lbs/gal}}{(8.34)}} \times 100$$

$$= \frac{2615 \text{ lbs/day Prim Sol.} + 1331 \text{ lbs/day Sec Sol.}}{47,538 \text{ lbs/day Prim. Slud.} + 35,028 \text{ lbs/day Sec. Slud.}} \times 100$$

$$= \frac{3946 \text{ lbs/day Solids}}{82,566 \text{ lbs/day Sludge}} \times 100$$

$$= \boxed{4.8\% \text{ Solids}}$$

Example 10: (Mixing Sludges)

❑ A 3.8% primary sludge flow of 6800 gpd is mixed with a 7% thickened secondary sludge flow of 4500 gpd. What is the percent solids of the combined sludge flow?

$$\text{\% Solids of Sludge Mixture} = \frac{\text{Prim. Sl. Sol., lbs/day} + \text{Sec. Sl. Sol., lbs/day}}{\text{Prim. Sludge, lbs/day} + \text{Sec. Sludge, lbs/day}} \times 100$$

$$= \frac{(6800 \text{ gpd}) \underset{\text{lbs/gal}}{(8.34)} \underset{100}{(\underline{3.8})} + (4500) \underset{\text{lbs/gal}}{(8.34)} \underset{100}{(\underline{7})}}{(6800 \text{ gpd}) (8.34 \text{ lbs/day}) + (4500) (8.34 \text{ lbs/day})} \times 100$$

$$= \frac{2155 \text{ lbs/day} + 2627 \text{ lbs/day}}{56,712 \text{ lbs/day} + 37,530 \text{ lbs/day}} \times 100$$

$$= \frac{4782 \text{ lbs/day Solids}}{94,242 \text{ lbs/day Sludge}} \times 100$$

$$= \boxed{5.1\% \text{ Solids}}$$

6.4 PERCENT VOLATILE SOLIDS

Sludge solids are comprised of organic matter (from plant or animal sources) and inorganic matter (material from mineral sources, such as sand and grit). The organic matter is called **volatile solids**, the inorganic matter is called **fixed solids**. Together, the volatile solids and fixed solids make up the **total solids**.

When calculating percent solids (also called percent total solids) and percent volatile solids, it is essential to focus on the general concept of percent:

$$\text{Percent} = \frac{\text{Part}}{\text{Whole}} \times 100$$

As illustrated in the diagrams to the right, **when calculating percent solids**, the "part" of interest is the weight of the total solids; the "whole" is the weight of the sludge:

$$\text{\% Solids} = \frac{\text{Wt. of Solids}}{\text{Wt. of Sludge}} \times 100$$

When calculating percent volatile solids, the "part" of interest is the weight of the volatile solids; the "whole" is the weight of total solids:

$$\text{\% VS} = \frac{\text{Wt. of VS}}{\text{Wt. of Tot. Solids}} \times 100$$

The calculation of volatile solids using laboratory data is described more thoroughly in Chapter 18.

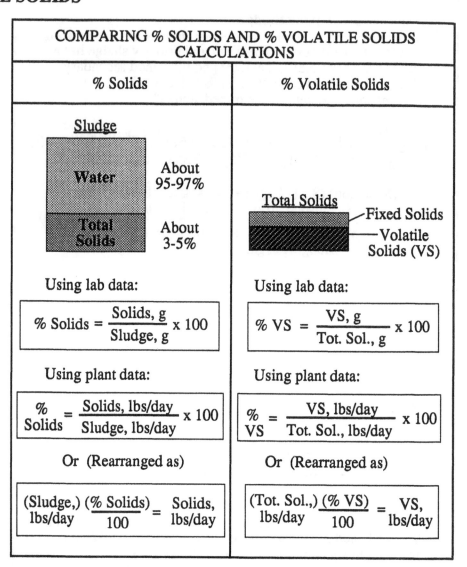

COMPARING % SOLIDS AND % VOLATILE SOLIDS CALCULATIONS

% Solids	% Volatile Solids

Sludge

Water — About 95-97%

Total Solids — About 3-5%

Total Solids — Fixed Solids / Volatile Solids (VS)

Using lab data:

$$\text{\% Solids} = \frac{\text{Solids, g}}{\text{Sludge, g}} \times 100$$

Using lab data:

$$\text{\% VS} = \frac{\text{VS, g}}{\text{Tot. Sol., g}} \times 100$$

Using plant data:

$$\text{\% Solids} = \frac{\text{Solids, lbs/day}}{\text{Sludge, lbs/day}} \times 100$$

Using plant data:

$$\text{\% VS} = \frac{\text{VS, lbs/day}}{\text{Tot. Sol., lbs/day}} \times 100$$

Or (Rearranged as)

$$(\text{Sludge,})_{\text{lbs/day}} \frac{(\text{\% Solids})}{100} = \text{Solids, lbs/day}$$

Or (Rearranged as)

$$(\text{Tot. Sol.,})_{\text{lbs/day}} \frac{(\text{\% VS})}{100} = \text{VS, lbs/day}$$

Example 11: (% Volatile Solids)
❑ If 1250 lbs/day solids are sent to the digester, with a volatile solids content of 72%, how many lbs/day volatile solids are sent to the digester?

Either the % Vol solids equation or the rearranged equation may be used to calculate the lbs/day volatile solids. The time factor (day) can be added to the Total Solids and Volatile Solids terms without affecting the calculation:

$$(\text{Total Solids})_{\text{lbs/day}} \frac{(\text{\% VS})}{100} = \text{Vol. Sol. lbs/day}$$

$$(1250 \text{ lbs/day}) \frac{(72)}{100} = \boxed{900 \text{ lbs/day Vol. Solids}}$$

Example 12: (% Volatile Solids)
❑ A total of 3000 gpd sludge is to be pumped to the digester. If the sludge has a 6% solids content with 70% volatile solids, how many lbs/day volatile solids are pumped to the digester?

$$\frac{(Sludge)}{lbs/day} \frac{(\%\ Solids)}{100} \frac{(\%Vol.\ Sol.)}{100} = \frac{Vol.\ Sol.}{lbs/day}$$

Since sludge is given in gpd, 8.34 lbs/gal must be added to the equation to convert gpd sludge to lbs/day sludge:*

lbs/day Sludge

$$\overbrace{\frac{(Sludge)}{gpd} \frac{(8.34)}{lbs/gal}} \frac{(\%\ Solids)}{100} \frac{(\%\ Vol.\ Sol.)}{100} = \frac{Vol.\ Sol.}{lbs/day}$$

$$\frac{(3000)}{gpd} \frac{(8.34)}{lbs/gal} \frac{(6)}{100} \frac{(70)}{100} = \boxed{\begin{array}{c} 1051\ lbs/day \\ Vol.\ Sol. \end{array}}$$

CALCULATING VOLATILE SOLIDS GIVEN SLUDGE DATA

Sometimes you will have lbs/day sludge information and will want to calculate lbs/day volatile solids. When this is the case, you must include the % solids factor in the equation as well, shown in the equation below. In effect, you are calculating lbs/day solids first, (using the % Solids factor), then the lbs/day Volatile Solids (using the % Volatile Solids factor):

$$\frac{(Sludge)}{lbs/day} \frac{(\%\ Sol.)}{100} \frac{(\%\ VS)}{100} = \frac{VS}{lbs/day}$$

Example 13: (% Volatile Solids)
❑ A 5% sludge has a volatile solids content of 68%. If 1200 lbs/day volatile solids are pumped to the digester (a) how many lbs/day sludge are pumped to the digester? and (b) how many gpd sludge are pumped to the digester? (Assume the sludge weighs 8.34 lbs/gal.)

(a)
$$\frac{(Sludge)}{lbs/day} \frac{(\%\ Sol.)}{100} \frac{(\%\ Vol.\ Sol.)}{100} = \frac{Vol.\ Sol.}{lbs/day}$$

$$(x\ lbs/day\ Sludge) \frac{(5)}{100} \frac{(68)}{100} = \begin{array}{c} 1200\ lbs/day \\ Vol.\ Sol. \end{array}$$

$$(x)\ (0.05)\ (0.68) = 1200$$

$$x = \frac{(1200)}{(0.05)\ (0.68)}$$

$$= \boxed{\begin{array}{c} 35,294\ lbs/day \\ Sludge \end{array}}$$

(b) Convert lbs/day to gpd:

$$\frac{35,294\ lbs/day}{8.34\ lbs/gal} = \boxed{\begin{array}{c} 4232\ gpd \\ Sludge \end{array}}$$

SOLVING FOR OTHER UNKNOWN FACTORS

The equations shown above can be used to solve for any one of the three or four variables shown. Use the same equation, fill-in the known information, and solve for the unknown variable. Example 13 illustrates this type of calculation.

* For a review of flow conversions, refer to Chapter 8 in *Basic Math Concepts.*

6.5 PERCENT SEED SLUDGE CALCULATIONS

There are many methods to determine seed sludge required to start a new digester. One method, discussed in Chapter 4, is to use a volatile solids loading ratio—lbs volatile solids added per lb volatile solids in the digester.

Another method is to calculate seed sludge required based on the volume of the digester. This method is not quite as sensitive to the volatile solids balance in the seed sludge and incoming sludge. Examples 1-4 illustrate this calculation.

Although most digesters have cone-shaped bottoms, for simplicity, it is assumed that the side water depth represents the average digester depth.

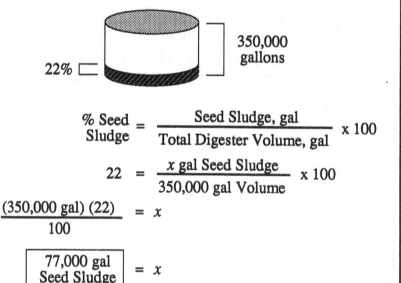

Example 1: (% Seed Sludge)
❏ A digester has a volume of 350,000 gallons. If the digester seed sludge is to be 22% of the digester volume, how many gallons of seed sludge will be required?*

22%   350,000 gallons

$$\frac{\% \text{ Seed}}{\text{Sludge}} = \frac{\text{Seed Sludge, gal}}{\text{Total Digester Volume, gal}} \times 100$$

$$22 = \frac{x \text{ gal Seed Sludge}}{350,000 \text{ gal Volume}} \times 100$$

$$\frac{(350,000 \text{ gal}) (22)}{100} = x$$

$$\boxed{\begin{array}{c} 77,000 \text{ gal} \\ \text{Seed Sludge} \end{array}} = x$$

Example 2: (% Seed Sludge)
❏ A 40-ft diameter digester has a typical water depth of 20 ft. If the seed sludge to be used is 20% of the tank volume, how many gallons of seed sludge will be required?

|← 40 ft →|

20%   20 ft

$$\frac{\% \text{ Seed}}{\text{Sludge}} = \frac{\text{Seed Sludge, gal}}{\text{Total Digester Volume, gal}} \times 100$$

$$20 = \frac{x \text{ gal Seed Sludge}}{(0.785) (40 \text{ ft}) (40 \text{ ft}) (20 \text{ ft}) (7.48 \text{ gal/cu ft})} \times 100$$

$$\frac{(0.785) (40 \text{ ft}) (40 \text{ ft}) (20 \text{ ft}) (7.48) (20)}{100} = x$$

$$\boxed{\begin{array}{c} 37,580 \text{ gal} \\ \text{Seed Sludge} \end{array}} = x$$

* For a review of volume calculations, refer to Chapter 11 in *Basic Math Concepts*.

Example 3: (% Seed Sludge)
❏ A digester 50 ft in diameter has a side water depth of 20 ft. If the digester seed sludge is to be 25% of the digester volume, how many gallons of seed sludge will be required?

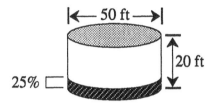

$$\frac{\% \text{ Seed}}{\text{Sludge}} = \frac{\text{Seed Sludge, gal}}{\text{Total Digester Volume, gal}} \times 100$$

$$25 = \frac{x \text{ gal Seed Sludge}}{(0.785)(50 \text{ ft})(50 \text{ ft})(20 \text{ ft})(7.48 \text{ gal/cu ft})} \times 100$$

$$\frac{(0.785)(50 \text{ ft})(50 \text{ ft})(20 \text{ ft})(7.48)(25)}{100} = x$$

$$\boxed{\begin{array}{c} 73,398 \text{ gal} \\ \text{Seed Sludge} \end{array}} = x$$

Example 4: (% Seed Sludge)
❏ A 40-ft diameter digester has a typical side water depth of 18 ft. If 45,700 gallons of seed sludge are to be used in starting up the digester, what percent of the digester volume will be seed sludge?

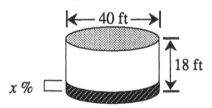

$$\frac{\% \text{ Seed}}{\text{Sludge}} = \frac{\text{Seed Sludge, gal}}{\text{Total Digester Volume, gal}} \times 100$$

$$x = \frac{45,700 \text{ gal Seed Sludge}}{(0.785)(40 \text{ ft})(40 \text{ ft})(18 \text{ ft})(7.48 \text{ gal/cu ft})} \times 100$$

$$x = \boxed{27\%}$$

CALCULATING OTHER UNKNOWN FACTORS

There are three variables in percent seed sludge calculations: percent seed sludge, gallons seed sludge, and total gallons digester volume.

In Examples 1-3, the unknown factor was seed sludge gallons. However, the same equation can be used to calculate either one of the other two variables. Example 4 is one such calculation.

6.6 PERCENT STRENGTH OF A SOLUTION CALCULATIONS

PERCENT STRENGTH USING DRY CHEMICALS

The strength of a solution is a measure of the amount of chemical (solute) dissolved in the solution. Since percent is calculated as "part over whole,"

$$\% = \frac{\text{Part}}{\text{Whole}} \times 100$$

percent strength is calculated as **part chemical**, in lbs, divided by the **whole solution**, in lbs:

$$\% \text{ Strength} = \frac{\text{Chem., lbs}}{\text{Sol'n, lbs}} \times 100$$

The denominator of the equation, lbs solution, includes both chemical (lbs) and water (lbs). Therefore, the equation can be written in expanded form as:

$$\frac{\%}{\text{Strength}} = \frac{\text{Chem., lbs}}{\underset{\text{lbs} \quad \text{lbs}}{\text{Water, + Chem.,}}} \times 100$$

As the two equations above illustrate, **the chemical added must be expressed in pounds.** If the chemical weight is expressed in ounces (as in Example 1) or grams (as in Example 2), it must first be converted to pounds (to correspond with the other lbs terms in the equation) before percent strength is calculated.

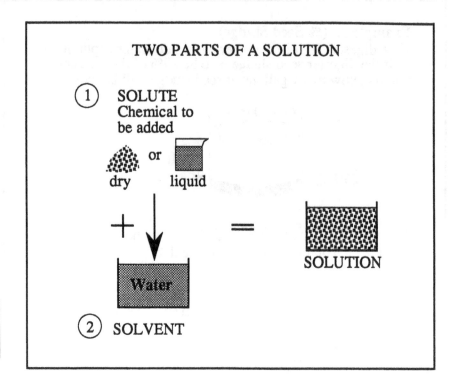

TWO PARTS OF A SOLUTION

① SOLUTE
Chemical to be added

dry   or   liquid

+

Water

② SOLVENT

= SOLUTION

Example 1: (Percent Strength)

❑ If a total of 8 ounces of dry polymer are added to 10 gallons of water, what is the percent strength (by weight) of the polymer solution?

Before calculating percent strength, the ounces chemical must be converted to lbs chemical:*

$$\frac{8 \text{ ounces}}{16 \text{ ounces/pound}} = 0.5 \text{ lbs chemical}$$

Now calculate percent strength:

$$\% \text{ Strength} = \frac{\text{Chemical, lbs}}{\text{Water, lbs + Chemical, lbs}} \times 100$$

$$= \frac{0.5 \text{ lbs Chemical}}{(10 \text{ gal}) (8.34 \text{ lbs/gal}) + 0.5 \text{ lbs}} \times 100$$

$$= \frac{0.5 \text{ lbs Chemical}}{83.9 \text{ lbs Solution}} \times 100$$

$$= \boxed{0.6\%}$$

---

* To review ounces to pounds conversions refer to Chapter 8 in *Basic Math Concepts*.

Example 2: (Percent Strength)
❑ If 100 grams of dry polymer are dissolved in 5 gallons of water, what percent strength is the solution? (1 g = 0.0022 lbs)

First, convert grams chemical to pounds chemical. Since 1 gram equals 0.0022 lbs, 100 grams is 100 times 0.0022 lbs:

$$\underset{\text{Chemical}}{(100 \text{ grams})} \ (0.0022 \text{ lbs/gram}) = \underset{\text{Chemical}}{0.22 \text{ lbs}}$$

Now calculate percent strength of the solution:

$$\text{\% Strength} = \frac{\text{lbs Chemical}}{\text{lbs Water} + \text{lbs Chemical}} \times 100$$

$$= \frac{0.22 \text{ lbs Chemical}}{(5 \text{ gal})(8.34 \text{ lbs/gal}) + 0.22 \text{ lbs}} \times 100$$

$$= \frac{0.22 \text{ lbs}}{41.92 \text{ lbs}} \times 100$$

$$= \boxed{0.52\%}$$

WHEN GRAMS CHEMICAL ARE USED

The chemical (solute) to be used in making a solution may be measured in grams rather than pounds or ounces. When this is the case, convert grams of chemical to pounds of chemical before calculating percent strength. The following conversion equations may be used:

$$\boxed{1 \ g = 0.0022 \text{ lbs}}$$

Or

$$\boxed{1 \text{ lb} = 454 \text{ grams **}}$$

Example 3: (Percent Strength)
❑ How many pounds of dry polymer must be added to 25 gallons of water to make a 1% polymer solution?

First, write the equation as usual and fill in the known information. Then solve for the unknown.*

$$\text{\% Strength} = \frac{\text{lbs Chemical}}{\text{lbs Water} + \text{lbs Chemical}} \times 100$$

$$1 = \frac{x \text{ lbs Chemical}}{(25 \text{ gal})(8.34 \text{ lbs/gal}) + x \text{ lbs Chemical}} \times 100$$

$$1 = \frac{100 \, x}{208.5 + x}$$

$$1(208.5 + x) = 100 \, x$$

$$208.5 = 100 \, x - 1 \, x$$

$$208.5 = 99 \, x$$

$$\boxed{2.1 \text{ lbs Chem.}} = x$$

SOLVING FOR OTHER UNKNOWN VARIABLES

In the percent strength equation there are three variables: percent strength, lbs chemical and lbs water. In Examples 1 and 2, the unknown value was percent strength. However, the same equation can be used to determine either of the other two variables. Example 3 illustrates this type of calculation.

Note that gallons water can also be the unknown variable in percent strength calculations. First set lbs water as the unknown variable in the equation. Then when you have calculated the lbs water required, you can then convert lbs water to gallons water using the 8.34 lbs/gal factor.

---

* To review solving for the unknown value, refer to Chapter 2 in *Basic Math Concepts*.
* * If the box method of conversions is used (see Chapter 8 in *Basic Math Concepts*), both numbers in the conversion equation must be greater than one.

6.7 MIXING DIFFERENT PERCENT STRENGTH SOLUTIONS CALCULATIONS

There are two types of solution mixture calculations. In one type of calculation, two solutions of different strengths are mixed, with no particular target solution strength. The calculation involves determining the percent strength of the solution mixture. These calculations are similar to the sludge mixture problems described in Section 6.3.

The second type of solution mixture calculation includes a desired or target strength. This calculation is described in Chapter 14, Section 5.

WHEN DIFFERENT % STRENGTH SOLUTIONS ARE MIXED

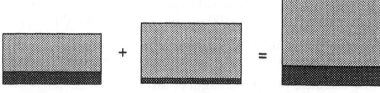

10% Strength Solution + 1% Strength Solution = Solution Mixture (% Strength somewhere between 10% and 1%)

Simplified Equation:

$$\frac{\% \text{ Strength}}{\text{of Mixture}} = \frac{\text{lbs Chemical in Mixture}}{\text{lbs Solution Mixture}} \times 100$$

Expanded Equations:

$$\frac{\% \text{ Strength}}{\text{of Mixture}} = \frac{\text{lbs Chem. from Solution 1} + \text{lbs Chem. from Solution 2}}{\text{lbs Solution 1} + \text{lbs Solution 2}} \times 100$$

$$\frac{\% \text{ Strength}}{\text{of Mixture}} = \frac{(\text{Sol'n 1}) \frac{(\% \text{ Strength})}{100} + (\text{Sol'n 2}) \frac{(\% \text{ Strength})}{100}}{\text{lbs Solution 1} + \text{lbs Solution 2}} \times 100$$

Example 7: (Solution Mixtures)

❑ If 20 lbs of a 10% strength solution are mixed with 50 lbs of 1% strength solution, what is the percent strength of the solution mixture?

$$\frac{\% \text{ Strength}}{\text{of Mixture}} = \frac{\text{lbs Chem. from Solution 1} + \text{lbs Chem. from Solution 2}}{\text{lbs Solution 1} + \text{lbs Solution 2}} \times 100$$

$$= \frac{(20 \text{ lbs}) \frac{(10)}{100} + (50 \text{ lbs}) \frac{(1)}{100}}{20 \text{ lbs} + 50 \text{ lbs}} \times 100$$

$$= \frac{2 \text{ lbs} + 0.5 \text{ lbs}}{70 \text{ lbs}} \times 100$$

$$= \frac{2.5 \text{ lbs}}{70 \text{ lbs}} \times 100$$

$$= \boxed{3.6\%}$$

Example 8: (Solution Mixtures)
❑ If 5 gallons of an 8% strength solution are mixed with 40 gallons of a 0.5% strength solution, what is the percent strength of the solution mixture? (Assume the 8% solution weighs 9.5 lbs/gal and the 0.5% solution weighs 8.34 lbs/gal.)

$$\frac{\% \text{ Strength}}{\text{of Mixture}} = \frac{\text{lbs Chem. from Solution 1} + \text{lbs Chem. from Solution 2}}{\text{lbs Solution 1} + \text{lbs Solution 2}}$$

$$= \frac{(5 \text{ gal}) (9.5 \text{ lbs/gal}) \frac{(8)}{100} + (40 \text{ gal}) (8.34 \text{ lbs/gal}) \frac{(0.5)}{100}}{(5 \text{ gal}) (9.5 \text{ lbs/gal}) + (40 \text{ gal}) (8.34 \text{ lbs/gal})} \times 100$$

$$= \frac{3.8 \text{ lbs Chem.} + 1.7 \text{ lbs Chem.}}{47.5 \text{ lbs Soln 1} + 333.6 \text{ lbs Soln 2}} \times 100$$

$$= \frac{5.5 \text{ lbs Chemical}}{381.1 \text{ lbs Solution}} \times 100$$

$$= \boxed{1.4\% \text{ Strength}}$$

USE DIFFERENT DENSITY FACTORS WHEN APPROPRIATE

Percent strength should be expressed in terms of **pounds chemical per pounds solution.** Therefore, when solutions are expressed in terms of gallons, the gallons should be expressed as pounds before continuing with the percent strength calculation. It is important to consider what density factor should be used to convert from gallons to pounds. If the solution has a density the same as water, 8.34 lbs/gal would be used. If, however, the solution has a higher density, such as for some polymer solutions, then a higher density factor should be used. When the density is unknown, sometimes it is possible to weigh the chemical solution to determine its density.

Example 9: (Solution Mixtures)
❑ If 15 gallons of a 10% strength solution are added to 50 gallons of 0.8% strength solution, what is the percent strength of the solution mixture? (Assume the 10% strength solution weighs 10.2 lbs/gal and the 0.8% strength solution weighs 8.8 lbs/gal.)

$$\frac{\% \text{ Strength}}{\text{of Mixture}} = \frac{\text{lbs Chem. from Solution 1} + \text{lbs Chem. from Solution 2}}{\text{lbs Solution 1} + \text{lbs Solution 2}} \times 100$$

$$= \frac{(15 \text{ gal}) (10.2 \text{ lbs/gal}) \frac{(10)}{100} + (50 \text{ gal}) (8.8 \text{ lbs/gal}) \frac{(0.8)}{100}}{(15 \text{ gal}) (10.2 \text{ lbs/gal}) + (50 \text{ gal}) (8.8 \text{ lbs/gal})} \times 100$$

$$= \frac{15.3 \text{ lbs Chem.} + 3.5 \text{ lbs Chem.}}{153 \text{ lbs Soln 1} + 440 \text{ lbs Soln 2}} \times 100$$

$$= \frac{18.8 \text{ lbs Chemical}}{593 \text{ lbs Solution}} \times 100$$

$$= \boxed{3.2\% \text{ Strength}}$$

6.8 PUMP AND MOTOR EFFICIENCY CALCULATIONS

Pump and motor efficiencies are a measure of horsepower output compared with the horsepower input to the pump or motor. Since percent is a calculation of "part over whole,"

$$\% = \frac{Part}{Whole} \times 100$$

in these efficiency calculations the "part" is represented by the hp output, and the "whole" is represented by the total hp supplied (or hp input), as shown in the general efficiency equation to the right.

PUMP AND MOTOR EFFICIENCIES ARE CALCULATIONS OF **PERCENT HORSEPOWER OUTPUT**

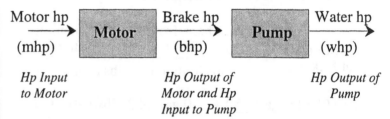

Motor hp		Brake hp		Water hp
(mhp)	**Motor**	(bhp)	**Pump**	(whp)
Hp Input to Motor		*Hp Output of Motor and Hp Input to Pump*		*Hp Output of Pump*

General Efficiency Equation:

$$\begin{matrix} \% \text{ Hp} \\ \text{Output} \end{matrix} = \frac{Hp\ Output}{Total\ hp\ Input} \times 100$$

Motor and Pump Efficiency Equations:

These equations can be written using the terms Hp Input and Hp Output or mhp and bhp, as shown below:

$$\begin{matrix} \% \text{ Motor} \\ \text{Efficiency} \end{matrix} = \frac{Hp\ Output}{Hp\ Input} \times 100 \quad \underline{Or} \quad \frac{bhp}{mhp} \times 100$$

$$\begin{matrix} \% \text{ Pump} \\ \text{Efficiency} \end{matrix} = \frac{Hp\ Output}{Hp\ Input} \times 100 \quad \underline{Or} \quad \frac{whp}{bhp} \times 100$$

Overall Efficiency Equation:

$$\begin{matrix} \% \text{ Overall} \\ \text{Efficiency} \end{matrix} = \frac{Hp\ Output}{Hp\ Input} \times 100 \quad \underline{Or} \quad \frac{whp}{mhp} \times 100$$

Example 1: (Pump and Motor Efficiency)
❏ The brake horsepower of a pump is 18 hp. If the water horsepower is 14 hp, what is the efficiency of the pump?

18 bhp → **Pump** → 14 whp

$$\begin{matrix} \% \text{ Pump} \\ \text{Efficiency} \end{matrix} = \frac{Hp\ Output}{Hp\ Input} \times 100$$

$$= \frac{14\ hp}{18\ hp} \times 100$$

$$= \boxed{78\%}$$

Example 2: (Pump and Motor Efficiency)
❏ If the motor horsepower is 25 hp and the brake horsepower is 22 hp, what is the efficiency of the motor?

25 mhp → | Motor | → 22 bhp

$$\frac{\% \text{ Motor}}{\text{Efficiency}} = \frac{\text{Hp Output}}{\text{Hp Input}} \times 100$$

$$= \frac{22 \text{ hp}}{25 \text{ hp}} \times 100$$

$$= \boxed{88\%}$$

Example 3: (Pump and Motor Efficiency)
❏ The brake horsepower is 13.5 hp. If the motor is 90% efficient, what is the motor horsepower?

x mhp → | Motor | → 13.5 bhp

$$90 = \frac{13.5 \text{ bhp}}{x \text{ mhp}} \times 100$$

$$x = \frac{13.5}{90} \times 100$$

$$= \boxed{15 \text{ mhp}}$$

CALCULATING OTHER UNKNOWN VALUES

Pump and motor efficiency calculations have three variables: efficiency, hp output, and hp input. In Examples 1 and 2, efficiency was the unknown term. However, **any one of the variables can be the unknown term** as long as data is given for the other two. Example 3 illustrates this type of calculation.

Example 4: (Pump and Motor Efficiency)
❏ A total of 20 hp is supplied to a motor. If the wire-to-water efficiency of the pump and motor is 65%, what will the whp be?

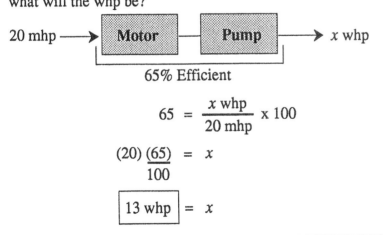

20 mhp → | Motor | | Pump | → x whp
65% Efficient

$$65 = \frac{x \text{ whp}}{20 \text{ mhp}} \times 100$$

$$(20)\frac{(65)}{100} = x$$

$$\boxed{13 \text{ whp}} = x$$

WIRE-TO-WATER EFFICIENCY

Wire-to-water efficiency is another name for **overall efficiency of the pump and motor**. In other words, the pump and motor are considered one unit. The hp input to the "unit" is the motor horsepower (mhp); the hp output to the "unit" is water horsepower (whp).

NOTES:

7 *Pumping Calculations*

SUMMARY

1. Density and Specific Gravity

The density of a substance is mass per volume (cu ft).

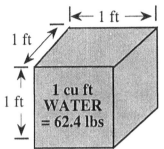

Density of Water
= 62.4 $\frac{\text{lbs}}{\text{cu ft}}$

$$\text{Density of any Substance} = \frac{\text{lbs}}{\text{cu ft}}$$

The density of liquids is commonly expressed as pounds per gallon:

$$\text{Density of a Liquid} = \frac{\text{lbs of Liquid}}{1 \text{ gal of Liquid}}$$

The specific gravity of a liquid is the density of the liquid compared to the density of water.

$$\text{Specific Gravity of a Liquid} = \frac{\text{Density of the Liquid}}{\text{Density of Water}}$$

If the specific gravity of a liquid is known, the density of that liquid may be calculated as:*

$$\begin{array}{l}\text{(Spec. Grav.) (Density)} \\ \text{of a liquid} \quad \text{of Water,} \\ \qquad \qquad 8.34 \text{ lbs/gal}\end{array} = \begin{array}{l}\text{Density of} \\ \text{the Liquid,} \\ \text{lbs/gal}\end{array}$$

The specific gravity of a gas is the density of the gas compared to the density of air.

$$\text{Specific Gravity of a Gas} = \frac{\text{Density of the Gas}}{\text{Density of Air}}$$

Pumping calculations include a variety of different types of problems including:

• Density and Specific Gravity,

• Pressure and Force,

• Head and Head Loss,

• Horsepower, and

• Pump Capacity.

* Note that this equation is simply the specific gravity equation with the terms rearranged.

SUMMARY

2. Pressure and Force

Pressure exerted by solid objects depends on contact area.

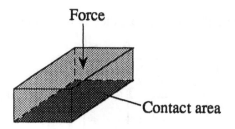

Force

Contact area

$$\text{Pressure} = \frac{\text{Force}}{\text{Area}}$$

Pressure exerted by a liquid depends on both liquid depth and density.

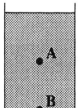

The pressure at Point B is greater than the pressure at Point A

$$\text{Pressure} = (\text{depth}) \, (\text{Density})$$

or

$$\text{Pressure} = dD$$

Since d and D may be confused, the depth is generally expressed as height, h:

$$\text{Pressure} = hD$$

SUMMARY—Cont'd

2. Pressure and Force—Cont'd

The total force on a surface is the sum of all unit pressures against it. The total force against the bottom of a tank is:

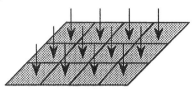

$$\text{Total Force, lbs} = \underset{\text{lbs/sq ft}}{(\text{Pressure})} \; \underset{\text{sq ft}}{(\text{Area})}$$

or

$$\text{Total Force, lbs} = \underset{\text{lbs/sq in.}}{(\text{Pressure})} \; \underset{\text{sq in.}}{(\text{Area})}$$

The total force against the side of a tank is:

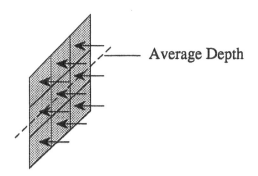

Average Depth

$$\text{Total Force, lbs} = \underset{\text{Depth, lbs/sq ft}}{(\text{Pressure at Average})} \; \underset{\text{sq ft}}{(\text{Total Area})}$$

or

$$\text{Total Force, lbs} = \underset{\text{Depth, lbs/sq in.}}{(\text{Pressure at Average})} \; \underset{\text{sq in.}}{(\text{Total Area})}$$

SUMMARY—Cont'd

2. Pressure and Force—Cont'd

The center of force on the side of a tank filled with water is located at a point two-thirds of the way down from the water surface:

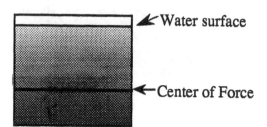

Front View of Wall

$$\text{Center of Force} = \frac{(2)}{3} \text{(Depth of) Water}$$

A hydraulic press operates on the principle of total force. The pressure applied at the smaller piston (Point A) is transferred through the liquid to the larger piston (Point B).

Point A

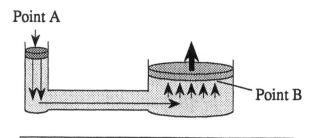

Point B

$$\text{Total Force at Point B} = \text{(Pressure) (Total Area)}\ \text{at Point B of Point B}$$

Gage pressures do not include atmospheric pressure. Absolute pressure includes both gage and atmospheric pressures.

$$\text{Absolute Pressure, psi} = \text{Gage Pressure, psi} + \text{Atmos. Pressure, psi}$$

SUMMARY—Cont'd

3. Head and Head Loss

When the two water surfaces are located above the pump, static head is the difference in water surface elevations:

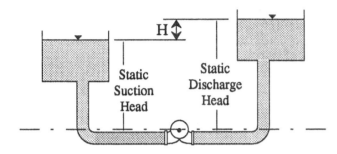

Total Static Head, ft	=	Higher Elevation, ft	−	Lower Elevation, ft

When the water surface on the suction side of the pump is below the pump centerline, the two distances must be added:

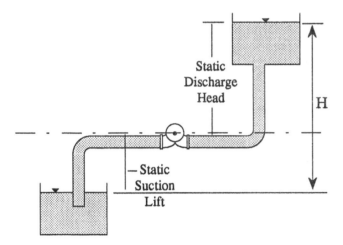

Total Static Head, ft	=	Height Above Pump Centerline, ft	+	Distance Below Pump Centerline, ft

SUMMARY—Cont'd

3. Head and Head Loss—Cont'd

Dynamic head is the static head plus friction and minor head losses:

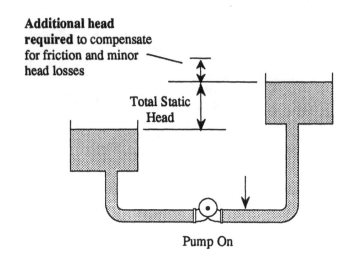

Additional head required to compensate for friction and minor head losses

Total Static Head

Pump On

$$\begin{array}{c} \text{Total Dynamic} \\ \text{Head, ft} \end{array} = \begin{array}{c} \text{Total Static} \\ \text{Head, ft} \end{array} + \begin{array}{c} \text{Head Losses} \\ \text{ft} \end{array}$$

Friction and minor head losses may be determined using hydraulics tables, such as those shown in this chapter.

4. Horsepower

The foundation of all horsepower problems is power—ft-lbs/min:

$$\text{ft} \times \frac{\text{lbs}}{\text{min}} = \frac{\text{ft-lbs}}{\text{min}}$$

$$\text{hp} = \frac{\text{ft-lbs/min}}{33,000 \text{ ft-lbs/min/hp}}$$

SUMMARY—Cont'd

4. Horsepower—Cont'd

Motor, brake, and water horsepower are terms used to indicate where the horsepower is measured.

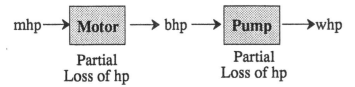

mhp ⟶ **Motor** ⟶ bhp ⟶ **Pump** ⟶ whp

Partial
Loss of hp

Partial
Loss of hp

The equations for motor, brake, and water horsepower are:

$$\text{Brake hp} = \frac{\text{Water hp}}{\dfrac{\text{Pump Effic.}}{100}}$$

$$\text{Motor hp} = \frac{\text{Brake hp}}{\dfrac{\text{Motor Effic.}}{100}}$$

$$\text{Motor hp} = \frac{\text{Water hp}}{\dfrac{(\text{Motor Effic.})}{100}\dfrac{(\text{Pump Effic.})}{100}}$$

Each of the three equations above may be rearranged as follows:

$$\text{Water hp} = \frac{(\text{Brake hp})(\text{Pump Effic.})}{100}$$

$$\text{Brake hp} = \frac{(\text{Motor hp})(\text{Motor Effic.})}{100}$$

$$\text{Water hp} = \frac{(\text{Motor hp})(\text{Motor})(\text{Pump})}{\dfrac{\text{Effic.}}{100}\dfrac{\text{Effic.}}{100}}$$

SUMMARY—Cont'd

4. Horsepower—Cont'd

When pumping liquids with a specific gravity different than that of water, the specific gravity factor must be added to the ft-lbs/min calculation:

$$\text{(ft)(lbs/min)(spec. grav)} = \begin{array}{l}\text{ft-lbs/min}\\\text{for different}\\\text{liquid}\end{array}$$

To calculate pumping costs, first calculate the kilowatt-hours (kWh) power consumption:

$$\begin{array}{l}\text{(kW)(Hrs of Pump)} = \text{kWh}\\\text{draw \ \ Operation} \quad \text{power}\\\qquad\qquad\qquad\qquad\text{consumed}\end{array}$$

Then determine pumping cost:

$$\begin{array}{l}\text{(kWh) \ (Cost/kWh)} = \text{Total}\\\text{Power Use \ \ \ Use} \qquad \text{Cost}\end{array}$$

5. Pump Capacity

These calculations are based on a volume of wastewater or sludge pumped during a specific time period. The two general types of pumping rate calculations include:

- Pumping into or out of a tank, and

- Positive displacement calculations

General Equation:

$$\begin{array}{l}\text{Pumping}\\\text{Rate, gpm}\end{array} = \frac{\text{gallons pumped}}{\text{minutes}}$$

SUMMARY—Cont'd

Specific Equations:

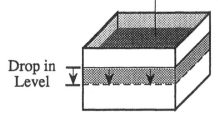

→Volume Pumped, gal

Drop in Level

When Influent Valve Is Closed—

$$\text{Pumping Rate, gpm} = \frac{\text{gallons pumped}}{\text{minutes}}$$

$$\text{Pumping Rate, gpm} = \frac{(\text{Length, ft}) (\text{Width, ft}) (\text{Drop, ft}) (7.48 \text{ gal/cu ft})}{\text{minutes}}$$

When Influent Valve Is Open—

The water level will drop if the pump is pumping at a rate greater than the influent flow:

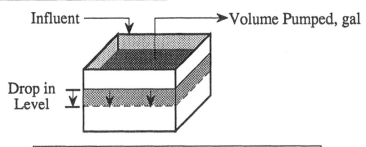

Influent ───→ →Volume Pumped, gal

Drop in Level

$$\text{Pumping Rate, gpm} = \text{Influent Flow, gpm} + \text{Drop in Level, gpm}$$

The water level will rise if the pump is pumping at a rate less than the influent flow:

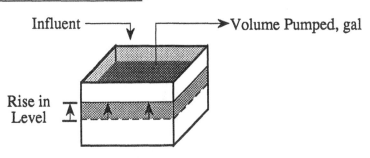

Influent ───→ →Volume Pumped, gal

Rise in Level

$$\text{Pumping Rate, gpm} = \text{Influent Flow, gpm} - \text{Rise in Level, gpm}$$

7.1 DENSITY AND SPECIFIC GRAVITY

DENSITY

The "lightness" or "heaviness" of an object is the layman's term for what scientists refer to as the density of an object. For example, petrified wood is heavy when compared to volcanic ash, and lead is heavy compared to aluminum. **Such comparisons presume a given volume.** That is, any given volume of lead is heavier than the same volume of aluminum. Without keeping volume constant, no comparison between objects or substances may be made.

The **density** of a substance is therefore the amount of matter or "mass" in a given volume of that substance. It is normally measured in lbs/cu ft. Tables listing the densities of a variety of substances are available in chemical and engineering handbooks. A listing of a few substances is given below.

Substance	Density (lbs/cu ft)
Water	62.4
Seawater	64
Gasoline	44
Aluminum	170
Lead	700
Wood, pine	30

In the water and wastewater field, the density of water and other liquids is commonly measured in lbs/gal. This is simply another expression of mass per unit volume.

DENSITY IS MEASURED IN LBS/CU FT OR LBS/GAL

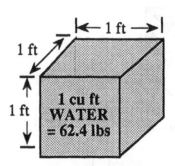

The density of water is 62.4 lbs/cu ft

The density of water is 8.34 lbs/gal

Example 1: (Density and Specific Gravity)
❑ A gallon of solution is weighed. After the weight of the container is subtracted, it is determined that the weight of the solution is 9.8 lbs. What is the density of the solution?

$$\text{Density} = \frac{\text{lbs of Solution}}{\text{gal of Solution}}$$

$$= \boxed{\frac{9.8 \text{ lbs}}{1 \text{ gal}}} \text{ or } \boxed{9.8 \text{ lbs/gal}}$$

Example 2: (Density and Specific Gravity)
❑ Suppose that only a half a gallon had been weighed in the example above, with a resulting weight of 4.9 lbs. How would density be calculated?

$$\text{Density} = \frac{\text{lbs of Solution}}{\text{gal of Solution}}$$

$$= \frac{4.9 \text{ lbs}}{0.5 \text{ gal}}$$

$$= \boxed{9.8 \text{ lbs/gal}}$$

Example 3: (Density and Specific Gravity)
❑ The density of a substance is given as 76.3 lbs/cu ft. What is this density expressed in lbs/gal?

To make a conversion in densities, first sketch the box diagram:

$$\boxed{\text{lbs/gal}} \xleftarrow{\;7.48\;} \boxed{\text{lbs/cu ft}}$$

Converting from lbs/cu ft to lbs/gal, the move is from a larger box to a smaller box. Therefore, division is indicated:

$$\frac{76.3 \text{ lbs/cu ft}}{\dfrac{7.48 \text{ lbs/cu ft}}{1 \text{ lb/gal}}} = \boxed{10.2 \text{ lbs/gal}}$$

Example 4: (Density and Specific Gravity)
❑ The density of a solution is 9.6 lbs/gal. What is the density expressed as lbs/cu ft?

First sketch the box diagram:

$$\boxed{\text{lbs/gal}} \xrightarrow{\;7.48\;} \boxed{\text{lbs/cu ft}}$$

Converting from lbs/gal to lbs/cu ft involves a move from the smaller box to the larger box. Therefore, multiplication is indicated:

$$\frac{(9.6 \text{ lbs/gal})(7.48 \text{ lbs/cu ft})}{1 \text{ lbs/gal}} = \boxed{71.8 \text{ lbs/cu ft}}$$

ESTIMATING THE DENSITY OF A SUBSTANCE

You can estimate the density of a substance by weighing a known volume of it. For example, to estimate the density of sludge being pumped, weigh a gallon sample of it. (Be sure to subtract the weight of the container.) You will then have the lbs/gal density of that sludge. The accuracy of your estimate, of course, depends on whether the sludge sample is a representative sample.

LBS/CU FT AND LBS/GAL EXPRESSIONS OF DENSITY

Occasionally you may know the density of a substance expressed in lbs/cu ft but need to know the density in lbs/gal, or vice versa. To make such a conversion, you may use the following box diagram:* (based on the conversion equation, 1 lb/gal = 7.48 lbs/cu ft)

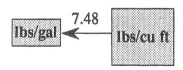

One aspect of this diagram is very different from other box diagrams. In other box diagram conversions, the same quantity is simply expressed in different units. For example, a quantity of water expressed as gallons is re-expressed in terms of cubic feet or pounds. The quantity of water has not changed, only how it is described.

In these density conversions, however, you are converting the weight of one quantity of water to the equivalent weight of a different quantity of water. Remember that the smaller box is associated with the smaller quantity of water (note <u>gallons</u> in the denominator of the smaller box), and the larger box is associated with the larger quantity of water (note <u>cubic feet</u> in the denominator of the larger box).

---

* For a review of the "box diagram" method of conversions, refer to Chapter 8 in *Basic Math Concepts*.

SPECIFIC GRAVITY

Two expressions of density have been mentioned thus far—lbs/cu ft and lbs/gal. There is a third measurement, a metric measurement, of density: grams per cubic centimeter (g/cm^3). With three different ways to express density, comparison of one density to another can be difficult. This problem is resolved by the use of specific gravity.

The density of water was established as the "standard" and all other densities are then compared to that of water. **The specific gravity of any liquid* is therefore the ratio or comparison of the density of a substance to the density of water.** Practically speaking, the specific gravity of a liquid may be determined by weighing a given volume of that liquid and then dividing that number by the weight of the same volume of water.

The specific gravity of water is one, since comparing the density of water to the density of water results in the following calculation:

$$\frac{62.4 \text{ lbs/cu ft}}{62.4 \text{ lbs/cu ft}} = 1.0$$

The specific gravity of seawater (with a density of 64 lbs/cu ft) would be:

$$\frac{64 \text{ lbs/cu ft}}{62.4 \text{ lbs/cu ft}} = 1.03$$

Any substance with a density greater than that of water will have a specific gravity greater than 1.0. And any substance with a density less than that of water will have a specific gravity less than 1.0.**

SPECIFIC GRAVITY IS A COMPARISON (OR RATIO) OF DENSITIES

The densities of all liquids are compared to the density of water.

$$\text{Specific Gravity of Liquids} = \frac{\text{Density of the Liquid}}{\text{Density of Water}}$$

The densities of all gases are compared to the density of air.

$$\text{Specific Gravity of Gases} = \frac{\text{Density of the Gas}}{\text{Density of Air}}$$

Example 5: (Density and Specific Gravity)
❑ Using a density of 44 lbs/cu ft for gasoline, what is the specific gravity of gasoline?

The specific gravity of gasoline is the comparison, or ratio, of the density of gasoline to that of water:

$$\text{Specific Gravity of Gasoline} = \frac{\text{Density of Gasoline}}{\text{Density of Water}}$$

$$= \frac{44 \text{ lbs/cu ft}}{62.4 \text{ lbs/cu ft}}$$

$$= \boxed{0.71}$$

(Note: both the density and specific gravity of gasoline indicate that gasoline will float in water.)

---

* The specific gravity of gases is based on a standard of air rather than water. These densities are very dependent on pressure and temperature. Therefore density listings of gases normally include pressure and temperature readings.

** The density (and specific gravity) of a substance indicates whether it will sink or float in water. If its density is greater than that of water, it will sink; if less, it will float.

Example 6: (Density and Specific Gravity)
❑ You wish to determine the specific gravity of a solution. After weighing a gallon of solution and subtracting the weight of the container, the solution is found to weigh 9.17 lbs. What is the specific gravity of the solution?

To determine the specific gravity of the solution, its density must be first be determined. Since one gallon of the solution weighed 9.17 lbs, the density is 9.17 lbs/gal.

Now compare the density of the solution to that of water to determine specific gravity:

$$\text{Specific Gravity} = \frac{\text{Density of the Solution}}{\text{Density of Water}}$$

$$= \frac{9.17 \text{ lbs/gal}}{8.34 \text{ lbs/gal}}$$

$$= \boxed{1.1}$$

DETERMINING THE SPECIFIC GRAVITY OF A SUBSTANCE

To determine the specific gravity of a substance, you must first determine its density (described on the previous two pages). Then the density of that substance is compared to the density of water. Example 6 illustrates such a problem.

Example 7: (Density and Specific Gravity)
❑ The specific gravity of a liquid is 0.95. What is the density of that liquid?(Density of water = 8.34 lbs/gal).

The specific gravity of a liquid can be used to determine its density. Multiply the specific gravity times the density of water:

(Spec. Grav.) (Density) = Density
of the Liq. of water   of the
                        Liquid

$$(0.95)(8.34 \text{ lbs/gal}) = \boxed{7.9 \text{ lbs/gal}}$$

WHEN SPECIFIC GRAVITY IS KNOWN AND DENSITY IS UNKNOWN

If you know the specific gravity of any substance, you can always determine its density by multiplying the specific gravity by the density of water:

(Specific)	(Density)	= Density
Gravity	of Water,	of the
of a	always	Liquid,
Liquid	8.34 lbs/gal	lbs/gal

Example 7 illustrates this type of calculation.

7.2 PRESSURE AND FORCE

Force is a push or pull measured in terms of weight, such as pounds or kilograms. The force on the bottom of an open tank, for example, is a measure of the weight of the water above it. The deeper the water, the more force on the bottom of the tank.

Although the force exerted against the entire tank bottom is an important calculation, we will first focus on another calculation related to force—that of pressure.

Pressure is a measure of the force or weight pushing against a specified area, usually a square inch or a square foot. Thus, pressure is normally expressed in pounds per square inch (lbs/sq in. or psi) or pounds per square foot (lbs/sq ft). The general equation used in calculations of pressure is:

$$\text{Pressure} = \frac{\text{Force}}{\text{Area}}$$

PRESSURE DEPENDS ON CONTACT AREA

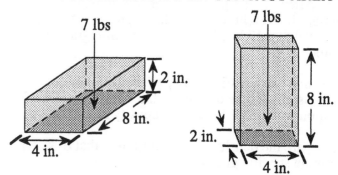

A brick is set on a table, as shown. If the 7 lbs of force (weight) is spread over the 32 sq in. of its side;* the pressure against the table at the point of contact is:

$$\frac{7 \text{ lbs}}{32 \text{ sq in.}} = \boxed{0.2 \text{ lbs/sq in.}}$$

A brick is placed on a table, as shown. Now the 7 lbs of force (weight) is spread over the 8 sq in. of its bottom. The pressure against the table at the point of contact is:

$$\frac{7 \text{ lbs}}{8 \text{ sq in.}} = \boxed{0.9 \text{ lbs/sq in.}}$$

Example 1: (Pressure and Force)

❑ The object shown below weighs 30 lbs. What is the lbs/sq in. pressure at the surface of contact?

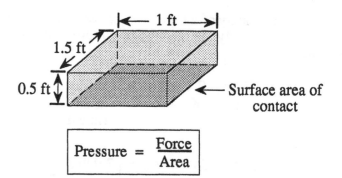

Surface area of contact

$$\text{Pressure} = \frac{\text{Force}}{\text{Area}}$$

Because lbs/sq in. pressure is desired, the dimensions will be expressed in inches rather than ft:

$$\text{Pressure} = \frac{30 \text{ lbs}}{(12 \text{ in.})(18 \text{ in.})}$$

$$= \boxed{0.14 \text{ lbs/sq in.}}$$

---

* For a review of area calculations, refer to Chapter 10, "Area Measurements", in *Basic Math Concepts*.

Example 2: (Pressure and Force)

❑ Compare the pressures, in lbs/sq ft, on the contact area for the two positions of the object shown below. The object weighs 300 lbs.

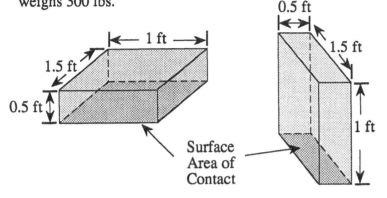

Surface Area of Contact

$$\text{Pressure} = \frac{\text{Force}}{\text{Area}} \qquad \text{Pressure} = \frac{\text{Force}}{\text{Area}}$$

$$= \frac{300 \text{ lbs}}{(1.5 \text{ ft})(1 \text{ ft})} \qquad = \frac{300 \text{ lbs}}{(0.5 \text{ ft})(1.5 \text{ ft})}$$

$$= \boxed{200 \text{ lbs/sq ft}} \qquad = \boxed{400 \text{ lbs/sq ft}}$$

Example 3: (Pressure and Force)

❑ An object rests on the floor. The pressure at the surface of contact is 0.5 lbs/sq in.. If the object is placed on another side that has only one-third the contact area, what is the new lbs/sq in. pressure?

Since the area and pressure are inversely related, with a decrease in contact area there will be an increase in pressure. From this, we know that the new pressure should be greater than 0.5 lbs/sq in.:

Area decreased to 1/3 ⟶ Pressure increased 3 times

$$(0.5 \text{ lbs/sq in.})(3) = \boxed{1.5 \text{ lbs/sq in.}}$$

---

* Inverse or indirect proportions are described in Chapter 7, "Ratios and Proportions", in *Basic Math Concepts*.

PRESSURE AND AREA ARE INVERSELY RELATED

As the pressure equation indicates, pressure reflects both the force exerted and the contact area. If the force is increased, the numerator is larger, resulting in a larger pressure reading. And if the force is decreased, the resulting pressure reading is also decreased. This is called a **direct proportion**—as force increases, pressure increases; and as force decreases, pressure decreases.

But what is the effect of changes in area? Assuming a constant force, if the area increases, the denominator of the fraction increases, resulting in a smaller pressure reading. Likewise, if the area is decreased, the denominator is thereby decreased, resulting in a larger pressure reading. This is called an **inverse or indirect proportion***—as one increases, the other decreases, and vice versa.

Apart from the mathematics, this reasoning makes sense. As the area decreases, the same weight is distributed over a smaller area. Therefore each square inch receives a greater force.

Assuming the force remains constant, if the area of contact is cut in half, the pressure is increased two-fold. If the area is reduced to one-quarter of its size, the pressure is increased fourfold. Note the relationship between these changes:

Area Decreased to	Pressure Increased by Factor of
1/2 size	2
1/4 size	4

Other numbers apply as well—1/8 and 8, etc. This inverse relationship also occurs when the area is increased from its original size.

Area Increased to	Pressure Decreased by Factor of
2 times size	1/2
4 times size	1/4

Example 3 illustrates a calculation using inverse relationships.

LIQUID PRESSURE

In a liquid at rest, such as in a container, tank, or reservoir, the pressure at any one point is exerted in all directions, not just toward the bottom contact surface. It is exerted toward the sides and top as well.

The amount of pressure at any point in the water depends on two factors:

• Depth (measured vertically) and

• Density.

The greater the water depth, the greater the pressure. At deeper and deeper levels of water, there is an increased weight of water above. As shown in the diagram at the top of the facing page, the pressure at 1 ft depth is 0.433 psi, while the pressure at 2.31 ft depth is 1 psi.*

Pressure is also dependent on the density of the liquid. For example, at point A   in the box to the right, the pressure would be less than that at point A   , because the liquid is less dense at A   than at A   . The equation used to determine pressure in a liquid is shown to the right.

Note how this equation results in the same units as the general equation for pressure:

$$P = hD$$

$$P = \frac{(ft)(lbs)}{ft^3}$$

$$P = \frac{lbs}{ft^2} \begin{matrix} \leftarrow(force) \\ \leftarrow(area) \end{matrix}$$

PRESSURE DEPENDS ON DEPTH

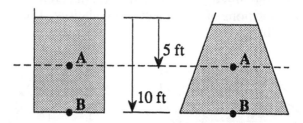

Twice the depth means twice the pressure, regardless of the shape of the container. The pressure at Point B is twice that at Point A.

PRESSURE DEPENDS ON DENSITY

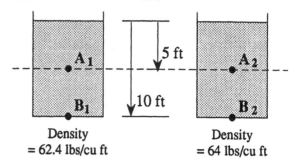

Density = 62.4 lbs/cu ft

Density = 64 lbs/cu ft

The pressures at A_1 and B_1 are less than corresponding pressures at A_2 and B_2. This is due to differences in the density of the liquids.

THE PRESSURE EQUATION INCLUDES BOTH DEPTH AND DENSITY

$$\boxed{Pressure = (depth)\ (Density)}$$

Due to possible confusion with abbreviation of terms, depth is replaced by height, *h*. Using *D* for density, the equation is written as:

$$\boxed{Pressure = hD}$$

Example 4: (Pressure and Force)
❑ What is the pressure (in lbs/sq ft) at a point 8 feet below the surface of the water? (The density of water is 62.4 lbs/cu ft.)

$$Pressure = hD$$

$$= (8\ ft)(62.4\ lbs/cu\ ft)$$

$$= \boxed{499\ lbs/sq\ ft}$$

---

* This is gage pressure, not absolute pressure. These terms are described later in this section.

PRESSURE AND PSI

1 ft = 0.433 psi

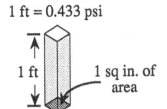

1 ft

1 sq in. of area

The weight bearing down on each square inch of area is about a half pound (0.433 lbs) for each foot of depth.

1 psi = 2.31 ft

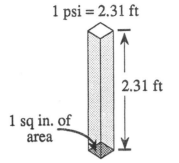

2.31 ft

1 sq in. of area

At a depth of 2.31 ft, there is one pound of weight bearing down on each square inch of area.

Example 5: (Pressure and Force)

❏ What is the pressure (in psi) at a point 12 ft below the surface?

Using the equation 1 psi = 2.31 ft, draw the box diagram:

$$\boxed{\text{psi}} \longleftarrow \overset{2.31}{\longleftarrow} \boxed{\text{ft}}$$

Converting from feet to psi, the move is from a larger box to a smaller box. Division by 2.31 is therefore indicated:

$$\frac{12 \text{ ft}}{2.31 \text{ ft/psi}} = \boxed{5.2 \text{ psi}}$$

Example 6: (Pressure and Force)

❏ At a point 3 ft below the liquid surface, what is the pressure in psi? (The specific gravity of the liquid is 0.95.)

First calculate psi as usual, using the box diagram:

$$\boxed{\text{psi}} \longleftarrow \overset{2.31}{\longleftarrow} \boxed{\text{ft}}$$

Division is indicated:

$$\frac{3 \text{ ft}}{2.31 \text{ ft/psi}} = \boxed{1.3 \text{ psi}}$$

The specific gravity may now be taken into account. A specific gravity less than that of water will result in a smaller psi reading for the same water depth:

Pressure at Different = (psi)(specific gravity)
Sp. Grav.

$$= (1.3 \text{ psi})(0.95)$$

$$= \boxed{1.2 \text{ psi}}$$

USING PSI AND FT

Since the density of water is a given in most water and wastewater calculations, the $P = hD$ equation can be shortened. The pressure at 1 ft depth is always 62.4 lbs/sq ft or 0.433 lbs/sq in.*

$$\boxed{1 \text{ ft} = 0.433 \text{ psi}}$$

When the "box method" of conversion is used, however, **both numbers of the equation must be greater than one.** The equation shown above may be easily converted to the desired form by dividing both sides of the equation by 0.433, as follows:

$$\frac{1 \text{ ft}}{0.433} = \frac{0.433 \text{ psi}}{0.433}$$

$$\boxed{2.31 \text{ ft} = 1 \text{ psi}}$$

Now rearrange the equation so the one is on the left side of the equation:

$$\boxed{1 \text{ psi} = 2.31 \text{ ft}}$$

This equation can be used in all conversions between feet and psi. It is recommended that this equation be memorized. Example 5 illustrates a conversion between feet and psi.

PRESSURE AND SPECIFIC GRAVITY

Specific gravity can be used to determine pressures within liquids of different densities. First, calculate the pressure in psi using the equation given above (1 psi = 2.13 ft). Then multiply the psi result by the specific gravity of the liquid. Example 6 illustrates this calculation.

---

* To verify this $P = (1 \text{ ft})(62.4 \text{ lbs/cu ft}) = 62.4$ lbs/sq ft. To convert to lbs/sq in., divide by 144 sq in./sq ft.
  For a discussion of square terms conversions, refer to Chapter 8 in *Basic Math Concepts*.
** The box method of conversions is described in Chapter 8 in *Basic Math Concepts*.

TOTAL FORCE

The total force of water against the side or bottom of a tank or wall is determined by multiplying the pressure at that depth times the entire area:

Total Force = (Pressure)(Area)

If the pressure is given as lbs/sq ft, the area must be expressed as square feet. And if the pressure is given in lbs/sq in., the area must be expressed as square inches.

Total = (Pressure)(Area)
Force   lbs/sq ft   sq ft

Total = (Pressure)(Area)
Force   lbs/sq in. sq in.

Calculating the total force against the bottom of a tank is a straight-forward calculation—simply determine the pressure at the bottom of the tank (based on the vertical distance beneath the surface of the water) and multiply this pressure by the area of the bottom.

Calculating the total force against the side of a tank is a little different—which pressure value should be used? The pressure at the water surface is zero*, and the pressures increase toward the bottom of the tank. The pressure to be used in these calculations is the **average pressure**. The average pressure occurs at half of the water depth, or what might be termed the **average depth**. Average depth may be determined as follows:

Average   Total Depth, ft
Depth, ft = ─────────────────
                     2

Example 8 illustrates this type of calculation.

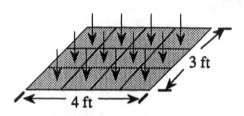

TOTAL FORCE AGAINST THE BOTTOM OF A TANK

3 ft

4 ft

Total = (Pressure) (Area)
Force   lbs/sq ft   sq ft

TOTAL FORCE AGAINST THE SIDE OF A TANK

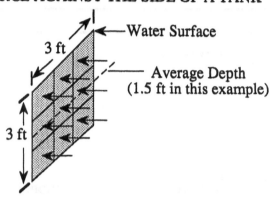

Water Surface

3 ft

Average Depth
(1.5 ft in this example)

3 ft

Total = (Pressure at Average) (Total Area)
Force   Depth, lbs/sq ft   sq ft

Example 7: (Pressure and Force)
❑ What is the total force against the bottom of a tank 30 ft long and 10 ft wide? The water depth is 8 ft.

First, calculate the pressure at the bottom of the tank. At 8 ft depth, the pressure is:

$$\boxed{\text{psi}} \xleftarrow{2.31} \boxed{\text{ft}}$$

$$\frac{8 \text{ ft}}{2.31 \text{ ft/psi}} = 3.46 \text{ psi}$$

Then calculate total force. Since pressure is given in psi, the dimensions must be reported in inches (30 ft x 12 in/ft = 360 in.; 10 ft x 12 in./ft = 120 in.)

Total Force = (Pressure, psi) (Area, sq in.)

= (3.46 psi) (360 in.) (120 in.)

= 149,472 lbs

*Referring to gage pressure. Absolute and gage pressures are described later in this section.

Example 8: (Pressure and Force)
❏ What is the total force exerted on the side of a tank if the tank is 16 ft wide and the water depth is 12 ft?

Since the water depth is 12 ft, the halfway point (or average water depth) is 6 ft. To calculate total force, the pressure at the average depth must be calculated. The pressure at a depth of 1 ft = 62.4 lbs/sq ft, so the pressure at 6 ft would be:

$$(6)(62.4 \text{ lbs/sq ft}) = 374 \text{ lbs/sq ft}$$
average pressure

Now the total force can be calculated:

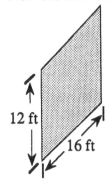

$$\begin{aligned}\text{Total Force} &= \text{(Pressure at Average Depth, lbs/sq ft) (Total Area sq ft)} \\ &= (374 \text{ lbs/sq ft}) (12 \text{ ft}) (16 \text{ ft}) \\ &= \boxed{71{,}808 \text{ lbs}}\end{aligned}$$

An equation sometimes given for the total force on the side of a tank is:

$$F = (31.2) (H^2) (L),$$

where F = force in lbs, H = head in ft, and L = length of wall in ft. This equation is simply an abbreviation of the typical total force equation, as shown below: (note the water depth is written as H, ft of head)

$$\begin{aligned}\text{Total Force, lbs} &= \text{(Pressure)(Area)} \\ &\qquad \text{lbs/sq ft} \quad \text{sq ft} \\[4pt] &= \text{(Aver. height)(Den.) (Area)} \\[4pt] &= \frac{(H)}{2} (62.4) (H) (L) \\[4pt] &= (31.2) (H^2) (L)\end{aligned}$$

TOTAL FORCE VS CENTER OF FORCE

Although the **total force** on a side wall is calculated using the average depth (or average pressure, since these are related), the center of force is not located at the halfway line—it is located along a line **two-thirds of the way down from the water surface.**

This is because the pressure against the wall increases with depth (**forming what is called a "triangular load"**). As shown in the diagram to the left, more of the force against the wall is located nearer the bottom. To calculate the center of force on a wall, simply multiply the water depth by 2/3:

THE CENTER OF FORCE IS LOCATED ALONG A LINE 2/3 FROM THE SURFACE

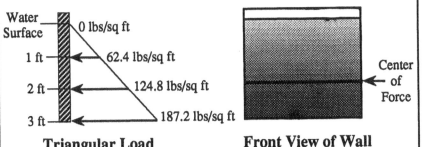

Triangular Load
(Side View of Wall)

Front View of Wall

In Example 8, the total force against the wall is 71,808 lbs. Where is the center of force located?

$$\begin{aligned}\text{Center of Force} &= \frac{(2)}{3} \text{(Depth of Water)} \\[4pt] &= \frac{(2)}{3} (12 \text{ ft}) \\[4pt] &= \boxed{8 \text{ ft}}\end{aligned}$$

$$\boxed{\text{Center of Force} = \frac{(2)}{3} \text{(Depth of Water)}}$$

HYDRAULIC PRESS

The operation of the **hydraulic press** or **hydraulic jack** is based on two primary principles:

• Force applied to a liquid is distributed equally within that liquid, and

• Total Force = (Pressure)(Area)

As illustrated in the graphic to the right, the force applied to the smaller cylinder is distributed evenly throughout the liquid. The larger cylinder has a greater surface area, so the total force applied is magnified several times.

To calculate the total force on the large cylinder, you must know the pressure against it (which is the same as that applied to the smaller cylinder) and the area. Examples 9 and 10 illustrate hydraulic press calculations.

THE HYDRAULIC PRESS OPERATES ON THE PRINCIPLE OF TOTAL FORCE

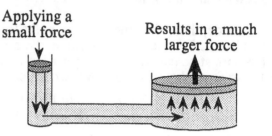

Applying a small force → Results in a much larger force

Example 9: (Pressure and Force)

❏ The force applied to the small cylinder (12-in. diameter) of a hydraulic jack is 35 lbs. If the diameter of the large cylinder is 3 ft, what is the total lifting force?

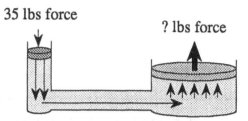

35 lbs force
? lbs force

First calculate the pressure applied to the small cylinder.* The pressure is calculated in lbs/sq ft since the cylinder diameters are given in ft:

$$\text{Pressure} = \frac{\text{Force, lbs}}{\text{Area, sq ft}}$$

$$= \frac{35 \text{ lbs}}{(0.785)(1 \text{ ft})(1 \text{ ft})}$$

$$= 45 \text{ lbs/sq ft}$$

The total force at the large cylinder can now be calculated:

$$\text{Total Force} = (\text{Pressure})(\text{Area})$$

$$= (45 \text{ lbs/sq ft})(0.785)(3 \text{ ft})(3 \text{ ft})$$

$$= \boxed{318 \text{ lbs}}$$

---

* For a review of circular area calculations, refer to Chapter 10 in *Basic Math Concepts*.

Example 10: (Pressure and Force)

❑ The force applied to the small cylinder of a hydraulic jack is 50 lbs. The diameter of the small cylinder is 18 inches. If the diameter of the large cylinder is 40 inches, what is the total lifting force?

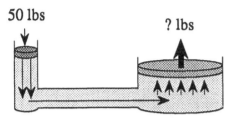

50 lbs

? lbs

First calculate the pressure applied to the small cylinder. Since the diameters of the cylinders are given in inches, the pressure will be calculated as lbs/sq in.:

$$\text{Pressure} = \frac{\text{Force, lbs}}{\text{Area, sq in.}}$$

$$= \frac{50 \text{ lbs}}{(0.785)(18 \text{ in.})(18 \text{ in.})}$$

$$= 0.2 \text{ lbs/sq in.}$$

The same pressure is transmitted to the larger cylinder. The total force at the large cylinder is calculated as:

$$\text{Total Force} = (\text{Pressure})(\text{Area})$$

$$= \frac{(0.2 \text{ lbs})(0.785) (40 \text{ in.}) (40 \text{ in.})}{\text{sq in.}}$$

$$= \boxed{251.2 \text{ lbs}}$$

Example 11: (Pressure and Force)

❑ A gage reading is 25 psi. What is the absolute pressure at the gage? (Assume sea level atmospheric pressure.)

$$\begin{matrix}\text{Absolute} \\ \text{Pressure, psi}\end{matrix} = \begin{matrix}\text{Gage Pressure} \\ \text{psi}\end{matrix} + \begin{matrix}\text{Atmospheric} \\ \text{Pressure, psi}\end{matrix}$$

$$= 25 \text{ psi} + 14.7 \text{ psi}$$

$$= \boxed{39.7 \text{ psi}}$$

GAGE VS. ABSOLUTE PRESSURES

When water pressures are measured in a tank or pipeline, they are measured by gages. These measurements are therefore called **gage pressures.** Gage pressures do not include all the pressures acting in the tank or pipeline—**they do not include atmospheric pressure.**

Atmospheric pressure is generally not considered in water and wastewater calculations, because atmospheric pressure is exerted both inside and outside the tank or pipeline. At sea level, the atmospheric pressure is 14.7 psi. The name given to the total pressure, including both gage and atmospheric pressures, is **absolute pressure.** Absolute pressure is calculated by adding the gage pressure and atmospheric pressure:

$$\begin{matrix}\text{Absolute} \\ \text{Press.,} \\ \text{psi}\end{matrix} = \begin{matrix}\text{Gage} \\ \text{Press.} \\ \text{psi}\end{matrix} + \begin{matrix}\text{Atmos.} \\ \text{Press.,} \\ \text{psi}\end{matrix}$$

Example 11 illustrates a calculation involving gage and absolute pressures.

7.3 HEAD AND HEAD LOSS

HEAD TERMINOLOGY

When describing the various types of head against which a pump must operate, several different terms may be used, depending on the side of the pump and whether or not the pump is operating.

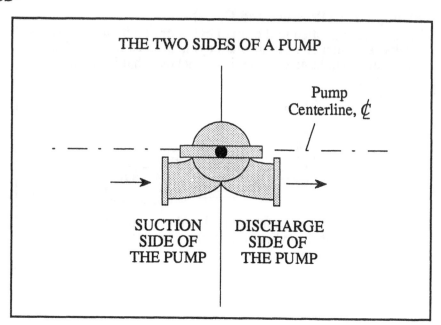

THE TWO SIDES OF A PUMP

SUCTION AND DISCHARGE HEADS

The terms **suction** and **discharge** indicate two different sides of the pump. As shown in the diagram to the right, the suction side of a pump is the inlet or low pressure side of the pump. The discharge side of a pump is the outlet or high pressure side of the pump.

Heads measured on the suction side of a pump are called **suction heads**. Heads measured on the discharge side of a pump are called **discharge heads**.

SUCTION HEAD, SUCTION LIFT, AND DISCHARGE HEAD

When the water feeding the pump is **above the pump**, this is called a **suction head**:

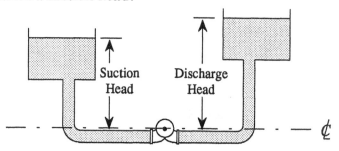

When the water feeding the pump is **below the pump**, this is called a **suction lift**:

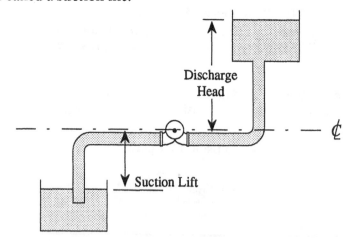

TOTAL STATIC HEAD IS THE VERTICAL DISTANCE BETWEEN THE TWO FREE WATER SURFACES

When the two water surfaces are located above the pump, the static suction head offsets part of the static discharge head. **The total static head is therefore the difference in height between the two heads:**

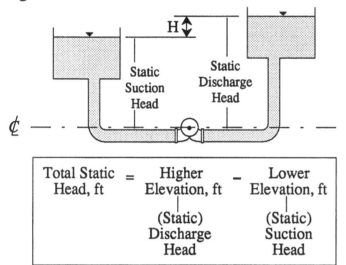

Total Static Head, ft	=	Higher Elevation, ft	–	Lower Elevation, ft
		(Static) Discharge Head		(Static) Suction Head

When the water surface on the suction side of the pump is below the pump centerline, there is no offsetting head on the suction side. **The total static head is therefore the sum of the two heads:**

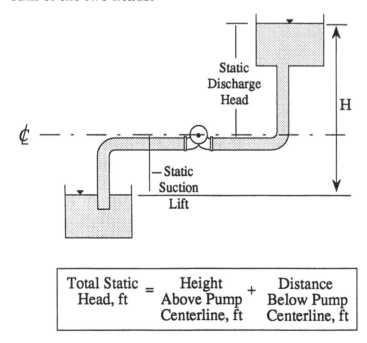

Total Static Head, ft	=	Height Above Pump Centerline, ft	+	Distance Below Pump Centerline, ft

STATIC HEAD

The total head against which a pump must operate is determined principally by two calculations:

- Total static head, and

- Friction and minor head losses

Head measurements taken on either side of the pump when the pump is off are called **static heads**. Both the static suction head and static discharge head are considered when determining the **total static head**, as illustrated in the diagrams to the left.

In simplified terms, however, total static head is a measure of the vertical distance between the two water surface elevations. It can be measured using elevations, vertical measurements, or gage pressures.

TOTAL DYNAMIC HEAD (TDH)

In addition to static head, the pump must work against **friction and minor head losses.** These are head losses resulting from friction as the water rubs against the pipeline and from friction and changes in direction as the water moves through valves and orifices.

For example, if the water is to be lifted 50 ft from Point A to Point B (as shown in the diagram to the right) and head losses are equal to 5 ft, the pump must produce a total of 55 ft of head to lift the water from Point A to Point B.

The total head against which the pump must operate is called the total dynamic head (TDH).

Total Dynamic Head, ft	=	Total Static Head, ft	+	Head Losses, ft

DYNAMIC HEAD IS THE STATIC HEAD PLUS FRICTION AND MINOR HEAD LOSSES

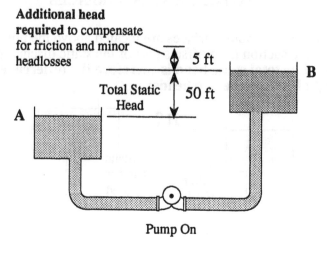

Additional head required to compensate for friction and minor headlosses — 5 ft

Total Static Head — 50 ft

A   B

Pump On

Example 1: (Head and Head Loss)

❏ The elevation of two water surfaces are given below. If the friction and minor head losses equal 9 ft, what is the total dynamic head (in ft)?

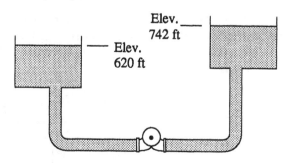

Elev. 742 ft

Elev. 620 ft

$$\text{Total Dynamic Head, ft} = \text{Total Static Head, ft} + \text{Head Losses, ft}$$

$$= (742 \text{ ft} - 620 \text{ ft}) + 9 \text{ ft}$$

$$= 122 \text{ ft} + 9 \text{ ft}$$

$$= \boxed{131 \text{ ft TDH}}$$

Example 2: (Head and Head Loss)
❑ The pump inlet and outlet pressure gage readings are given below. (Pump is off.) If the friction and minor head losses are equal to 12 ft, what is the total dynamic head (in ft)?

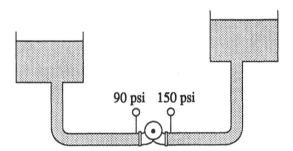

90 psi   150 psi

First calculate the static head in psi, then convert head in psi to head in ft. The static head is the difference in inlet and outlet pump pressures: 150 psi – 90 psi = 60 psi.

Next, convert 60 psi to ft*: (1 psi = 2.31 ft)

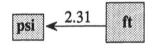

(60 psi) (2.31 ft/psi) = ⌈ 139 ft Static Head ⌉

Then calculate the total dynamic head:

$$TDH \;=\; Static\ Head \;+\; Head\ Losses$$

$$= \; 139\ ft \;+\; 12\ ft$$

$$= \; \boxed{151\ ft}$$

If the pressure is expressed in psi, it can be converted to feet, if desired.* Example 2 illustrates such a calculation.

Example 3: (Head and Head Loss)
❑ Readings taken from the inlet and outlet pressure gages of a pump while it is in operation are as follows: 87 psi and 143 psi, respectively. What is the TDH, in ft?

The difference in pressure readings is 143 psi – 87 psi = 56 psi. Convert this difference in psi to ft:

(56 psi)(2.31 ft/psi) = ⌈ 129 ft TDH ⌉

When a pump is operating, it operates against the heads and head losses described above. Inlet and outlet gage pressures **taken when the pump is operating**, therefore, are a good estimate of the total dynamic head. Example 3 illustrates this calculation.

* Refer to Section 7.2 for a discussion of psi and ft conversions.

FRICTION HEAD LOSS

Friction head losses within a pipeline depend on several factors:

- Velocity or rate of flow,

- Diameter of pipe,

- Length of pipe, and

- Pipe roughness.

Friction losses are determined by using tables such as that shown on the facing page. Values on this table are calculated using the Hazen-Williams formula, which includes a roughness coefficient, C. The smoother the pipe, the higher the C-value.

The table on the facing page is based on a C-value of 100. To determine friction loss:

1. Enter the table at the known pipe diameter. (Diameters are shown across the top of the table.)

2. Follow the column down until it is opposite the given flow rate, gpm.

3. Read the corresponding friction loss. The loss is given in feet and represents the friction loss for every 100-ft section of pipe.

4. Calculate the total friction loss for the pipeline by multiplying the friction loss by the number of 100-ft segments of pipe.

A wide variety of pipes in use today have C-values from 130 to 140 and greater. Reference books are available that include friction loss tables for various other C-values.

Note that these tables **do not apply when pumping sludges.**

Example 4: (Head and Head Loss)

❑ A 6-inch diameter pipe has a C-value of 100. When the flow rate is 240 gpm, what is the friction loss for a 2000-ft length of pipe?

Enter the top column at "6-in. Pipe". Come down the column to the entry across from 240 gpm. The friction loss shown is 0.87 ft per 100-ft segments.

$$2000. \longrightarrow \text{There are 20 100-ft segments of pipe}$$

hundreds place
decimal

$$\begin{array}{c}\text{Total Friction}\\ \text{Loss, ft}\end{array} = \begin{array}{c}(0.87\ \text{ft})(20\ \text{segments})\\ \text{of 100-ft}\end{array}$$

$$= \boxed{17.4\ \text{ft}}$$

Example 5: (Head and Head Loss)

❑ Flow through an 8-inch pipeline is 1300 gpm. The C-value is 100. What is the friction loss through a 3500-ft section of pipe ?

Enter the table at the 8-inch diameter heading and follow the column down until you are across from the 1300 gpm value. The friction loss indicated is 4.85 ft per 100-ft segment of pipeline.

$$3500. \longrightarrow \text{There are 35 100-ft segments of pipe}$$

hundreds place
decimal

$$\begin{array}{c}\text{Total Friction}\\ \text{Loss, ft}\end{array} = \begin{array}{c}(4.85)(35\ \text{segments})\\ \text{of 100-ft}\end{array}$$

$$= \boxed{169.8\ \text{ft}}$$

FRICTION LOSS IN FEET PER 100-FT LENGTH OF PIPE
(Based on Williams & Hazen Formula Using $C = 100$)

In the lower portion of the table, the leftmost columns are re-used to show larger pipe diameters, as marked by the inline labels. The mapping is: ½-in. column → **8″ Pipe**, ¾-in. → **10″ Pipe**, 1-in. → **12″ Pipe**, 1¼-in. → **14″ Pipe**, 1½-in. → **16″ Pipe**, 2-in. → **20″ Pipe**, 2½-in. → **24″ Pipe**, 3-in. → **30″ Pipe**. "Vel" = velocity in ft per sec; "Loss" = loss in ft.

US gal per min	½-in Vel	½-in Loss	¾-in Vel	¾-in Loss	1-in Vel	1-in Loss	1¼-in Vel	1¼-in Loss	1½-in Vel	1½-in Loss	2-in Vel	2-in Loss	2½-in Vel	2½-in Loss	3-in Vel	3-in Loss	4-in Vel	4-in Loss	5-in Vel	5-in Loss	6-in Vel	6-in Loss	US gal per min
2	2.10	7.4	1.20	1.9																			2
4	4.21	27.0	2.41	7.0	1.49	2.14																	4
6	6.31	57.0	3.61	14.7	2.23	4.55	1.29	1.20	.94	.56	.61	.20											6
8	8.42	98.0	4.81	25.0	2.98	7.8	1.72	2.03	1.26	.95	.82	.33	.52	.11									8
10	10.52	147.0	6.02	38.0	3.72	11.7	2.14	3.05	1.57	1.43	1.02	.50	.65	.17	.45	.07							10
12			7.22	53.0	4.46	16.4	2.57	4.3	1.89	2.01	1.23	.79	.78	.23	.54	.10							12
15			9.02	80.0	5.60	25.0	3.21	6.5	2.36	3.00	1.53	1.08	.98	.36	.68	.15							15
18			10.84	108.2	6.69	35.0	3.86	9.1	2.83	4.24	1.84	1.49	1.18	.50	.82	.21							18
20			12.03	136.0	7.44	42.0	4.29	11.1	3.15	5.20	2.04	1.82	1.31	.61	.91	.25	.51	.06					20
25					9.30	64.0	5.36	16.6	3.80	7.30	2.55	2.73	1.63	.92	1.13	.38	.64	.09					25
30					11.15	89.0	6.43	23.0	4.72	11.0	3.06	3.84	1.96	1.29	1.36	.54	.77	.13	.49	.04			30
35					13.02	119.0	7.51	31.2	5.51	14.7	3.57	5.10	2.29	1.72	1.59	.71	.89	.17	.57	.06			35
40					14.88	152.0	8.58	40.0	6.30	18.8	4.08	6.6	2.61	2.20	1.82	.91	1.02	.22	.65	.08			40
45							9.65	50.0	7.08	23.2	4.60	8.2	2.94	2.80	2.04	1.15	1.15	.28	.73	.09			45
50							10.72	60.0	7.87	28.4	5.11	9.9	3.27	3.32	2.27	1.38	1.28	.34	.82	.11	.57	.04	50
55							11.78	72.0	8.66	34.0	5.62	11.8	3.59	4.01	2.45	1.58	1.41	.41	.90	.14	.62	.05	55
60							12.87	85.0	9.44	39.6	6.13	13.9	3.92	4.65	2.72	1.92	1.53	.47	.98	.16	.68	.06	60
65							13.92	99.7	10.23	45.9	6.64	16.1	4.24	5.4	2.89	2.16	1.66	.53	1.06	.19	.74	.076	65
70							15.01	113.0	11.02	53.0	7.15	18.4	4.58	6.2	3.18	2.57	1.79	.63	1.14	.21	.79	.08	70
75							16.06	129.0	11.80	60.0	7.66	20.9	4.91	7.1	3.33	3.00	1.91	.73	1.22	.24	.85	.10	75
80							17.16	145.0	12.59	68.0	8.17	23.7	5.23	7.9	3.63	3.28	2.04	.81	1.31	.27	.91	.11	80
85							18.21	163.8	13.38	75.0	8.68	26.5	5.56	8.1	3.78	3.54	2.17	.91	1.39	.31	.96	.12	85
90							19.30	180.0	14.71	84.0	9.19	29.4	5.88	9.8	4.08	4.08	2.30	1.00	1.47	.34	1.02	.14	90
95									14.95	93.0	9.70	32.6	6.21	10.8	4.22	4.33	2.42	1.12	1.55	.38	1.08	.15	95
100									15.74	102.0	10.21	35.8	6.54	12.0	4.54	4.96	2.55	1.22	1.63	.41	1.13	.17	100
110									17.31	122.0	11.23	42.9	7.18	14.5	5.00	6.0	2.81	1.46	1.79	.49	1.25	.21	110
120	*8″ Pipe*								18.89	143.0	12.25	50.0	7.84	16.8	5.45	7.0	3.06	1.71	1.96	.58	1.36	.24	120
130									20.46	166.0	13.28	58.0	8.48	18.7	5.91	8.1	3.31	1.97	2.12	.67	1.47	.27	130
140	.90	.08							22.04	190.0	14.30	67.0	9.15	22.3	6.35	9.2	3.57	2.28	2.29	.76	1.59	.32	140
150	.96	.09									15.32	76.0	9.81	25.5	6.82	10.5	3.82	2.62	2.45	.88	1.70	.36	150
160	1.02	.10									16.34	86.0	10.46	29.0	7.26	11.8	4.08	2.91	2.61	.98	1.82	.40	160
170	1.08	.11									17.36	96.0	11.11	34.1	7.71	13.3	4.33	3.26	2.77	1.08	1.92	.45	170
180	1.15	.13	*10″ Pipe*								18.38	107.0	11.76	35.7	8.17	14.0	4.60	3.61	2.94	1.22	2.04	.50	180
190	1.21	.14									19.40	118.0	12.42	39.6	8.63	15.5	4.84	4.01	3.10	1.35	2.16	.55	190
200	1.28	.15									20.42	129.0	13.07	43.1	9.08	17.8	5.11	4.4	3.27	1.48	2.27	.62	200
220	1.40	.18	.90	.06							22.47	154.0	14.38	52.0	9.99	21.3	5.62	5.2	3.59	1.77	2.50	.73	220
240	1.53	.22	.98	.07							24.51	182.0	15.69	61.0	10.89	25.1	6.13	6.2	3.92	2.08	2.72	.87	240
260	1.66	.25	1.06	.08							26.55	211.0	16.99	70.0	11.80	29.1	6.64	7.2	4.25	2.41	2.95	1.00	260
280	1.79	.28	1.15	.09									18.30	81.0	12.71	33.4	7.15	8.2	4.58	2.77	3.18	1.14	280
300	1.91	.32	1.22	.11									19.61	92.0	13.62	38.0	7.66	9.3	4.90	3.14	3.40	1.32	300
320	2.05	.37	1.31	.12									20.92	103.0	14.52	42.8	8.17	10.5	5.23	3.54	3.64	1.47	320
340	2.18	.41	1.39	.14	*12″ Pipe*								22.22	116.0	15.43	47.9	8.68	11.7	5.54	3.97	3.84	1.62	340
360	2.30	.45	1.47	.15	1.08	.069							23.53	128.0	16.34	53.0	9.19	13.1	5.87	4.41	4.08	1.83	360
380	2.43	.50	1.55	.17	1.14	.075							24.84	142.0	17.25	59.0	9.69	14.0	6.19	4.86	4.31	2.00	380
400	2.60	.54	1.63	.19			*14″ Pipe*						26.14	156.0	18.16	65.0	10.21	16.0	6.54	5.4	4.55	2.20	400
450	2.92	.68	1.84	.23	1.28	.095									20.40	78.0	11.49	19.8	7.35	6.7	5.11	2.74	450
500	3.19	.82	2.04	.28	1.42	.113	1.04	.06							22.70	98.0	12.77	24.0	8.17	8.1	5.68	2.90	500
550	3.52	.97	2.24	.33	1.56	.135	1.15	.07							24.96	117.0	14.04	28.7	8.99	9.6	6.25	3.96	550
600	3.84	1.14	2.45	.39	1.70	.159	1.25	.08							27.23	137.0	15.32	33.7	9.80	11.3	6.81	4.65	600
650	4.16	1.34	2.65	.45	1.84	.19	1.37	.09									16.59	39.0	10.62	13.2	7.38	5.40	650
700	4.46	1.54	2.86	.52	1.99	.22	1.46	.10									17.87	44.9	11.44	15.1	7.95	6.21	700
750	4.80	1.74	3.06	.59	2.13	.24	1.58	.11	*16″ Pipe*								19.15	51.0	12.26	17.2	8.50	7.12	750
800	5.10	1.90	3.26	.66	2.27	.27	1.67	.13									20.42	57.0	13.07	19.4	9.08	7.96	800
850	5.48	2.20	3.47	.75	2.41	.31	1.79	.14	1.36	.08							21.70	64.0	13.89	21.7	9.65	8.95	850
900	5.75	2.46	3.67	.83	2.56	.34	1.88	.16	1.44	.084	*20″ Pipe*						22.98	71.0	14.71	24.0	10.20	10.11	900
950	6.06	2.87	3.88	.91	2.70	.38	2.00	.18	1.52	.095									15.52	26.7	10.77	11.20	950
1000	6.38	2.97	4.08	1.03	2.84	.41	2.10	.19	1.60	.10	1.02	.04							16.34	29.2	11.34	12.04	1000
1100	7.03	3.52	4.49	1.19	3.13	.49	2.31	.23	1.76	.12	1.12	.04							17.97	34.9	12.48	14.55	1100
1200	7.66	4.17	4.90	1.40	3.41	.58	2.52	.27	1.92	.14	1.23	.05							19.61	40.9	13.61	17.10	1200
1300	8.30	4.85	5.31	1.62	3.69	.67	2.71	.32	2.08	.17	1.33	.06									14.72	18.4	1300
1400	8.95	5.50	5.71	1.87	3.98	.78	2.92	.36	2.24	.19	1.43	.064									15.90	22.60	1400
1500	9.58	6.24	6.12	2.13	4.26	.89	3.15	.41	2.39	.21	1.53	.07	*24″ Pipe*								17.02	25.60	1500
1600	10.21	7.00	6.53	2.39	4.55	.98	3.34	.47	2.56	.24	1.63	.08									18.10	26.9	1600
1800	11.50	8.78	7.35	2.95	5.11	1.21	3.75	.58	2.87	.30	1.84	.10	1.28	.04									1800
2000	12.78	10.71	8.16	3.59	5.68	1.49	4.17	.71	3.19	.37	2.04	.12	1.42	.05									2000
2200	14.05	12.78	8.98	4.24	6.25	1.81	4.59	.84	3.51	.44	2.25	.15	1.56	.06	*30″ Pipe*								2200
2400	15.32	14.2	9.80	5.04	6.81	2.08	5.00	.99	3.83	.52	2.45	.17	1.70	.07	1.09	.02							2400
2600			10.61	5.81	7.38	2.43	5.47	1.17	4.15	.60	2.66	.20	1.84	.08	1.16	.027							2600
2800			11.41	6.70	7.95	2.75	5.84	1.32	4.47	.68	2.86	.23	1.98	.09	1.27	.03							2800
3000			12.24	7.62	8.52	3.15	6.01	1.49	4.79	.78	3.08	.27	2.13	.10	1.37	.037							3000
3200			13.05	7.8	9.10	3.51	6.68	1.67	5.12	.88	3.27	.30	2.26	.12	1.46	.041							3200
3500			14.30	10.08	9.95	4.16	7.30	1.97	5.59	1.04	3.59	.35	2.49	.14	1.56	.047							3500
3800			15.51	13.4	10.80	4.90	7.98	2.36	6.07	1.20	3.88	.41	2.69	.17	1.73	.05							3800
4200					11.92	5.88	8.76	2.77	6.70	1.44	4.29	.49	2.99	.20	1.91	.07							4200
4500					12.78	6.90	9.45	3.22	7.18	1.64	4.60	.56	3.20	.22	2.04	.08							4500
5000					14.20	8.40	10.50	3.92	8.01	2.03	5.13	.68	3.54	.27	2.26	.09							5000
5500							11.55	4.65	8.78	2.39	5.64	.82	3.90	.33	2.50	.11							5500
6000							12.60	5.50	9.58	2.79	6.13	.94	4.25	.38	2.73	.13							6000
6500							13.65	6.45	10.39	3.32	6.64	1.10	4.61	.45	2.96	.15							6500
7000							14.60	7.08	11.18	3.70	7.15	1.25	4.97	.52	3.18	.17							7000
8000									12.78	4.74	8.17	1.61	5.68	.66	3.63	.23							8000
9000									14.37	5.90	9.20	2.01	6.35	.81	4.08	.28							9000
10000									15.96	7.19	10.20	2.44	7.07	.98	4.54	.33							10000
12000											12.25	3.41	8.50	1.40	5.46	.48							12000

MINOR HEAD LOSS

Minor head losses are a result of water moving through valves and orifices, causing rapid changes in velocity and direction of flow.

These losses may be estimated using the nomograph* given on the facing page.

To read the nomograph, follow these steps:

1. Place one end of a straightedge at the known pipe diameter on the right scale.

2. Align the other end of the straight edge with the point designated for the desired fitting or orifice on the left scale.

3. Draw a line from the left scale to the right scale and read the head loss value on the middle scale.

The values given in the table represent **equivalent length of straight pipe**. These values are to be **added to actual pipe length** for calculation of friction losses, described on the previous page.

Minor losses are normally just that—minor, when compared to friction loss values. However, the smaller the runs of pipe, the more significant are these minor head losses.

Examples 6 and 7 illustrate the use of the nomograph in determining minor friction losses.

Example 6: (Head and Head Loss)
❏ Determine the "equivalent length of pipe" for a flow through a swing check valve, fully open, if the diameter of the pipe is 6 inches.

Align one end of the straightedge with the 6-inch (nominal diameter) mark on the left side of the scale shown to the right.

Align the other end of the straightedge with the point indicated for the "Swing Check Valve, Fully Open."

Draw a line from the left scale to the right scale.

The "equivalent length" value can be read on the middle scale:

> 40 ft equivalent length of pipe

Example 7: (Head and Head Loss)
❏ What is the head loss through a gate valve (1/2 closed) for a 10-inch diameter pipeline?

Align one end of the straightedge with 10-inches (nominal diameter) on the scale to the right.

Align the other end of the straightedge with the point corresponding with "Gate Valve—1/2 Closed" on the scale to the left.

Draw a line between the two outside scales and read the head loss value from the middle scale. The approximate reading is:

> 170 ft equivalent length of pipe

---

* For a review of reading nomographs, refer to Chapter 12 in *Basic Math Concepts*.

RESISTANCE OF VALVES AND FITTINGS TO FLOW

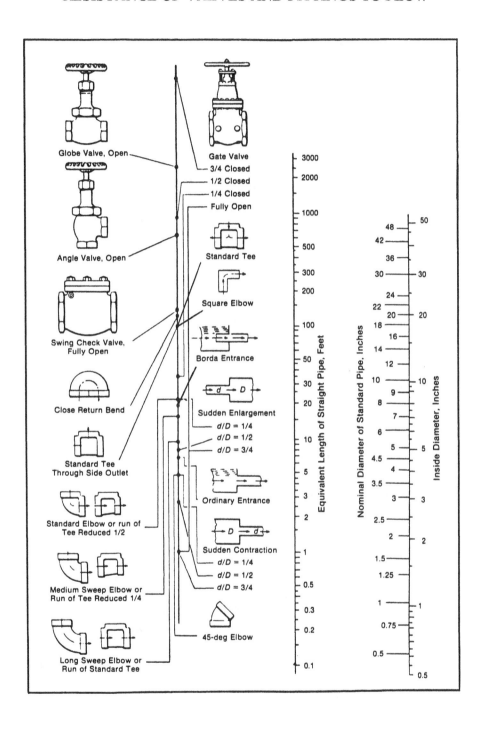

Globe Valve, Open

Angle Valve, Open

Swing Check Valve, Fully Open

Close Return Bend

Standard Tee Through Side Outlet

Standard Elbow or run of Tee Reduced 1/2

Medium Sweep Elbow or Run of Tee Reduced 1/4

Long Sweep Elbow or Run of Standard Tee

Gate Valve
3/4 Closed
1/2 Closed
1/4 Closed
Fully Open

Standard Tee

Square Elbow

Borda Entrance

Sudden Enlargement
d/D = 1/4
d/D = 1/2
d/D = 3/4

Ordinary Entrance

Sudden Contraction
d/D = 1/4
d/D = 1/2
d/D = 3/4

45-deg Elbow

Equivalent Length of Straight Pipe, Feet

Nominal Diameter of Standard Pipe, Inches

Inside Diameter, Inches

Reprinted with permission of Crane Valves

7.4 HORSEPOWER

The selection of a pump or combination of pumps with an adequate pumping capacity depends upon the flow rate desired and the effective height* to which the flow must be pumped.

Calculations of horsepower and head are made in conjunction with many treatment plant operations. The basic concept from which the horsepower calculation is derived is the concept of work.

Work involves the operation of a force (lbs) over a specific distance (ft). The **amount of work** accomplished is measured in foot-pounds:

$$(ft)(lbs) = ft\text{-}lbs$$

The **rate of doing work** is called **power**. The time factor in which the work occurs now becomes important. James Watt was the first to use the term **horsepower**. He used it to compare the power of a horse to that of the steam engine. The rate at which a horse could work was determined to be about 550 ft-lbs/sec (or expressed as 33,000 ft-lbs/min). This rate has become the definition of the standard unit called horsepower:

$$1 \text{ hp} = 33,000 \text{ ft-lbs/min}$$

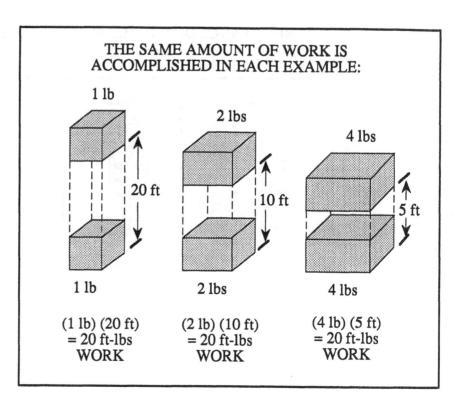

THE SAME AMOUNT OF WORK IS ACCOMPLISHED IN EACH EXAMPLE:

(1 lb) (20 ft) = 20 ft-lbs WORK

(2 lb) (10 ft) = 20 ft-lbs WORK

(4 lb) (5 ft) = 20 ft-lbs WORK

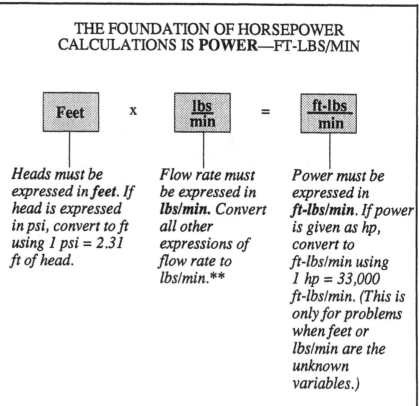

THE FOUNDATION OF HORSEPOWER CALCULATIONS IS **POWER—FT-LBS/MIN**

$$\text{Feet} \quad \text{x} \quad \frac{\text{lbs}}{\text{min}} \quad = \quad \frac{\text{ft-lbs}}{\text{min}}$$

*Heads must be expressed in **feet**. If head is expressed in psi, convert to ft using 1 psi = 2.31 ft of head.*

*Flow rate must be expressed in **lbs/min**. Convert all other expressions of flow rate to lbs/min.***

*Power must be expressed in **ft-lbs/min**. If power is given as hp, convert to ft-lbs/min using 1 hp = 33,000 ft-lbs/min. (This is only for problems when feet or lbs/min are the unknown variables.)*

---

* "Effective height" refers to the feet of head against which the pump must pump.

** To review flow rate conversions, refer to Chapter 8 in *Basic Math Concepts*.

Example 1: (Horsepower)
❑ A pump must pump 2000 gpm against a total head of 20 ft. What horsepower is required for this work?

First calculate ft - lbs/min required:

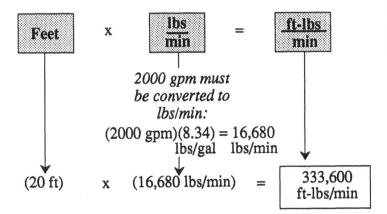

2000 gpm must be converted to lbs/min:

$$(2000 \text{ gpm})(8.34) = 16,680$$
$$\text{lbs/gal} \quad \text{lbs/min}$$

$$(20 \text{ ft}) \quad x \quad (16,680 \text{ lbs/min}) = \boxed{\begin{array}{c} 333,600 \\ \text{ft-lbs/min} \end{array}}$$

Then convert ft-lbs/min to hp:

$$\frac{333,600 \text{ ft-lbs/min}}{33,000 \text{ ft-lbs/min/hp}} = \boxed{10 \text{ hp}}$$

Example 2: (Horsepower)
❑ A flow of 8 MGD must be pumped against a total dynamic head (TDH) of 25 ft. What horsepower is required for this work?

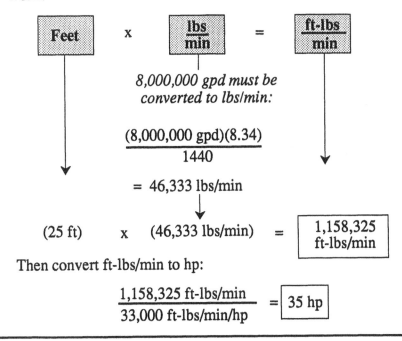

8,000,000 gpd must be converted to lbs/min:

$$\frac{(8,000,000 \text{ gpd})(8.34)}{1440}$$

$$= 46,333 \text{ lbs/min}$$

$$(25 \text{ ft}) \quad x \quad (46,333 \text{ lbs/min}) = \boxed{\begin{array}{c} 1,158,325 \\ \text{ft-lbs/min} \end{array}}$$

Then convert ft-lbs/min to hp:

$$\frac{1,158,325 \text{ ft-lbs/min}}{33,000 \text{ ft-lbs/min/hp}} = \boxed{35 \text{ hp}}$$

CALCULATING HORSEPOWER

When calculating horsepower requirements:

1. Determine the ft-lbs/min power required:

$$\boxed{(\text{ft})(\text{lbs/min}) = \text{ft-lbs/min}}$$

2. Once ft-lbs/min power has been calculated, horsepower can be determined using the equation 1 hp = 33,000 ft-lbs/min.

$$\boxed{\text{hp} = \frac{\text{ft-lbs/min}}{33,000 \text{ ft-lbs/min/hp}}}$$

AN ALTERNATE EQUATION

An equation frequently given for horsepower calculations is:

$$\boxed{\text{whp} = \frac{(\overset{\text{flow rate}}{\underset{\text{gpm}}{}}) (\overset{\text{total head}}{\underset{\text{ft}}{}})}{3960}}$$

This equation is derived from the horsepower equation described above:

$$\text{hp} = \frac{\text{ft-lbs/min}}{33,000 \text{ ft-lbs/min/hp}}$$

It is then adjusted to reflect gpm flow rate, rather than lbs/min flow rate. The advantage of this equation is that gpm may be used directly, without conversions. The disadvantage is that it is somewhat "cut off from its roots":—the concept of power, ft-lbs/min. Because of this, there is often a lack in flexible application of the equation. It tends to become an equation memorized but not fully understood.

In the examples thus far, we have calculated the horsepower required to accomplish a particular pumping job. Due to motor and pump inefficiencies, however, more horsepower must be supplied in order to deliver the desired horsepower. To illustrate, in Example 2 it was calculated that 35 hp would be needed to pump 8 MGD against a TDH of 25 ft. Due to pump and motor inefficiencies, however, about 50 hp would have to be supplied to the motor in order to deliver the 35 hp from the pump.

HORSEPOWER TERMINOLOGY

Three different horsepower terms are used to distinguish the type of horsepower being referred to in any particular calculation:

- Motor horsepower,

- Brake horsepower, and

- Water horsepower.

Motor horsepower (mhp) refers to the horsepower supplied to the motor in the form of electrical current. Some of this horsepower is lost due to the conversion of electrical energy to mechanical energy. The efficiency of most motors ranges from 80-95%, and is listed in manufacturer's literature.

Brake horsepower (bhp) refers to the horsepower supplied to the pump from the motor. As the power moves through the pump, additional horsepower is lost, resulting from slippage and friction of the shaft and other factors. Pump efficiencies generally range between 50-85%.

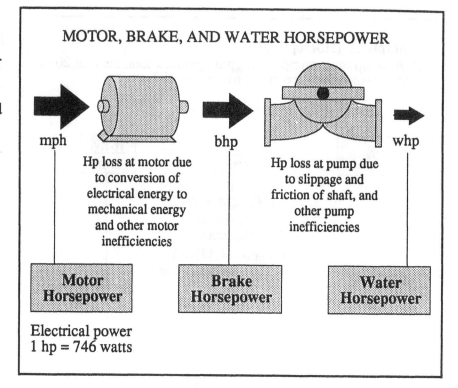

MOTOR, BRAKE, AND WATER HORSEPOWER

mph — Hp loss at motor due to conversion of electrical energy to mechanical energy and other motor inefficiencies

bhp — Hp loss at pump due to slippage and friction of shaft, and other pump inefficiencies

whp

Motor Horsepower **Brake Horsepower** **Water Horsepower**

Electrical power
1 hp = 746 watts

Example 3: (Horsepower)
❑ If 12 hp is supplied to a motor (mhp), what is the bhp and whp if the motor is 90% efficient and the pump is 85% efficient?

It is helpful to diagram the information given and desired:

12 mhp ⟶ **Motor** $\xrightarrow{\text{? bhp}}$ **Pump** $\xrightarrow{\text{? whp}}$

(90% Effic.)   (85% Effic.)

Mhp is 12 hp. Bhp and whp will be smaller numbers. To calculate bhp and mhp, multiply by the motor and pump efficiencies, as indicated below.

Calculate brake horsepower:

$$(12 \text{ mhp}) \frac{(90)}{100} = \boxed{10.8 \text{ bhp}}$$

Calculate water horsepower:

$$(10.8 \text{ bhp}) \frac{(85)}{100} = \boxed{9.2 \text{ whp}}$$

Note: Always check your answers. Note that bhp and whp are smaller numbers than mhp.

Example 4: (Horsepower)
❏ 40 hp is supplied to a motor. How many horsepower will be available for actual pumping loads, if the motor is 92% efficient and the pump is 85% efficient?

In this problem, the unknown hp is <u>whp</u> ("available for actual pumping loads"). First diagram the problem:

$$40 \text{ mhp} \longrightarrow \boxed{\textbf{Motor}} \longrightarrow \boxed{\textbf{Pump}} \xrightarrow{\text{? whp}}$$
$$\text{(92\% Effic.)} \qquad \text{(85\% Effic.)}$$

To calculate whp, **multiply** mhp by both the motor and pump efficiencies:

$$(40 \text{ mhp}) \frac{(92)}{100} \frac{(85)}{100} = \boxed{31.3 \text{ whp}}$$

Example 5: (Horsepower)
❏ A total of 28 hp is required for a particular pumping application. If the pump efficiency is 75% and the motor efficiency is 85%, what horsepower must be supplied to the motor?

In this problem, the unknown term is mhp. Remember that the answer must be a **larger number** than the whp (28 hp).

First diagram the information in the problem:

$$\text{? mhp} \longrightarrow \boxed{\textbf{Motor}} \longrightarrow \boxed{\textbf{Pump}} \longrightarrow 28 \text{ whp}$$
$$\text{(85\% Effic.)} \qquad \text{(75\% Effic.)}$$

To calculate mhp, divide by pump and motor efficiencies:

$$\frac{28 \text{ whp}}{(0.85 (0.75)} = \boxed{43.9 \text{ mhp}}$$

Note that the mhp calculated is in fact a larger number than whp.

Water horsepower (whp) refers to the actual horsepower available to pump the water. Examples 1 and 2 in this section were actually calculations of water horsepower.

When making calculations of motor, brake, and water horsepower, it is important to remember that motor horsepower will be the largest number, brake horsepower next largest, followed by water horsepower. Knowing the relative sizes of these terms will help you know if your answers are reasonable.

The conversion between horsepower terms can be calculated as follows:

- When converting from a smaller term to a larger term* (such as whp to bhp, bhp to mhp, or whp to mhp), divide by the efficiency of the pump or motor:**

$$\text{Brake hp} = \frac{\text{Water hp}}{\text{Pump Effic.}}$$

$$\text{Motor hp} = \frac{\text{Brake hp}}{\text{Motor Effic.}}$$

$$\text{Motor hp} = \frac{\text{Water hp}}{(\text{Motor Effic.})(\text{Pump Effic.})}$$

- When converting from a larger term to a smaller term (such as mhp to bhp, bhp to whp, or mhp to whp), multiply by the efficiency of the pump or motor:**

Brake hp = (Motor)(Motor)
       hp   Effic.

Water hp = (Brake)(Pump)
       hp   Effic.

Water hp = (Motor)(Motor)(Pump)
       hp   Effic.   Effic.

---

* Normally, when converting from a smaller term to a larger term, multiplication is indicated. However, that is only true when multiplying by a number <u>greater than one</u>. When a number less than one (such as pump or motor efficiency) is multiplied times a number , the answer is a smaller number; when a number less than one is used to divide a number, the resulting answer is larger number.

** Efficiency is written as %/100. For example 80% efficiency is written as 80/100. This can be simplified as 0.80.

HORSEPOWER AND SPECIFIC GRAVITY

In Examples 1-5, the horsepower calculations were based on pumping water. If another liquid is to be pumped, the specific gravity* of the liquid must be considered.

The specific gravity of a liquid is an indication of its density, or generally its weight, compared to that of water.

To account for differences in specific gravity, include the specific gravity factor when calculating ft-lbs/min pumping requirements:

> (ft)(lbs/min)(sp. gr.) = ft-lbs/min for different liquid

Example 6 illustrates such a calculation.

MHP AND KILOWATT REQUIREMENTS

Motor horsepower requirements can be converted to watts and then kilowatts requirements using the following equation:**

> 1 hp = 746 watts

Once watts requirements are determined, kilowatts are easily determined by a metric system conversion.

Example 6: (Horsepower)

❑ A pump must pump against a total dynamic head of 50 ft at a flow rate of 1300 gpm. The liquid to be pumped has a specific gravity of 1.3. What is the water horsepower requirement for this pumping application?

Water horsepower is essentially a calculation of ft-lbs/min. A specific gravity factor must be included in this calculation:

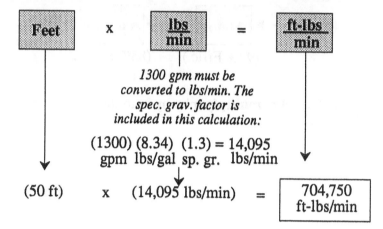

1300 gpm must be converted to lbs/min. The spec. grav. factor is included in this calculation:

$$(1300)(8.34)(1.3) = 14,095$$
gpm   lbs/gal   sp. gr.   lbs/min

$$(50 \text{ ft}) \times (14,095 \text{ lbs/min}) = \boxed{704,750 \text{ ft-lbs/min}}$$

Now convert ft-lbs/min to hp:

$$\frac{704,750 \text{ ft-lbs/min}}{33,000 \text{ ft-lbs/min/hp}} = \boxed{21.4 \text{ whp}}$$

Example 7: (Horsepower)

❑ The motor horsepower requirement has been calculated to be 35 hp. How many kilowatts electric power does this represent?

First calculate the watts required using the equation 1 hp = 746 watts. The box method of conversions may be used:

$$\boxed{hp} \xrightarrow{746} \boxed{watts}$$

Multiplication by 746 is indicated:

$$(35 \text{ hp})(746 \text{ watts/hp}) = 26,110 \text{ watts}$$

Then: $$\frac{26,110 \text{ watts}}{1000 \text{ watts/kW}} = \boxed{26.1 \text{ kW}}$$

---

* Specific gravity is discussed in Section 7.1 of this chapter.

** This equation is preferred to the equation 1 hp = 0.746 kW, since the box method of conversion works only if both numbers are greater than one. Refer to Chapter 8 in *Basic Math Concepts*.

Example 8: (Horsepower)
❑ 22 mhp is required for a pumping application. If the cost of power is $0.0526/kWh, and the pump is in operation 24 hrs/day, what is the daily pump cost?

To calculate kWh pump operation, you must know the kW power requirements of the motor and hours of operation. First convert 22 mhp to kW: (1 hp = 746 watts)

$$(22 \text{ mhp}) (746 \text{ watts/hp}) = 16,412 \text{ watts}$$

$$\text{or} = 16.4 \text{ kW}$$

The kWh of power consumption can now be determined:

$$(16.4 \text{ kW}) (24 \text{ hrs/day}) = \boxed{\begin{array}{l} 393.6 \text{ kWh} \\ \text{daily} \end{array}}$$

Now complete the cost calculation:

$$(393.6 \text{ kWh/day}) (\$0.0526/\text{kWh}) = \boxed{\begin{array}{l} \$20.70 \\ \text{daily} \end{array}}$$

Example 9: (Horsepower)
❑ The motor horsepower requirement has been calculated to be 50 mhp. During the week, the pump is in operation a total of 148 hours. Using a power cost of $0.09439/kWh, what would be the power cost that week for the pumping?

First convert 50 mhp to kW so that kWh can be calculated: (1 hp = 746 watts)

$$(50 \text{ mhp}) (746 \text{ watts/hp}) = 37,300 \text{ watts}$$

$$\text{or} = 37.3 \text{ kW}$$

Next calculate kWh of power consumed:

$$(37.3 \text{ kW}) (148 \text{ hrs}) = \boxed{5520 \text{ kWh}}$$

Then powers costs may be calculated:

$$(5520 \text{ kWh/wk}) \frac{(\$0.09439)}{\text{kWh}} = \boxed{\begin{array}{l} \$521.03 \\ \text{cost for} \\ \text{the week} \end{array}}$$

PUMPING COST CALCULATIONS

Pumps costs are determined on the basis of two primary considerations:

• Kilowatt-hours of pump operation, and

• Power cost per kilowatt-hour.

Kilowatt-hours of pump operation are determined by multiplying power drawn by the pump (kW) by the hours of operation (hrs):

$$\boxed{(\text{kW}) (\text{hrs}) = \begin{array}{l} \text{kWh} \\ \text{used} \end{array}}$$

Once the kilowatt-hours of power use has been determined, then determine the cost of that power use using the cost factor:

$$\boxed{\begin{array}{lll} (\text{kWh}) & (\text{Cost/kWh}) & = \text{Total} \\ \text{Power} & \text{use} & \text{Cost} \\ \text{Use} & & \end{array}}$$

Examples 8 and 9 are pumping cost calculations.

7.5 PUMP CAPACITY

PUMP CAPACITY TESTING

Pump capacity may be determined by timing the pumping into or out of a tank of known size. Assuming water is not entering or leaving from any source other than the pump being tested, then—

- **When pumping into a tank** the rise in water level will correspond with the pumping rate.

- **When pumping out of a tank**, the drop in water level will correspond with the pumping rate.

To calculate the pumping capacity or rate, determine the gallons rise or fall and divide by the time of the pump test:

$$\text{Pumping Rate, gpm} = \frac{\text{gal rise or fall}}{\text{minutes of test}}$$

Or

(In expanded form for a rectangular tank)

$$\text{Pumping Rate, gpm} = \frac{(l)\,(w)\,(d)\,(7.48\text{ gal/cu ft})}{\text{minutes}}$$

WHEN PUMPING INTO AN EMPTY TANK, **THE RISE IN WATER LEVEL** INDICATES PUMPING RATE

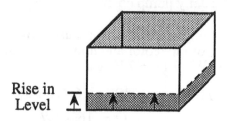

Rise in Level

WHEN PUMPING FROM A TANK, (WITH INFLUENT VALVE CLOSED) **THE DROP IN WATER LEVEL** INDICATES PUMPING RATE

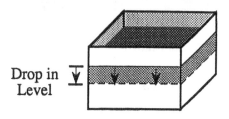

Drop in Level

Example 1: (Pump Capacity)

❏ A wet well is 15 ft long and 12 ft wide. The influent valve to the wet well is closed. If a pump lowers the water level 1.25 ft during a 5-minute pumping test, what is the gpm pumping rate?

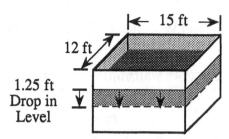

15 ft

12 ft

1.25 ft Drop in Level

$$\text{Pumping Rate, gpm} = \frac{(\text{Length, ft})\,(\text{Width, ft})\,(\text{Drop, ft})\,(7.48\text{ gal/cu ft})}{\text{Test time, min}}$$

$$= \frac{(15\text{ ft})\,(12\text{ ft})\,(1.25\text{ ft})\,(7.48\text{ gal/cu ft})}{5\text{ minutes}}$$

$$= \boxed{337\text{ gpm}}$$

Example 2: (Pump Capacity)
❑ A pump is discharged into a 55-gallon barrel. If it takes 35 seconds to fill the barrel, what is the pumping rate?

Since gallons <u>per minute</u> are desired, first, convert 35 seconds to minutes,

$$\frac{35 \text{ sec}}{60 \text{ sec/min}} = 0.58 \text{ min}$$

then calculate the gpm pumping rate. The equation using tank dimensions is not needed since the gallons pumped (55 gallons) is already known. The general equation may be used:

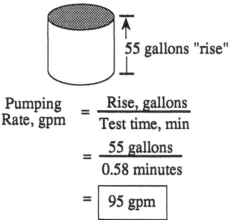

55 gallons "rise"

$$\text{Pumping Rate, gpm} = \frac{\text{Rise, gallons}}{\text{Test time, min}}$$

$$= \frac{55 \text{ gallons}}{0.58 \text{ minutes}}$$

$$= \boxed{95 \text{ gpm}}$$

Example 3: (Pump Capacity)
❑ A wet well pump is rated at 300 gpm. A pump test is conducted for 3 minutes. What is the actual gpm pumping rate if the wet well is 10 ft long and 8 ft wide and the water level drops 1.33 ft during the pump test?

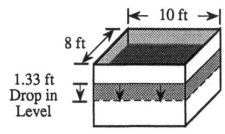

10 ft

8 ft

1.33 ft Drop in Level

$$\text{Pumping Rate, gpm} = \frac{(\text{Length, ft}) (\text{Width, ft}) (\text{Drop, ft}) (7.48 \text{ gal/cu ft})}{\text{Test time, min}}$$

$$= \frac{(10 \text{ ft}) (8 \text{ ft}) (1.33 \text{ ft}) (7.48 \text{ gal/cu ft})}{3 \text{ minutes}}$$

$$= \boxed{265 \text{ gpm}}$$

CAPACITY TESTING WHEN INFLUENT VALVE IS OPEN

In the previous two pages, pump capacity was determined for two conditions:

- Pumping into a tank with known dimensions, and

- Pumping from a tank with the influent valve closed.

However, it is possible to determine pumping capacity (or pumping rate) even when the influent valve is open. Examples 4 and 5 illustrate how to calculate pumping rate out of a tank when there is influent entering the tank.

WHEN INFLUENT VALVE IS OPEN INFLUENT FLOW (GPM) MUST BE INCLUDED IN THE CALCULATION

When the water level remains the same, the pumping rate is <u>equal to the influent rate</u>:

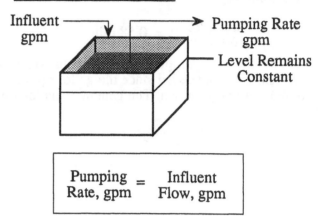

$$\text{Pumping Rate, gpm} = \text{Influent Flow, gpm}$$

When the water level drops, the pumping rate is <u>greater than the influent rate</u>:

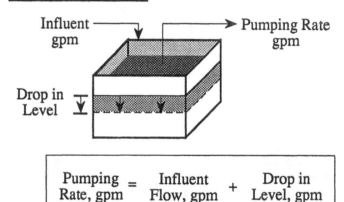

$$\text{Pumping Rate, gpm} = \text{Influent Flow, gpm} + \text{Drop in Level, gpm}$$

When the water level rises, the pumping rate is <u>less than the influent rate</u>:

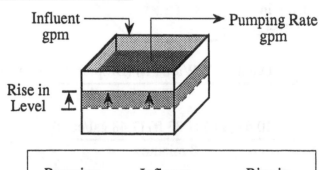

$$\text{Pumping Rate, gpm} = \text{Influent Flow, gpm} - \text{Rise in Level, gpm}$$

Example 4: (Pump Capacity)
❏ A tank is 6 ft wide and 10 ft long. During a 3-minute pumping test, the influent valve remains open. If the water level drops 6 inches during the pump test, what is the pumping rate in gpm? The influent flow is 0.8 MGD.

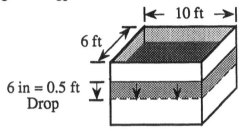

First calculate the gpm corresponding to the drop in water level: (6 in = 0.5 ft)

$$\frac{\text{Drop,}}{\text{gpm}} = \frac{(10 \text{ ft}) (6 \text{ ft}) (0.5 \text{ ft}) (7.48) \text{ gallons}}{3 \text{ minutes}}$$

$$= 75 \text{ gpm}$$

Now calculate pumping rate:

$$\frac{\text{Pumping}}{\text{Rate, gpm}} = \frac{\text{Influent}}{\text{Flow, gpm}} + \frac{\text{Drop in}}{\text{Level, gpm}}$$

$$= \frac{800,000 \text{ gpd}}{1440 \text{ min/day}} + 75 \text{ gpm}$$

$$= 556 \text{ gpm} + 75 \text{ gpm}$$

$$= \boxed{631 \text{ gpm}}$$

Example 5: (Pump Capacity)
❏ A pump test is conducted for 5 minutes while influent flow continues. During the test, the water level rises 3 inches. If the tank is 10 ft by 10 ft and the influent flow is 750,000 gpd, what is the pumping rate in gpm?

First calculate the gpm corresponding to the rise in level: (3 inches = 0.25 ft)

$$\frac{\text{Rise,}}{\text{gpm}} = \frac{(10 \text{ ft}) (10 \text{ ft}) (0.25 \text{ ft}) (7.48) \text{ gallons}}{5 \text{ minutes}}$$

$$= 37 \text{ gpm}$$

Now calculate the pumping rate:

$$\frac{\text{Pumping}}{\text{Rate, gpm}} = \frac{\text{Influent}}{\text{Flow, gpm}} - \frac{\text{Rise in}}{\text{Level, gpm}}$$

$$= \frac{750,000 \text{ gpd}}{1440 \text{ min/day}} - 37 \text{ gpm}$$

$$= 521 \text{ gpm} - 37 \text{ gpm}$$

$$= \boxed{484 \text{ gpm}}$$

CAPACITY FOR POSITIVE DISPLACEMENT PUMPS

One of the most common types of sludge pumps is the piston pump.* This type of pump operates on the principle of positive displacement. This means that it displaces, or pushes out, a volume of sludge equal to the volume of the piston. The length of the piston, called the stroke, can be adjusted (lengthened or shortened) to increase or decrease the gpm sludge delivered by the pump. Normally, the piston pump is operated no faster than about 50 gpm.

EACH STROKE OF A PISTON PUMP "DISPLACES" OR PUSHES OUT SLUDGE

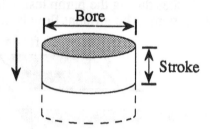

Simplified Equation:

$$\text{Volume of Sludge Pumped (gal/min)} = \frac{(\text{Gallons pumped})}{\text{Stroke}} \frac{(\text{No. of Strokes})}{\text{Minute}}$$

Expanded Equation:

$$\text{Volume of Sludge Pumped (gal/min)} = \left[(0.785)(D^2) \frac{(\text{Stroke})}{\text{Length}} \frac{(7.48)}{\text{gal/cu ft}} \right] \left[\frac{\text{No. of}}{\text{Strokes/min}} \right]$$

Example 6: (Pump Capacities)

❑ A piston pump discharges a total of 0.8 gallons per stroke (or revolution). If the pump operates at 20 revolutions per minute, what is the gpm pumping rate? (Assume the piston is 100% efficient and displaces 100% of its volume each stroke)

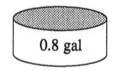

0.8 gal

$$\text{Vol. of Sludge Pumped} = \frac{(\text{Gallons pumped})}{\text{Stroke}} \frac{(\text{No. of Strokes})}{\text{Minute}}$$

$$= \frac{(0.8 \text{ gal})}{\text{stroke}} \frac{(20 \text{ strokes})}{\text{min}}$$

$$= \boxed{16 \text{ gpm}}$$

---

* This type pump is also known as a plunger-type pump or positive displacement pump.

Example 7: (Pump Capacities)
❑ A sludge pump has a bore of 8 inches and a stroke length of 3 inches. If the pump operates at 50 strokes (or revolutions) per minute, how many gpm are pumped? (Assume the piston is 100% efficient and displaces 100% of its volume each stroke.)

|← 0.67 ft →|

$$\frac{8 \text{ in.}}{12 \text{ in./ft}} = 0.67 \text{ ft}$$

$$\frac{3 \text{ in.}}{12 \text{ in./ft}} = 0.25 \text{ ft}$$

$$\begin{array}{c} \text{Vol. of Sludge} \\ \text{Pumped} \end{array} = \frac{(\text{Gallons pumped})}{\text{Stroke}} \frac{(\text{No. of Strokes})}{\text{Minute}}$$

$$= \left[(0.785) (D^2) \begin{array}{c}(\text{Stroke}) \\ \text{Length}\end{array} \begin{array}{c}(7.48) \\ \text{gal/cu ft}\end{array} \right] \left[\text{Strokes/min} \right]$$

$$= \left[(0.785) (0.67 \text{ ft}) (0.67 \text{ ft}) (0.25 \text{ ft}) (7.48 \frac{\text{gal}}{\text{cu ft}}) \right] \left[50 \frac{\text{Strokes}}{\text{min}} \right]$$

$$= \frac{(0.66 \text{ gal})}{\text{stroke}} \frac{(50 \text{ strokes})}{\text{min}}$$

$$= \boxed{33 \text{ gpm}}$$

Example 8: (Pump Capacities)
❑ A sludge pump has a bore of 6 inches and a stroke setting of 3 inches. The pump operates at 45 revolutions per minute. If the pump operates a total of 80 minutes during a 24-hour period, what is the gpd pumping rate? (Assume the piston is 100% efficient.)

|← 0.5 ft →|

$$\frac{6 \text{ in.}}{12 \text{ in./ft}} = 0.5 \text{ ft}$$

$$\frac{3 \text{ in.}}{12 \text{ in./ft}} = 0.25 \text{ ft}$$

First calculate the gpm pumping rate:

$$\begin{array}{c} \text{Vol. Pumped} \\ \text{gpm} \end{array} = \frac{(\text{Gallons pumped})}{\text{Stroke}} \frac{(\text{No. of Strokes})}{\text{Minute}}$$

$$= \left[(0.785) (0.5 \text{ ft}) (0.5 \text{ ft}) (0.25 \text{ ft}) (7.48 \frac{\text{gal}}{\text{cu ft}}) \right] \left[45 \frac{\text{Strokes}}{\text{min}} \right]$$

$$= \frac{(0.37 \text{ gal})}{\text{stroke}} \frac{(45 \text{ strokes})}{\text{min}}$$

$$= 16.7 \text{ gpm}$$

Then convert gpm to gpd pumping rate, based on total minutes pumped during 24-hours:

$$(16.7 \text{ gpm}) \frac{(80 \text{ min})}{\text{day}} = \boxed{1336 \text{ gpd}}$$

CALCULATING GPD PUMPED

There are two methods to determine gpd pumping rate:

- Calculate the gpm pumping rate, then multiply by the total minutes operation during the 24-hour period:

$$\begin{array}{c} \text{Pumping} \\ \text{Rate, gpd} \end{array} = \begin{array}{c} (\text{Pumping}) \\ \text{Rate, gpm} \end{array} \begin{array}{c} (\text{Total min}) \\ \text{pumping} \\ \text{in 24 hrs} \end{array}$$

- Calculate the gallons pumped each revolution, then multiply by the total revolutions during the 24-hour period:

$$\begin{array}{c} \text{Pumping} \\ \text{Rate, gpd} \end{array} = \frac{(\text{Gallons})}{\text{Revolution}} \frac{(\text{Total Revol.})}{\text{day}}$$

NOTES:

8 *Water Sources and Storage*

SUMMARY

1. Well Drawdown

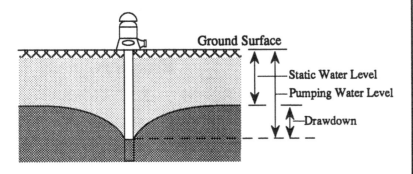

Drawdown, ft	= Pumping Water Level, ft	− Static Water Level, ft

2. Well Yield

$$\text{Well Yield, gpm} = \frac{\text{Flow, gallons}}{\text{Duration of Test, min}}$$

3. Specific Yield

$$\text{Specific Yield, gpm/ft} = \frac{\text{Well Yield, gpm}}{\text{Drawdown, ft}}$$

4. Well Casing Disinfection

First calculate chlorine required, lbs:

$$(\text{mg}/L\ Cl_2)(\text{Water-filled Casing Volume, MG})(8.34\ \text{lbs/gal}) = \text{lbs}\ Cl_2$$

Then calculate chlorine compound required, if applicable:

$$\text{Chlorine Compound, lbs} = \frac{\text{Chlorine, lbs}}{\dfrac{\%\ \text{Avail.}\ Cl_2}{100}}$$

SUMMARY

5. Deep-Well Turbine Pump Calculations

• Head calculations

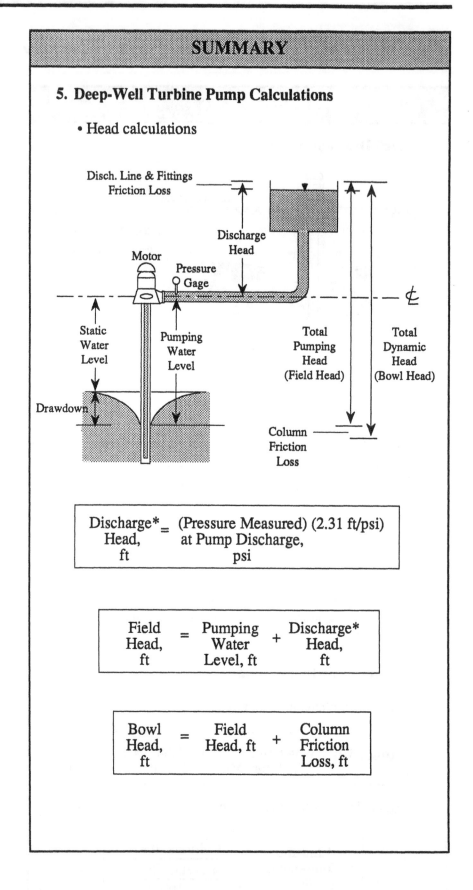

$$\begin{array}{c} \text{Discharge*} \\ \text{Head,} \\ \text{ft} \end{array} = \begin{array}{c} \text{(Pressure Measured)} \\ \text{at Pump Discharge,} \\ \text{psi} \end{array} \text{(2.31 ft/psi)}$$

$$\begin{array}{c} \text{Field} \\ \text{Head,} \\ \text{ft} \end{array} = \begin{array}{c} \text{Pumping} \\ \text{Water} \\ \text{Level, ft} \end{array} + \begin{array}{c} \text{Discharge*} \\ \text{Head,} \\ \text{ft} \end{array}$$

$$\begin{array}{c} \text{Bowl} \\ \text{Head,} \\ \text{ft} \end{array} = \begin{array}{c} \text{Field} \\ \text{Head, ft} \end{array} + \begin{array}{c} \text{Column} \\ \text{Friction} \\ \text{Loss, ft} \end{array}$$

* The discharge head includes friction losses through the discharge line and any fittings.

- Horsepower Calculations

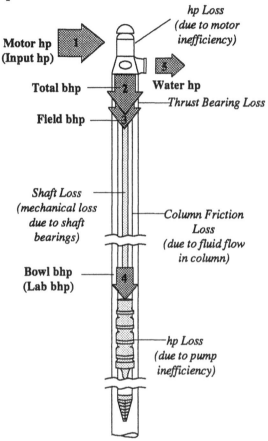

$$\text{Motor hp (Input hp)} = \frac{\text{Total bhp}}{\dfrac{\text{Motor Effic.}}{100}}$$

$$\text{Total bhp} = \text{Field bhp} + \text{Thrust Bearing Loss, hp}$$

$$\text{Field bhp} = \text{Bowl bhp} + \text{Shaft Loss, hp}$$

$$\text{Bowl bhp (Lab bhp)} = \frac{(\text{Bowl Head, ft})(\text{Capacity, gpm})}{\dfrac{(3960)(\text{Bowl Effic.})}{100}}$$

$$\text{Water hp} = \frac{(\text{Field Head, ft})(\text{Capac., gpm})}{3960}$$

SUMMARY

• Efficiencies

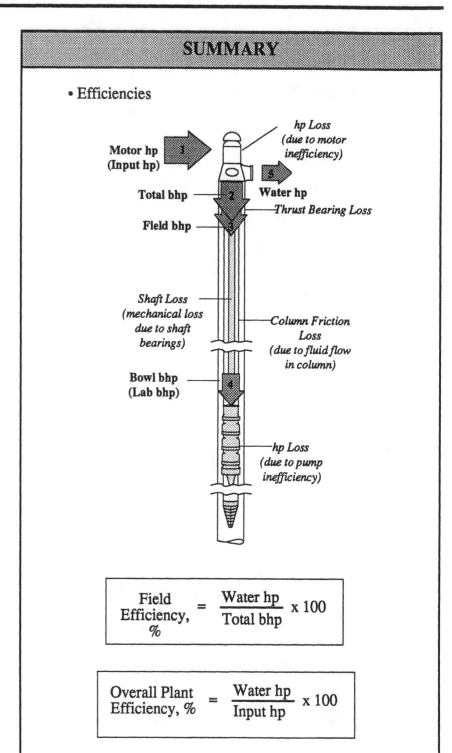

$$\text{Field Efficiency, \%} = \frac{\text{Water hp}}{\text{Total bhp}} \times 100$$

$$\text{Overall Plant Efficiency, \%} = \frac{\text{Water hp}}{\text{Input hp}} \times 100$$

Or

$$\text{Overall Plant Efficiency, \%} = \frac{(\text{Motor Effic., \%})(\text{Pump Effic., \%})}{100}$$

6. Pond or Small Lake Storage Capacity

(Top View of Lake)

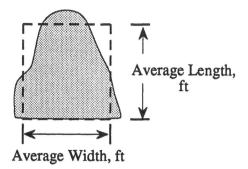

Average Length, ft

Average Width, ft

Pond or Lake Storage Capacity, gal	=	(Average) Length, ft	(Average) Width, ft	(Average) Depth, ft	(7.48) gal/cu ft

7. Copper Sulfate Dosing

The desired copper sulfate dosage may be expressed in three different ways, as indicated below:

• <u>Given mg/*L* Copper</u>

First calculate the lbs copper (Cu) required:

$$\frac{(mg/L\ Cu)\ (MG\ Vol.)\ (8.34)}{of\ Reserv.\ lbs/gal} = lbs\ Cu$$

Then determine the lbs of copper sulfate pentahydrate ($CuSO_4 \cdot 5H_2O$): (This compound contains 25% copper.)

$$\frac{lbs\ Cu}{\dfrac{\%\ Copper}{100}} = lbs\ CuSO_4 \cdot 5H_2O$$

• <u>Given lbs Copper Sulfate/ac-ft</u>
(Using 0.9 lbs $CuSO_4$ /ac-ft as an example)

$$Copper\ Sulfate,\ lbs = \frac{(0.9\ lbs\ CuSO_4)\ (Actual\ ac\text{-}ft)}{1\ ac\text{-}ft}$$

• <u>Given lbs Copper Sulfate/ac</u>
(Using 5.4 lbs $CuSO_4$ /ac as an example)

$$Copper\ Sulfate,\ lbs = \frac{(5.4\ lbs\ CuSO_4)\ (Actual\ ac)}{1\ ac}$$

8.1   WELL DRAWDOWN

When a well is pumped, the water in the vicinity of the well is pulled into the pump intake and pumped to the surface. This results in a lowered water level near the well, as shown to the right. **Drawdown** is a measure of how much the water level lowers during pumping.

To determine drawdown, you must know the water level <u>before pumping</u> (called the **static water level**) and the water level <u>during pumping</u> (called the **pumping water level**). The pumping water level is not measured until the pump has been operating at a given capacity for a specified time.

Once the static and pumping water levels are known, the drawdown can be calculated as follows:

Drawdown, ft = Pumping Water Level, ft − Static Water Level, ft

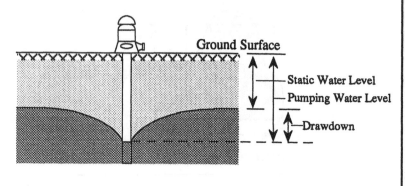

WELL DRAWDOWN IS THE DIFFERENCE BETWEEN THE PUMPING WATER LEVEL AND THE STATIC WATER LEVEL

Drawdown, ft = Pumping Water Level, ft − Static Water Level, ft

Example 1: (Well Drawdown)

❑ The static water level for a well is 65 ft. If the pumping water level is 90 ft, what is the drawdown?

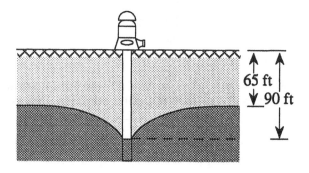

Drawdown, ft = Pumping Water Level, ft − Static Water Level, ft

= 90 ft − 65 ft

= 25 ft

Example 2: (Well Drawdown)
❏ Before the pump is started the water level is measured as 140 ft. The pump is then started. If the pumping water level is determined to be 167 ft, what is the drawdown?

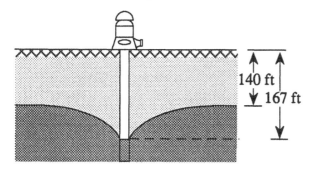

$$\begin{array}{ccc} \text{Drawdown,} & \text{Pumping Water} & \text{Static Water} \\ \text{ft} & = \text{Level, ft} & - \text{Level, ft} \end{array}$$

$$= 167 \text{ ft} - 140 \text{ ft}$$

$$= \boxed{27 \text{ ft}}$$

Example 3: (Well Drawdown)
❏ The static water level of a well is 128 ft. The pumping water level is determined using the sounding line. The air pressure applied to the sounding line is 3.5 psi and the length of the sounding line is 170 ft. What is the drawdown?

First calculate the water depth in the sounding line (1) and the pumping water level (2):

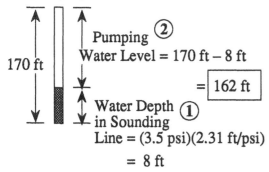

Pumping ② Water Level = 170 ft – 8 ft

$$= \boxed{162 \text{ ft}}$$

Water Depth ① in Sounding Line = (3.5 psi)(2.31 ft/psi)

$$= 8 \text{ ft}$$

Then calculate drawdown as usual:

$$\begin{array}{ccc} \text{Drawdown,} & \text{Pumping Water} & \text{Static Water} \\ \text{ft} & = \text{Level, ft} & - \text{Level, ft} \end{array}$$

$$= 162 \text{ ft} - 128 \text{ ft}$$

$$= \boxed{34 \text{ ft}}$$

The water level in a well may be measured by inserting a disinfected rope or measuring tape into the well sounding tube or casing vent and lowering it to the water level.

A water level sounding line may also be used to determine water level. In this case, air pressure is applied to the sounding line to force all the water out of the line. The amount of pressure (in psi) required to force out the water is directly related to the water depth in the sounding line. When the length of the sounding line is known and the water depth in the sounding line is determined, the water surface level in the well can be calculated, as follows:

1. Calculate the water depth in the sounding line by converting psi pressure to ft:*

$$\boxed{(\text{Air Press., psi})(2.31 \text{ ft/psi}) = \begin{array}{c} \text{Water} \\ \text{Depth, ft} \end{array}}$$

2. Calculate the water surface level by subtracting the depth of water in the sounding line from the total length of the sounding line:

$$\boxed{\begin{array}{ccccc} \text{Length of} & & \text{Water} & & \text{Water} \\ \text{Sounding} & - & \text{Depth in} & = & \text{Surface} \\ \text{Line, ft} & & \text{Sounding} & & \text{Level,} \\ & & \text{Line, ft} & & \text{ft} \end{array}}$$

Example 3 illustrates the calculation of water surface level using air pressure.

* For a review of psi and head conversions, refer to Chapter 7, Section 7.2.

8.2 WELL YIELD

Well yield is a measure of how much water the well will produce, often expressed as gallons per minute (gpm). It may also be expressed as gph, gpd, or gallons per year. Several different methods may be used to determine well yield, including:

- the bailing test method,

- the air blow test method,

- the variable rate method,

- the constant rate method, and

- the step-continuous composite method.

Determining well yield is an important part of the well construction process and this information is therefore obtained from the well driller.

All four methods of determining well yield measure essentially the same thing: the number of gallons produced during a specified time.

The **bailer test method** is used most often for low-yield wells. The "bailer" is an elongated bucket or tube with a check valve in the lower end. A driller uses a drilling rig to lower the bailer into the well, fills it with water, then lifts it quickly to the surface and empties the water into a trench. The time period for each round trip into and out of the well is noted.

Example 1: (Well Yield)
❏ Once the drawdown level of a well stabilized, it was determined that the well produced 380 gallons during a five-minute test. What is the yield of the well expressed in gpm?

$$\text{Well Yield, gpm} = \frac{\text{Gallons Produced}}{\text{Duration of Test, min}}$$

$$= \frac{380 \text{ gallons}}{5 \text{ minutes}}$$

$$= \boxed{76 \text{ gpm}}$$

Example 2: (Well Yield)
❏ During a 5-minute test for well yield, a total of 740 gallons are removed from the well. What is the well yield in gpm? in gph?

$$\text{Well Yield, gpm} = \frac{\text{Gallons Removed}}{\text{Duration of Test, min}}$$

$$= \frac{740 \text{ gallons}}{5 \text{ minutes}}$$

$$= \boxed{148 \text{ gpm}}$$

Then convert gpm flow to gph flow:*

$$\frac{(148 \text{ gal})}{\text{min}} \frac{(60 \text{ min})}{\text{hr}} = \boxed{8880 \text{ gph}}$$

---

* For a review of flow conversions, refer to Chapter 8 in *Basic Math Concepts*.

Example 3: (Well Yield)
❑ During a test for well yield, a total of 950 gallons were pumped during the 5-minute test. What is the well yield expressed as gpm?

$$\text{Well Yield,} \atop \text{gpm} = \frac{\text{Gallons Produced}}{\text{Duration of Test, min}}$$

$$= \frac{950 \text{ gallons}}{5 \text{ minutes}}$$

$$= \boxed{190 \text{ gpm}}$$

Example 4: (Well Yield)
❑ A bailer is used to determine the approximate yield of a well. The bailer is 10 ft long and has a diameter of 10 inches. If the bailer is placed in the well and removed a total of nine times during the 5-minute test, what is the well yield in gpm?*

10 in. = 0.83 ft

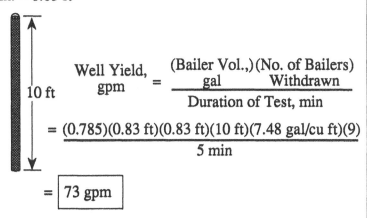

$$\text{Well Yield,} \atop \text{gpm} = \frac{(\text{Bailer Vol.,})(\text{No. of Bailers})}{\text{gal} \qquad \text{Withdrawn}} \atop \text{Duration of Test, min}$$

$$= \frac{(0.785)(0.83 \text{ ft})(0.83 \text{ ft})(10 \text{ ft})(7.48 \text{ gal/cu ft})(9)}{5 \text{ min}}$$

$$= \boxed{73 \text{ gpm}}$$

---

* For a review of volume calculations, refer to Chapter 11 in *Basic Math Concepts*.

By repeating this procedure, well yield can be determined in one of two ways:

1. Once the drawdown level has stabilized, count the <u>total number of buckets withdrawn</u> during a specified period of time. The well yield is then determined as:

<u>Simplified Equation</u>

$$\text{Well Yield,} \atop \text{gpm} = \frac{\text{Gallons Removed}}{\text{Duration of Test, min}}$$

<u>Expanded Equation*</u>

$$\text{Well} \atop \text{Yield,} = {(\text{Bailer Vol.,})(\text{No. of Bailers}) \atop \text{gal} \qquad \text{Withdrawn} \over \text{Duration of Test, min}} \atop \text{gpm}$$

2. Once the drawdown level has stabilized (the same volume of water is withdrawn for equal time periods), the volume of water withdrawn <u>each round trip</u> can be used to calculate well yield:

$$\text{Well Yield,} \atop \text{gpm} = \frac{\text{Gal. Withdr. in One} \atop \text{Round Trip}}{\text{Time for One} \atop \text{Round Trip, min}}$$

Many times the bailer method is used to determine a general well yield and then a test pump is used to determine a more accurate well yield.

The **air blow test method** is similar to the bailer method except that water is blown out of the well rather than bailed. A deflector at the top of the well captures the water which can then be measured. The well yield is then calculated using the gallons removed from the well and the duration of the test.

The **variable rate, constant rate,** and **step-continuous composite** methods use pumping rates compared with drawdown levels to determine well yield.

8.3 SPECIFIC YIELD

In the previous section, well yield was expressed as the discharge capacity of the well, usually reported in terms of gpm.

Another way to express well yield is the discharge capacity of the well per foot of drawdown. This is called the **specific yield*** of the well. It is calculated using the equation:

$$\text{Specific Yield, gpm/ft} = \frac{\text{Well Yield, gpm}}{\text{Drawdown, ft}}$$

Example 1: (Specific Yield)
❑ A well produces 250 gpm. If the drawdown for the well is 22 ft, what is the specific yield in gpm/ft of drawdown?

$$\text{Specific Yield, gpm/ft} = \frac{\text{Well Yield, gpm}}{\text{Drawdown, ft}}$$

$$= \frac{250 \text{ gpm}}{22 \text{ ft}}$$

$$= \boxed{11.4 \text{ gpm/ft}}$$

The specific yield may range from 1 gpm/ft drawdown in a tight aquifer to more than 100 gpm/ft drawdown for a properly developed well in a highly permeable aquifer.

Example 2: (Specific Yield)
❑ The discharge capacity of a well is 195 gpm. If the drawdown is 18 ft, what is the specific yield of the well in gpm/ft of drawdown?

$$\text{Specific Yield, gpm/ft} = \frac{\text{Well Yield, gpm}}{\text{Drawdown, ft}}$$

$$= \frac{195 \text{ gpm}}{18 \text{ ft}}$$

$$= \boxed{10.8 \text{ gpm/ft}}$$

---

* It is sometimes referred to as the *specific capacity* of the well.

Example 3: (Specific Yield)

❏ The yield for a particular well is 300 gpm. If the drawdown for this well is 35 ft, what is the specific yield in gpm/ft of drawdown?

$$\text{Specific Yield, gpm/ft} = \frac{\text{Well Yield, gpm}}{\text{Drawdown, ft}}$$

$$= \frac{300 \text{ gpm}}{35 \text{ ft}}$$

$$= \boxed{8.6 \text{ gpm/ft}}$$

Example 4: (Specific Yield)

❏ The specific yield of a well is listed on a report as 30.46 gpm/ft of drawdown. If the drawdown of the well is 43.3 ft, what is the well yield in gpm?

$$\text{Specific Yield, gpm/ft} = \frac{\text{Well Yield, gpm}}{\text{Drawdown, ft}}$$

$$30.46 \text{ gpm/ft} = \frac{x \text{ gpm}}{43.3 \text{ ft}}$$

Now solve for the unknown value:*

$$(30.46)(43.3) = x$$

$$\boxed{1319 \text{ gpm}} = x$$

CALCULATING OTHER UNKNOWN FACTORS

If you know any two of the three variables in the specific yield equation, you can calculate the third variable. Example 4 illustrates such a calculation.

---

* Refer to Chapter 2 in Basic Math Concepts for a review of solving for the unknown value.

8.4 WELL CASING DISINFECTION

Wells are disinfected during construction to ensure that they have not been contaminated by the drilling tools, mud, gravel, makeup water, etc. A chlorine dosage of 50 mg/L in the water-filled casing is normally provided.

Existing wells are disinfected after well or pump repairs. Since deposits of slime or bacterial growth may be dislodged or introduced to the well column during repair work, a higher chlorine dosage, 100 mg/L, is used for existing well disinfection.

The following equation is used to calculate the pounds of chlorine required for disinfection:*

$$\frac{(mg/L) \; (Water\text{-}filled) \; (8.34)}{Cl_2 \quad Casing \; Vol, \quad lbs/gal} = lbs \; Cl_2$$
$$MG$$

Examples 1-4 illustrate this calculation.

Typically, a well is chlorinated using some type of chlorine compound such as a hypochlorite or chloride of lime. The available chlorine in these compounds ranges from 5.25% to 65%, depending on the compound used. Examples 5-8 illustrate how to determine the amount of chlorine compound required for a given desired chlorine dosage.

Example 1: (Well Casing Disinfection)

❏ A new well is to be disinfected with chlorine at a dosage of 50 mg/L. If the well casing diameter is 6 inches and the length of the water-filled casing is 120 ft, how many pounds of chlorine will be required?

First calculate the volume of the water-filled casing:

$$(0.785)(0.5 \; ft)(0.5 \; ft)(120 \; ft)(7.48 \; \frac{gal}{cu \; ft}) = 176 \; gallons$$

Then determine the pounds of chlorine required using the mg/L to lbs equation:

$$(mg/L \; Cl_2) \; (MG \; Vol.) \; (8.34 \; lbs/gal) = lbs \; Cl_2$$

$$(50 \; mg/L) \; (0.000176 \; MG) \; (8.34 \; lbs/gal) = \boxed{0.07 \; lbs \; Cl_2}$$

Example 2: (Well Casing Disinfection)

❏ A new well with a casing diameter of 10 inches is to be disinfected. The desired chlorine dosage is 50 mg/L. If the casing is 190 ft long and the water level in the well is 70 ft from the top of the well, how many pounds of chlorine will be required?

First calculate the volume of the water-filled casing:
(Length of water-filled casing = 190 ft – 70 ft = 120 ft)

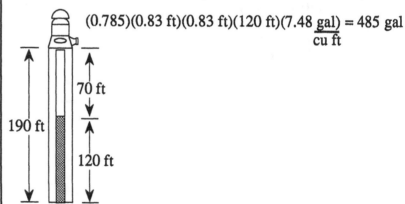

$$(0.785)(0.83 \; ft)(0.83 \; ft)(120 \; ft)(7.48 \; \frac{gal}{cu \; ft}) = 485 \; gal$$

Then determine the pounds of chlorine required:

$$(mg/L \; Cl_2) \; (MG \; Vol.) \; (8.34 \; lbs/gal) = lbs \; Cl_2$$

$$(50 \; mg/L) \; (0.000485 \; MG) \; (8.34) = \boxed{0.2 \; lbs \; Cl_2}$$

---

* For a review of mg/L to lbs calculations, refer to Chapter 3 in this text. For a review of volume calculations, refer to Chapter 1 in this text and Chapter 11 in *Basic Math Concepts*.

Example 3: (Well Casing Disinfection)

❑ An existing well has a total casing length of 210 ft. The top 160 ft of casing has a 12-inch diameter and the bottom 50 ft of casing has an 8-inch casing. The water level is 65 ft from the top of the well. How many pounds of chlorine will be required if a chlorine dosage of 100 mg/*L* is desired? (Refer to diagram shown to the right.)

First calculate the volume of water in the water-filled casing:

The <u>12-in. diameter casing</u> has a length of 160 ft – 65 ft = 95 ft of water-filled casing—

$$(0.785) \ (1 \ ft) \ (1 \ ft) \ (95 \ ft) \ \left(7.48 \ \frac{gal}{cu \ ft}\right) = 558 \ gal$$

The <u>8-in. diameter casing</u> has a water-filled length of 50 ft. The volume of water is—

$$(0.785) \ (0.67 \ ft) \ (0.67 \ ft) \ (50 \ ft) \ \left(7.48 \ \frac{gal}{cu \ ft}\right) = 132 \ gal$$

The <u>total volume</u> of water-filled casing is 558 gal + 132 gal = 690 gal

Now calculate the lbs chlorine required:

$$(100 \ mg/L) \ (0.00069 \ MG) \ (8.34 \ lbs/gal) = \boxed{0.58 \ lbs \ Cl_2}$$

<u>Diagram for Example 3:</u>

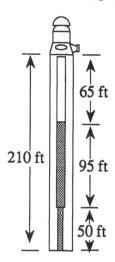

65 ft

210 ft   95 ft

50 ft

Example 4: (Well Casing Disinfection)

❑ The water-filled casing of a well has a volume of 550 gallons. If 0.5 pounds of chlorine were used in disinfection, what was the chlorine dosage in mg/*L*?

Write the equation and fill in known information.

$$(mg/L \ Cl_2) \ (MG \ Vol.) \ (8.34 \ lbs/gal) = lbs \ Cl_2$$

$$(x \ mg/L) \ (0.000550 \ gal) \ (8.34 \ lbs/gal) = 0.5 \ lbs \ Cl_2$$

Then solve for the unknown value:*

$$x = \frac{0.5}{(0.000550) \ (8.34)}$$

$$x = \boxed{109 \ mg/L}$$

CALCULATING OTHER UNKNOWN VALUES

The same equation can be used to calculate any of the factors in the equation. For this type calculation, write the equation as usual and fill in the known values. Then solve for the unknown value.* Example 4 illustrates one such calculation.

---

* For a review of solving for the unknown value, refer to Chapter 2 in *Basic Math Concepts.*

USING CHLORINE COMPOUNDS

In Examples 1-4, the required pounds of chlorine were calculated. These calculations are based on 100% available chlorine. In practice, however, chlorine compounds are often used for well disinfection. These compounds have an available chlorine content ranging from 5.25% to 65%, depending on the compound used.

The chlorine compounds and their percent available chlorine are listed below:

Compound	% Avail. Chlorine
• Calcium hypochlorite (HTH or Perchloron)	65%
• Chloride of Lime	25%
• Sodium hypochlorite* (such as Purex or Clorox)	5.25%

To calculate the required pounds of chlorine compound, simply divide the lbs of chlorine required by the percent available chlorine, as follows:**

Simplified Equation

$$\text{Chlorine Compound, lbs} = \frac{\text{Chlorine Req'd, lbs}}{\frac{\% \text{ Avail. Cl}_2}{100}}$$

Expanded Equation

$$\text{Chlorine Compound, lbs} = \frac{(\text{mg/}L)(\text{MG})(8.34) \quad \text{Cl}_2 \quad \text{Vol.} \quad \text{lbs/gal}}{\frac{\% \text{ Avail. Cl}_2}{100}}$$

Example 5: (Well Casing Disinfection)
❑ Use the answer to Example 1 (0.07 lbs chlorine required) and calculate the pounds of sodium hypochlorite required (5.25% available chlorine).

$$\text{Sodium Hypochlorite, lbs} = \frac{\text{Chlorine Req'd, lbs}}{\frac{\% \text{ Avail. Cl}_2}{100}}$$

$$= \frac{0.07 \text{ lbs}}{\frac{5.25}{100}}$$

$$= \boxed{1.3 \text{ lbs sodium hypochlorite}}$$

Note that <u>more pounds</u> of chlorine compound are required to achieve the same disinfection level.

Example 6: (Well Casing Disinfection)
❑ A new well is to be disinfected with sodium hypochlorite (5.25% available chlorine). The well casing diameter is 6 inches and the length of the water-filled casing is 125 ft. If the desired chlorine dosage is 50 mg/*L*, how many pounds of sodium hypochlorite are required?

First calculate the volume of the water-filled casing:

$$(0.785)(0.5 \text{ ft})(0.5 \text{ ft})(125 \text{ ft})(7.48 \text{ gal/cu ft}) = 183 \text{ gal}$$

Then determine the lbs sodium hypochlorite required:

$$\text{Sodium Hypochlorite, lbs} = \frac{(\text{mg/}L \text{ Cl}_2)(\text{MG Vol.})(8.34 \text{ lbs/gal})}{\frac{\% \text{ Avail. Cl}_2}{100}}$$

$$= \frac{(50 \text{ mg/}L)(0.000183 \text{ gal})(8.34 \text{ lbs/gal})}{\frac{5.25}{100}}$$

$$= \boxed{1.5 \text{ lbs}}$$

---

* There is a commercial-strength sodium hypochlorite with an available chlorine of about 11-12%.

** For a review of percents, refer to Chapter 5 in *Basic Math Concepts*.

Example 7: (Well Casing Disinfection)
❏ How many pounds of calcium hypochlorite (65% available chlorine) is required to disinfect a well if the casing is 18 inches in diameter and 140 ft long, with a water level at 60 ft from the top of the well? The desired chlorine dose is 100 mg/L.

First calculate the volume of the water-filled casing:
(Length of water-filled casing = 140 ft – 60 ft = 80 ft)

$$(0.785)\ (1.5\ ft)\ (1.5\ ft)\ (80\ ft)\ (7.48\ gal/cu\ ft) = 1057\ gal$$

Then determine the lbs of calcium hypochlorite required:

$$\frac{(100\ mg/L\ Cl_2)\ (0.001057\ MG)\ (8.34\ lbs/gal)}{\dfrac{65}{100}} = \begin{array}{l}\text{Calcium}\\ \text{hypochlorite,}\\ \text{lbs}\end{array}$$

$$= \boxed{1.4\ lbs}$$

Example 8: (Well Casing Disinfection)
❏ A well casing contains 210 gallons of water. How many fluid ounces of sodium hypochlorite (5.25% available chlorine) are required to disinfect the well if a chlorine concentration of 50 mg/L is desired?

First calculate the lbs sodium hypochlorite required:

$$\frac{(50\ mg/L\ Cl_2)\ (0.000210\ MG)\ (8.34\ lbs/gal)}{\dfrac{5.25}{100}} = 1.7\ lbs$$

Next calculate the gallons sodium hypochlorite required:*

$$\frac{1.7\ lbs}{8.34\ lbs/gal} = 0.2\ gal$$

Then convert gallons to fluid ounces:

$$\frac{(0.2\ gal)\ (128\ fl\ oz)}{gal} = \boxed{25.6\ fl\ oz}$$

CALCULATING OTHER MEASURES OF CHLORINE COMPOUND

In Examples 5-7, the pounds of chlorine compound are calculated. To determine the dry ounces, fluid ounces or other measures of chlorine compound required, the following identities will be needed.

16 oz = 1 lb (dry weight)
16 fluid oz = 1 pint
2 pints = 1 quart
4 quarts = 1 gallon
128 fluid oz = 1 gallon

Example 8 illustrates the calculation of fluid ounces of chlorine compound.

---

* For a review of lbs to gallons conversions, refer to Chapter 8 in *Basic Math Concepts*.

8.5 DEEP-WELL TURBINE PUMP CALCULATIONS

The calculations pertaining to well pumps can be grouped into three types:

- Head calculations,
- Horsepower calculations, and
- Efficiency calculations.

Each of these three types is discussed in this section.

HEAD CALCULATIONS

As shown in the diagram to the right, there are three types of head in well pump calculations:

- Discharge head

- Field head or "total pumping head", and

- Bowl assembly head or "lab head" or "total dynamic head".

Discharge head is measured at the pressure gage located close to the pump discharge flange. The pressure, expressed in psi, can be converted to feet of head using the equation:*

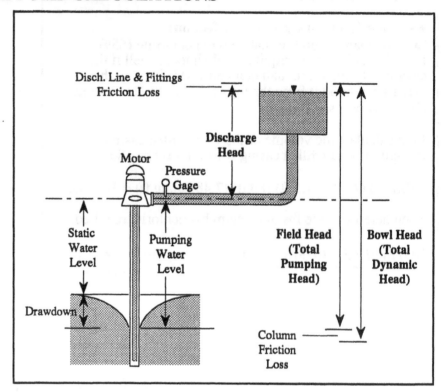

$$(\text{Press., psi})(2.31\ \text{ft/psi}) = \begin{array}{c}\text{Discharge}\\ \text{Head, ft}\end{array}$$

Field head is a measure of the lift <u>below</u> the discharge head centerline (pumping water level) and the head <u>above</u> the discharge head centerline (discharge head). Field head is therefore calculated as:

$$\begin{array}{c}\text{Field Head,}\\ \text{ft}\end{array} = \begin{array}{c}\text{Pumping}\\ \text{Water}\\ \text{Level, ft}\end{array} + \begin{array}{c}\text{Discharge}\\ \text{Head, ft}\end{array}$$

Field head is sometimes referred to as "total pumping head".

COLUMN FRICTION LOSS (IN FT) PER 100 FT OF COLUMN											
Column Size	8"			10"				12"			
Tube Size	2"	2½"	3"	2"	2½"	3"	3½"	2"	2½"	3"	3½"
gpm											
1000	3.2	3.9	5.4	0.85	0.97	1.2	1.4	0.34	0.38	0.44	0.50
1200	4.5	5.4	7.6	1.2	1.4	1.6	2.0	0.47	0.54	0.62	0.71
1400	6.0	7.2	10.0	1.6	1.8	2.2	2.7	0.62	0.71	0.82	0.94
1600	7.6	9.1	13.0	2.0	2.3	2.8	3.4	0.80	0.90	1.1	1.2

Source: Floway Turbine Pumps; Peabody Floway, Inc., Fresno, CA.

Example 1: (Vertical Turbine Pumps)
❑ The pressure gage reading at a pump discharge head is 4.4 psi. What is this discharge head expressed in feet?

$$(4.4\ \text{psi})\ (2.31\ \text{ft/psi}) = \boxed{10.2\ \text{ft}}$$

---

* For a review of psi and feet conversions, refer to Chapter 7, Section 7.2. The gage reading is psi x 2.31 plus any correction to the center line of the discharge head.

Example 2: (Vertical Turbine Pumps)
❑ The static water level of a pump is 90 ft. The well drawdown is 29 ft. If the gage reading at the pump discharge head is 3.9 psi, what is the field head?

The field head (or total pumping head) is a total of the lift <u>below</u> the discharge head centerline plus the head <u>above</u> the discharge head centerline:

$$
\begin{array}{ll}
\text{Field Head,} & = \text{Pumping} + \text{Discharge} \\
\text{ft} & \quad \text{Water} \quad \text{Head, ft} \\
& \quad \text{Level, ft}
\end{array}
$$

$$= (90 \text{ ft} + 29 \text{ ft}) + (3.9 \text{ psi}) (2.31 \text{ ft/psi})$$

$$= 119 \text{ ft} + 9 \text{ ft}$$

$$= \boxed{128 \text{ ft}}$$

Example 3: (Vertical Turbine Pumps)
❑ The field head for a deep-well vertical turbine pump is 140 ft. The 8-inch diameter column is 180 ft long with a shaft enclosure tube 2 inches in diameter. If the flow through the column is 1200 gpm, what is the lab head for the pump?

$$
\begin{array}{ll}
\text{Lab Head,} & = \text{Field Head,} + \text{Column} \\
\text{ft} & \quad\quad \text{ft} \quad\quad \text{Friction} \\
& \quad\quad\quad\quad\quad \text{Loss, ft}
\end{array}
$$

From the column friction loss table on the opposite page, the friction loss per 100 ft for a 8-inch diameter column, 2-inch tube, and 1200 gpm flow is 4.5 ft.

The friction loss for the entire 140 ft of column is:

$$\frac{(4.5 \text{ ft loss}) (140 \text{ ft})}{100 \text{ ft}} = 6.3 \text{ ft}$$

Lab head can now be determined:

$$
\begin{array}{ll}
\text{Lab Head,} & = 140 \text{ ft} + 6.3 \text{ ft} \\
\text{ft} & \\
& = \boxed{146.3 \text{ ft}}
\end{array}
$$

Bowl assembly head, usually referred to simply as **bowl head,** is also called total dynamic head. It is a measure of all heads against which the pump must operate. In addition to lift below and head above the discharge head centerline (field head), the friction loss due to water flow through the well column must be included. To calculate bowl head, therefore, add column friction loss (in ft) to field head:

$$
\begin{array}{ll}
\text{Bowl Head,} & = \text{Field Head,} + \text{Column} \\
\text{ft} & \quad\quad \text{ft} \quad\quad \text{Friction} \\
& \quad\quad\quad\quad\quad \text{Loss, ft}
\end{array}
$$

To determine the column friction loss, a table such as that shown on the opposite page is used. The friction loss is generally given as loss per 100 ft of column. Once you know the column diameter, the shaft enclosure tube diameter, and gpm flow rate through the column, the column friction loss can be determined. Example 3 illustrates a calculation of this type.

HORSEPOWER CALCULATIONS

The required horsepower for a particular pumping application depends on several factors such as:*

- total head against which the pump must operate (total dynamic head)

- The flow rate to be pumped (usually expressed as gpm)

- The efficiency of the pump, and

- the efficiency of the motor.

There are five types of horsepower calculations for deep-well turbine pumps, as illustrated in the diagram to the right. Before reviewing the calculations, however, it is important to have a general understanding of the various terms.

Motor horsepower or **input horsepower** refers to the horsepower supplied to the motor. Next, there is a loss of horsepower at the motor due to motor inefficiency. Note that the arrow representing horsepower output of the motor, **total brake horsepower** (arrow 2), is smaller than the input arrow to reflect the horsepower loss. There is a horsepower loss at the motor resulting from thrust bearing loss. The horsepower after this loss has been considered is called **field horsepower** (arrow 3) in the diagram). Field horsepower is the horsepower required at the top of pump shaft (not including thrust bearing losses).

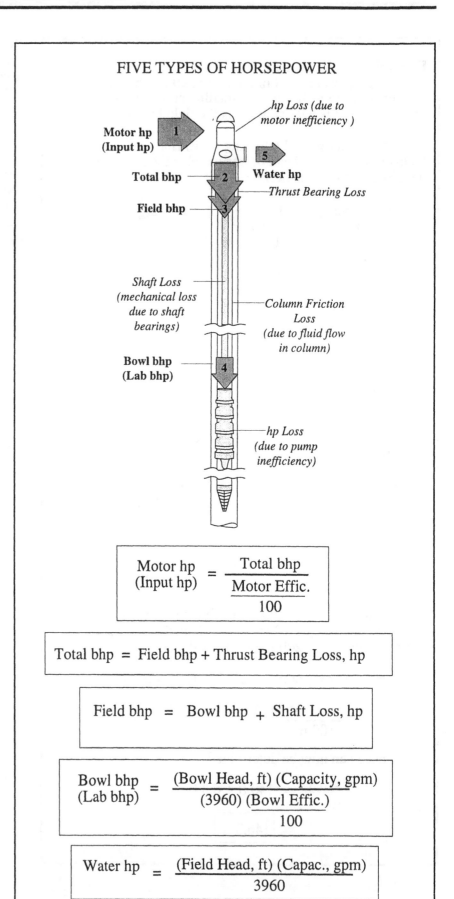

FIVE TYPES OF HORSEPOWER

$$\frac{\text{Motor hp}}{\text{(Input hp)}} = \frac{\text{Total bhp}}{\dfrac{\text{Motor Effic.}}{100}}$$

$$\text{Total bhp} = \text{Field bhp} + \text{Thrust Bearing Loss, hp}$$

$$\text{Field bhp} = \text{Bowl bhp} + \text{Shaft Loss, hp}$$

$$\frac{\text{Bowl bhp}}{\text{(Lab bhp)}} = \frac{(\text{Bowl Head, ft}) (\text{Capacity, gpm})}{(3960) \dfrac{(\text{Bowl Effic.})}{100}}$$

$$\text{Water hp} = \frac{(\text{Field Head, ft}) (\text{Capac., gpm})}{3960}$$

* Additional pumping calculations are described in Chapter 7.

Example 4: (Vertical Turbine Pumps)
❏ The pumping water level for a well pump is 160 ft and the discharge pressure measured at the pump discharge centerline is 3.6 psi. If the flow rate from the pump is 800 gpm, what is the water horsepower? (Use the <u>first equation</u> shown in the text column to the right.)

First calculate the field head.* The discharge head must be converted from psi to ft:

$$(3.6 \text{ psi}) (2.31 \text{ ft/psi}) = 8.3 \text{ ft}$$

The field head is therefore:

$$160 \text{ ft} + 8.3 \text{ ft} = 168.3 \text{ ft}$$

The water horsepower can now be determined:

$$whp = \frac{(\text{Field Head,}) (\text{Capac.,}) (8.34)}{33,000 \text{ ft-lbs/min}}$$
$$= \frac{(168.3 \text{ ft}) (800 \text{ gpm}) (8.34 \text{ lbs/gal})}{33,000 \text{ ft-lbs/min}}$$
$$= \boxed{34 \text{ whp}}$$

Example 5: (Vertical Turbine Pumps)
❏ The pumping water level for a pump is 190 ft. The discharge pressure measured at the pump discharge head is 4.1 psi. If the pump flow rate is 900 gpm, what is the water horsepower? (Use the <u>second equation</u> shown to the right.)

The field head must first be determined. In order to determine field head, the discharge head must be converted from psi to ft:

$$(4.1 \text{ psi}) (2.31 \text{ ft/psi}) = 9.5 \text{ ft}$$

The field head can now be calculated:

$$190 \text{ ft} + 9.5 \text{ ft} = 199.5 \text{ ft}$$

And then the water horsepower can be calculated:

$$whp = \frac{(199.5 \text{ ft}) (900 \text{ gpm})}{3960}$$
$$= \boxed{45 \text{ whp}}$$

* Head calculations are described at the beginning of this section.

From field horsepower, traveling down the shaft to the top of the pump bowls, there is a horsepower loss due to the mechanical friction of the line shaft bearings. The horsepower at the entry to the pump bowls is called **bowl horsepower** or **laboratory horsepower** (arrow 4). As the power passes through the pump bowls there is a loss of horsepower resulting from pump inefficiency. Then, as the water flows upward through the well column, there is another loss due to fluid flow through the column.

WATER HORSEPOWER

The horsepower at the pump discharge is called water horsepower (arrow 5). It is the horsepower that must be delivered at the pump discharge head to accomplish the desired pumping job.

When describing the various horsepower calculations, it is advantageous to begin with water horsepower. The water horsepower (whp) required depends on:

• the **field head** (total pumping head) against which the pump must operate, and

• the **flow rate** or capacity to be pumped, in gpm.

There are two equations that may be used to calculate water horsepower:

$$whp = \frac{(\text{Field Head,}) (\text{Capac.,}) (8.34)}{33,000 \text{ ft-lbs/min}}$$

Or

$$whp = \frac{(\text{Field Head,}) (\text{Capacity})}{3960}$$

These two equations are equivalent. To derive the second equation, simply divide the numerator and denominator of the first equation by 8.34.

BOWL HORSEPOWER

Bowl horsepower (bowl bhp or lab bhp) is the horsepower required at the input to the bowl assembly. Since horsepower input to a pump is generally referred to as brake horsepower (bhp), the bowl horsepower is often abbreviated as bowl bhp.

In addition to the water horsepower, bowl bhp must also incorporate additional horsepower to account for:

• the horsepower loss due to the pump inefficiency as well as

• the friction loss in the well column

Thus, **the bowl horsepower will always be greater than the water horsepower**.

Dividing water hp by pump efficiency accounts for horsepower lost as a result of pump inefficiency.* But column friction loss must also be considered. Thus, instead of using field head to determine horsepower requirements, lab head is used. (Lab head is field head plus column friction loss.)

As with water horsepower, there are two equations that may be used to calculate bowl bhp:

$$\text{Bowl bhp} = \frac{(\text{Bowl Head, ft})(\text{Capac., gpm})(8.34 \text{ lbs/gal})}{\dfrac{(33,000 \text{ ft-lbs/min/hp})(\text{Bowl Effic.})}{100}}$$

Or

$$\text{Bowl bhp} = \frac{(\text{Bowl Head, ft})(\text{Capac., gpm})}{\dfrac{(3960)(\text{Bowl Effic.})}{100}}$$

Example 6: (Vertical Turbine Pumps)

❏ A deep-well vertical turbine pump delivers 700 gpm. If the lab head is 195 feet and the bowl efficiency is 83.5%, what is the bowl horsepower? (Use the second equation given in the column to the left.)

$$\text{Bowl bhp} = \frac{(\text{Bowl Head, ft})(\text{Capac., gpm})}{\dfrac{(3960)(\text{Bowl effic.})}{100}}$$

$$= \frac{(195 \text{ ft})(700 \text{ gpm})}{\dfrac{(3960)(83.5)}{100}}$$

$$= \frac{(195)(700)}{(3960)(0.835)}$$

$$= \boxed{41.3 \text{ bowl bhp}}$$

Example 7: (Vertical Turbine Pumps)

❏ A deep-well vertical pump delivers 900 gpm. The lab head is 200 ft and the bowl efficiency is 85%. What is the bowl horsepower?

$$\text{Bowl bhp} = \frac{(\text{Bowl Head, ft})(\text{Capac., gpm})}{\dfrac{(3960)(\text{Bowl effic.})}{100}}$$

$$= \frac{(200 \text{ ft})(900 \text{ gpm})}{\dfrac{(3960)(85)}{100}}$$

$$= \frac{(200)(900)}{(3960)(0.85)}$$

$$= \boxed{53.5 \text{ bowl bhp}}$$

---

* When horsepower is divided by efficiency (written as a decimal number) the result is always a <u>larger horsepower</u>.

SHAFT FRICTION LOSS IN HP PER 100-FT OF SHAFT

Shaft Diam. (inches)	RPM of Shaft				
	2900	1760	1450	960	860
3/4	0.51	0.31	0.26	0.17	–
1	0.87	0.53	0.44	0.29	0.26
1-1/4	1.33	0.79	0.67	0.44	0.39
1-1/2	1.90	1.14	0.96	0.63	0.56
1-3/4	2.50	1.50	1.25	0.83	0.74
2	–	1.90	1.60	1.05	0.95
1-1/4	–	2.40	2.00	1.35	1.20

Source: Floway Turbine Pumps; Peabody Floway, Inc., Fresno, CA.

FIELD HORSEPOWER

Field horsepower (field bhp) is the horsepower required at the top of the pump shaft. It is the bowl bhp plus the shaft loss, expressed as horsepower:

Field bhp = Bowl bhp + Shaft Loss, hp

The shaft loss depends on the shaft diameter and length, and is read from a table such as that shown to the left.

TOTAL BRAKE HORSEPOWER

Total brake horsepower includes field horsepower plus thrust bearing loss. Thrust bearing loss data is calculated using the equation shown below.

$$\text{Thrust Bearing Loss, hp} = \frac{(0.0075)(rpm)}{100}\frac{(\text{Total Thrust, lbs})}{1000}$$

Total thrust, lbs, depends on the shaft weights and area and the thrust factor for the pump. The equation for calculating total lbs thrust is given in Appendix 1.

MOTOR OR INPUT HORSEPOWER

Motor horsepower or input horsepower must include total brake horsepower plus an additional horsepower to account for motor inefficiency. Since motor horsepower will always be greater than total brake horsepower, it can be calculated by dividing the total brake horsepower by the motor efficiency:

$$mhp = \frac{\text{Total bhp}}{\frac{\text{Motor Effic.}}{100}}$$

Example 8: (Vertical Turbine Pumps)
❏ In Example 7 it was calculated that the bowl bhp was 53.5 bhp. If the 1-inch diameter shaft is 160 ft long and is rotating at 1450 rpm, what is the field bhp?

Before field bhp can be calculated, the shaft loss (in hp) must be determined using the table above. For a 1-inch diameter shaft rotating at 1450 rpm, the friction loss is 0.44 hp per 100 ft. Therefore, for a total of 160 ft of shaft, the friction loss is:

$$\frac{(0.44 \text{ hp loss})(160 \text{ ft})}{100 \text{ ft}} = 0.7 \text{ hp loss}$$

Now the field bhp can be determined:

Field bhp = Bowl bhp + Shaft Loss, hp

= 53.5 bhp + 0.7 hp

= 54.2 bhp

Example 9: (Vertical Turbine Pumps)
❏ The field horsepower for a deep-well turbine pump is 61.8 bhp. If the thrust bearing loss is 0.6 hp and the motor efficiency is 90% (from motor manufacturer), what is the input horsepower to the motor?

$$mhp = \frac{\text{Total bhp}}{\frac{\text{Motor Effic.}}{100}}$$

$$= \frac{61.8 \text{ bhp} + 0.6 \text{ hp}}{0.90}$$

$$= 69.3 \text{ mhp}$$

EFFICIENCY CALCULATIONS

There are four types of efficiencies considered with respect to well installations:

- Bowl efficiency (or Lab efficiency),

- Field efficiency,

- Motor efficiency, and

- Overall efficiency (or wire-to-water efficiency).

Each calculation of efficiency, whether calculated by the operator or determined by the equipment manufacturer, is a comparison of horsepower output of the unit or system with horsepower input to that unit or system. The general equation used in calculating percent efficiency (shown in the box to the right) follows the general percent equation:*

$$\% = \frac{Part}{Whole} \times 100$$

This subsection discusses each of the four types of efficiencies.

Bowl efficiency (sometimes called "lab efficiency") is a comparison of horsepower input and output of the bowl assembly. Bowl efficiency is determined by the pump manufacturer using precise testing procedures. The results are then used to prepare pump performance curves such as that shown on the opposite page.

EFFICIENCY COMPARES HP OUTPUT WITH HP INPUT

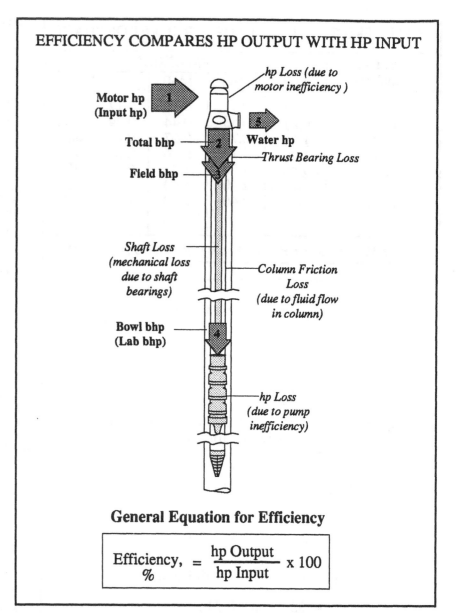

General Equation for Efficiency

$$\text{Efficiency, } \% = \frac{\text{hp Output}}{\text{hp Input}} \times 100$$

Example 10: (Vertical Turbine Pumps)
❏ Using the pump performance curve shown on the next page, determine the pump efficiency when the bowl head per stage is 52 hp and the capacity is 1000 gpm.

Find 1000 gpm along the bottom scale and follow the line up to the bowl head curves. Next, locate 52 hp along the left scale. (Note that the left scale has five divisions between every 20-point span. Thus, each division represents 20/5 = 4 points.) Draw a horizontal line from 52 hp until it intersects with the 1000-gpm line.

The intersection is at approximately $\boxed{82.5\%}$

---

* For a review of percent calculations, refer to Chapter 6 in this text and Chapter 5 in *Basic Math Concepts*.

PUMP PERFORMANCE CURVE

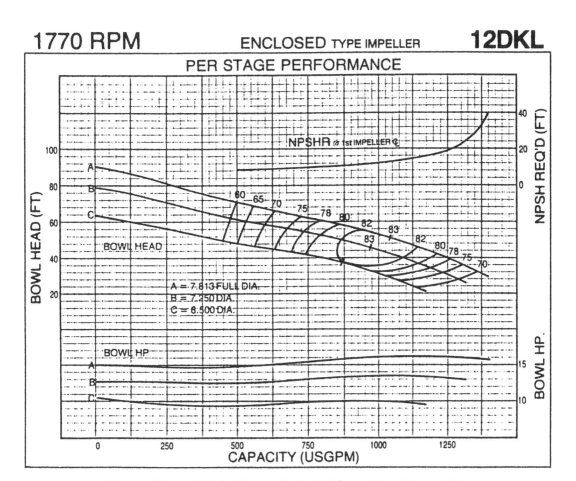

Source: Floway Turbine Pumps; Peabody Floway, Inc., Fresno, CA.

Example 11: (Vertical Turbine Pumps)
❑ Using the pump performance curve, determine the bowl efficiency when the bowl head per stage is 60 ft and the capacity is 800 gpm.

First, find 800 gpm along the bottom scale and draw a vertical line (Note that this scale divides every 250 points into 10 divisions. Therefore each division represents 250/10 = 25 points.)

Next, find 60 ft along the left scale and draw a horizontal line.

The point of intersection represents the efficiency:

78% Efficiency

To read the bowl efficiency on the pump performance curve provided by the pump manufacturer, you will need to know:

• The bowl head per stage

$$\text{Bowl Head/Stage} = \frac{\text{Bowl Head, ft}}{\text{No. of Stages}}$$

• The pump capacity, gpm

To read the pump performance curve, find the given capacity along the bottom scale and draw a vertical line at that point. Find the bowl head per stage along the left scale and draw a horizontal line to the point of intersection with the vertical capacity line. Read the efficiency at that point.

Field efficiency is a comparison of the horsepower of leaving the pump unit (at the discharge centerline) with that entering the pump unit (at the top of the impeller shaft). The horsepower output of the pump unit is the water hp (whp) and horsepower input to the pump unit is the total bhp. Thus, the equations used to calculate field efficiency are:

Simplified Equation

$$\text{Field Effic.,} \% = \frac{\text{Hp Output}}{\text{Hp Input}} \times 100$$

Expanded Equations

$$\text{Field Effic.,} \% = \frac{\text{Whp}}{\text{Total Input}} \times 100$$

Whp in the numerator can be replaced by:

$$\text{Field Effic.,} \% = \frac{\dfrac{(\text{Field head,)} \ (\text{Capac.,)}}{\text{ft} \qquad \text{gpm}}}{3960}}{\text{Total bhp}} \times 100$$

This equation is usually rewritten as:

$$\text{Field Effic.,} \% = \frac{(\text{Field head,) (Capac.,)}}{\text{ft} \qquad \text{gpm}}{(3960)(\text{Total bhp})} \times 100$$

Example 12: (Vertical Turbine Pumps)
❑ The total bhp for a deep-well turbine pump is 61.2 bhp. If the water horsepower is 50.8 hp, what is the field efficiency?

$$\text{Field Effic.,} \% = \frac{\text{Whp}}{\text{Total Input}} \times 100$$

$$= \frac{50.8 \text{ hp}}{61.2 \text{ bhp}} \times 100$$

$$= \boxed{83\%}$$

Example 13: (Vertical Turbine Pumps)
❑ Given the data below, calculate the field efficiency of the deep-well turbine pump.

Field head—195 ft
Capacity—950 gpm
Total bhp—59.2 bhp

Field efficiency is a comparison of water horsepower and total brake horsepower. Since water horsepower has not been given, the expanded equation must be used:

$$\text{Field Effic.,} \% = \frac{(\text{Field head,) (Capac.,)}}{\text{ft} \qquad \text{gpm}}{(3960) \ (\text{Total bhp})} \times 100$$

$$= \frac{(195 \text{ ft}) \ (950 \text{ gpm})}{(3960) \ (59.2 \text{ bhp})} \times 100$$

$$= \boxed{79\%}$$

Example 14: (Vertical Turbine Pumps)
❑ The total bhp for a pump is 58.2 bhp. If the motor is 90% efficient and the water horsepower is 45.9 whp, what is the overall efficiency of the unit?

$$\frac{\text{Overall}}{\text{Efficiency, \%}} = \frac{\text{whp}}{\text{Input hp}} \times 100$$

Before overall efficiency can be determined, the input hp must be calculated:

$$\text{Input hp} = \frac{\text{Total bhp}}{\dfrac{\text{Motor Effic., \%}}{100}}$$

$$= \frac{58.2 \text{ bhp}}{0.90}$$

$$= 64.7 \text{ hp Input}$$

Now the overall efficiency can be calculated:

$$\frac{\text{Overall}}{\text{Efficiency, \%}} = \frac{45.9 \text{ whp}}{64.7 \text{ Input hp}} \times 100$$

$$= \boxed{71\%}$$

Example 15: (Vertical Turbine Pumps)
❑ The efficiency of a motor is 90%. If the field efficiency is 81%, what is the overall efficiency of the unit?

$$\frac{\text{Overall}}{\text{Efficiency, \%}} = \frac{(\text{Field Effic., \%}) (\text{Motor Effic., \%})}{100}$$

$$= \frac{(81) (90)}{100}$$

$$= \boxed{72.9\%}$$

Motor efficiency data is provided by the manufacturer.

Overall efficiency or **wire-to-water efficiency** is a comparison of the horsepower output of the system (water horsepower) with that entering the system (motor horsepower or input horsepower). The equations used to calculate overall efficiency are:

Simplified Equation

$$\frac{\text{Overall}}{\text{Effic., \%}} = \frac{\text{hp Output}}{\text{hp Input}} \times 100$$

Expanded Equations

$$\frac{\text{Overall}}{\text{Effic., \%}} = \frac{\text{whp}}{\text{Input hp}} \times 100$$

Overall efficiency can also be calculated by multiplying the motor and field efficiencies, as follows:

$$\frac{\text{Overall}}{\text{Effic., \%}} = \frac{(\text{Field}) (\text{Motor})}{100} \frac{\text{Effic.,\%}}{100} \frac{\text{Effic.,\%}}{100} \times 100$$

This equation can be simplified by dividing out a 100 in the numerator and denominator:

$$\frac{\text{Overall}}{\text{Effic., \%}} = \frac{(\text{Field}) (\text{Motor})}{100} \frac{\text{Effic.,\%} \cdot \text{Effic.,\%}}{100}$$

8.6 POND OR SMALL LAKE STORAGE CAPACITY

The storage capacity of a pond or small lake can be estimated using average length and width dimensions and an estimated average depth.

The two equations that are used to calculate the gallons and acre-feet volume are shown to the right.*

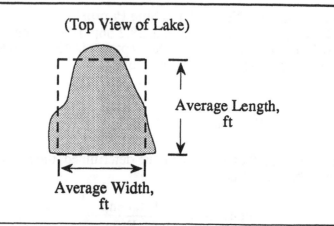

(Top View of Lake)

Average Length, ft

Average Width, ft

$$\begin{array}{c}\text{Pond or Lake}\\\text{Storage Capacity,}\\\text{gal}\end{array} = \begin{array}{c}\text{(Average)}\\\text{Length, ft}\end{array}\begin{array}{c}\text{(Average)}\\\text{Width, ft}\end{array}\begin{array}{c}\text{(Average)}\\\text{Depth, ft}\end{array}\begin{array}{c}\text{(7.48)}\\\text{gal/cu ft}\end{array}$$

$$\begin{array}{c}\text{Volume,}\\\text{ac-ft}\end{array} = \frac{\text{Volume, cu ft}}{43,560 \text{ cu ft/ac-ft}}$$

Example 1: (Storage Capacity)
❑ A pond has an average length of 250 ft, an average width of 90 ft, and an estimated average depth of 10 ft. What is the estimated volume of the pond in gallons?

$$\begin{array}{c}\text{Volume,}\\\text{gal}\end{array} = \begin{array}{c}\text{(Average)}\\\text{Length, ft}\end{array}\begin{array}{c}\text{(Average)}\\\text{Width, ft}\end{array}\begin{array}{c}\text{(Average)}\\\text{Depth, ft}\end{array}\begin{array}{c}\text{(7.48)}\\\text{gal/cu ft}\end{array}$$

$$= (250 \text{ ft}) (90 \text{ ft}) (10 \text{ ft}) \left(\frac{7.48 \text{ gal}}{\text{cu ft}}\right)$$

$$= \boxed{1,683,000 \text{ gal}}$$

---

* For a review of volume calculations, refer to Chapter 1 in this text or Chapter 11 in *Basic Math Concepts*.

Example 2: (Storage Capacity)

❏ A small lake has an average length of 310 ft and an average width of 115 ft. If the maximum depth of the lake is 18 ft, what is the estimated gallons volume of the lake?

First, the average depth of the lake must be estimated:

$$\text{Estimated Aver. Depth, ft} = (\text{Greatest}) (0.4) \text{ Depth, ft}$$

$$= (18 \text{ ft}) (0.4)$$

$$= 7.2 \text{ ft}$$

Then the lake volume can be determined:

$$\text{Volume, gal} = (\text{Average}) (\text{Average}) (\text{Average}) (7.48) \\ \text{Length, ft Width, ft Depth, ft gal/cu ft}$$

$$= (310 \text{ ft}) (115 \text{ ft}) (7.2 \text{ ft}) (7.48 \text{ gal/cu ft})$$

$$= \boxed{1,919,966 \text{ gal}}$$

ESTIMATING THE AVERAGE DEPTH

For small ponds and lakes, the average depth is generally about 0.4 times the greatest depth. Therefore, to estimate the average depth, measure the greatest depth, then multiply that number by 0.4:

$$\boxed{\text{Estimated Aver. Depth, ft} = (\text{Greatest}) (0.4) \text{ Depth, ft}}$$

Example 3: (Storage Capacity)

❏ A pond has an average length of 160 ft, an average width of 93 ft, and an average depth of 9 ft. What is the acre-feet volume of the pond?

$$\text{Volume, ac-ft} = \frac{(\text{Average}) (\text{Average}) (\text{Average})}{\text{Length, ft Width, ft Depth, ft}} \Big/ {43,560 \text{ cu ft/ac-ft}}$$

$$= \frac{(160 \text{ ft}) (93 \text{ ft}) (9 \text{ ft})}{43,560 \text{ cu ft/ac-ft}}$$

$$= \boxed{3.1 \text{ ac-ft}}$$

ACRE-FEET VOLUME

In many calculations pertaining to lakes, ponds, and reservoirs, the volume must be expressed in terms of ac-ft. The equation to be used is:

$$\boxed{\text{Volume, ac-ft} = \frac{\text{Volume, cu ft}}{43,560 \text{ cu ft/ac-ft}}}$$

8.7 COPPER SULFATE DOSING

Copper sulfate is used for algae control in lakes and ponds. The copper sulfate compound generally used is called copper sulfate pentahydrate ($CuSO_4 \cdot 5\ H_2O$), or "bluestone". The copper ions in the water kill the algae.

The three principal water quality characteristics that affect the effectiveness of copper sulfate in algae control are:

• alkalinity,

• turbidity or suspended matter, and

• water temperature.

Of these three, the alkalinity of the water is particularly important. The recommended copper sulfate dose is therefore often based on the alkalinity of the water.

The desired copper sulfate dosage may be expressed in several different ways:

• mg/L Copper

• lbs Copper Sulfate/ac-ft

• lbs Copper Sulfate/ac

When the desired dose is expressed as **mg/L copper**, the following equation is used to calculate lbs copper sulfate required:

Simplified Equation

$$\text{Copper Sulfate, lbs} = \dfrac{\text{Copper, lbs}}{\dfrac{\%\ \text{Avail. Cu}}{100}}$$

Expanded Equation*

$$\text{Copper Sulfate, lbs} = \dfrac{\dfrac{(\text{mg}/L)(\text{MG})(8.34}{\text{Cu}\quad \text{Vol.}\quad \text{lbs/gal})}{\dfrac{\%\ \text{Avail. Cu}}{100}}$$

Examples 1 and 2 illustrate this type of calculation.

Example 1: (Copper Sulfate Dosing)
❑ For algae control in a small lake, a dosage of 0.5 mg/L copper is desired. The lake has a volume of 20 MG. How many pounds of copper sulfate pentahydrate ($CuSO_4 \cdot 5\ H_2O$) will be required? (Copper sulfate pentahydrate contains 25% available copper.)

$$\text{Copper Sulfate, lbs} = \dfrac{(\text{mg}/L\ \text{Cu})(\text{Vol., MG})(8.34\ \text{lbs/gal})}{\dfrac{\%\ \text{Avail. Cu}}{100}}$$

$$= \dfrac{(0.5\ \text{mg}/L)(20\ \text{MG})(8.34\ \text{lbs/gal})}{\dfrac{25}{100}}$$

$$= \boxed{\begin{array}{l}334\ \text{lbs}\\ \text{Copper Sulfate}\end{array}}$$

Example 2: (Copper Sulfate Dosing)
❑ The desired copper dosage at a reservoir is 0.5 mg/L. The reservoir has a volume of 65 ac-ft. How many pounds of copper sulfate pentahydrate (25% available copper) will be required?

Before the equation can be used, the reservoir volume must be expressed in terms of million gallons. First convert ac-ft to cu ft:**

$$(65\ \text{ac-ft})\left(43{,}560\ \dfrac{\text{cu ft}}{\text{ac-ft}}\right) = 2{,}831{,}400\ \text{cu ft}$$

Then to gallons:

$$(2{,}831{,}400\ \text{cu ft})(7.48\ \text{gal/cu ft}) = 21{,}178{,}872\ \text{gal}$$

$$\text{or} = 21.2\ \text{MG}$$

The lbs copper sulfate can now be calculated:

$$\text{Copper Sulfate, lbs} = \dfrac{(\text{mg}/L\ \text{Cu})(\text{Vol., MG})(8.34\ \text{lbs/gal})}{\dfrac{\%\ \text{Avail. Cu}}{100}}$$

$$= \dfrac{(0.5\ \text{mg}/L)(21.2\ \text{MG})(8.34\ \text{lbs/gal})}{0.25}$$

$$= \boxed{\begin{array}{l}354\ \text{lbs}\\ \text{Copper Sulfate}\end{array}}$$

---

* These calculations are similar to hypochlorite problems. Refer to Chapter 3, Section 3.1.
** For a review of conversions, refer to Chapter 8 in *Basic Math Concepts*.

Example 3: (Copper Sulfate Dosing)
❏ A pond has a volume of 40 ac-ft. If the desired copper sulfate dose is 0.9 lbs per ac-ft, how many lbs of copper sulfate will be required?

Either of the two equations shown to the right may be used:

$$\frac{0.9 \text{ lbs CuSO}_4}{1 \text{ ac-ft}} = \frac{x \text{ lbs CuSO}_4}{40 \text{ ac-ft}}$$

Then solve for x:

$$(0.9)(40) = x$$

$$\boxed{36 \text{ lbs CuSO}_4} = x$$

Example 4: (Copper Sulfate Dosing)
❏ A small lake has a surface area of 6.2 acres. If the desired copper sulfate dose is 5.4 lbs/ac, how many pounds of copper sulfate are required?

Again, either of two equations shown to the right may be used. In this example, the first equation will be used:

$$\text{Copper Sulfate, lbs} = \frac{(5.4 \text{ lbs CuSO}_4)(6.2 \text{ ac})}{1 \text{ ac}}$$

$$= \boxed{33.5 \text{ lbs CuSO}_4}$$

Sometimes the desired copper sulfate dose is expressed as **lbs copper sulfate/ac-ft.** This is a dosage of lbs chemical per <u>volume</u> of water. Either of the two following equations may be used to make this calculation (assume the desired copper sulfate dosage is 0.9 lbs/ac-ft):

$$\text{Copper Sulfate, lbs} = \frac{(0.9 \text{ lbs CuSO}_4)(\text{ac-ft})}{1 \text{ ac-ft}}$$

or, written as a proportion:*

Desired Dosage	Actual Dosage

$$\frac{0.9 \text{ lbs CuSO}_4}{1 \text{ ac-ft}} = \frac{x \text{ lbs CuSO}_4}{\text{Actual Vol., ac ft}}$$

Example 3 illustrates this type of calculation. For large reservoirs sometimes only a depth of perhaps 20 ft or the depth down to the thermocline is included in the calculation of ac-ft volume.

The desired copper sulfate dosage may also be expressed in terms of **lbs copper sulfate/ac.** This is a dosage of lbs chemical per <u>area</u> of water. This expression of dosage is generally used for waters with a relatively high alkalinity. Under these conditions, most of the algal control is limited to a relatively shallow area near the surface (due to interference of "competing ions"), and thus the overall volume of water is irrelevant.

Either of the two following equations may be used to calculate the desired lbs copper sulfate (assume a desired dose of 5.4 lbs $CuSO_4$ /ac):

$$\text{Copper Sulfate, lbs} = \frac{(5.4 \text{ lbs CuSO}_4)(\text{ac})}{1 \text{ ac}}$$

or, written as a proportion:

Desired Dosage	Actual Dosage

$$\frac{5.4 \text{ lbs CuSO}_4}{1 \text{ ac}} = \frac{x \text{ lbs CuSO}_4}{\text{Actual Area, ac}}$$

---

* To review proportions, refer to Chapter 7 in *Basic Math Concepts.*

** Refer to Chapter 2 in Basic Math Concepts for a review of solving for the unknown value.

NOTES:

9 *Coagulation and Flocculation*

SUMMARY

1. Chamber or Basin Volume

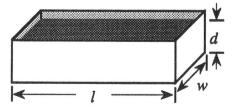

$$\boxed{\text{Vol, cu ft} = (\text{length, ft}) (\text{width, ft}) (\text{depth, ft})}$$

Or

$$\boxed{\text{Vol, gal} = (\text{length, ft}) (\text{width, ft}) (\text{depth, ft}) (7.48 \text{ gal/cu ft})}$$

2. Detention Time

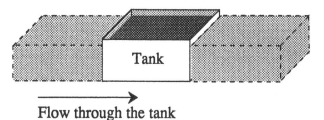

Flow through the tank

For Flash Mix Chambers:

$$\boxed{\frac{\text{Detention}}{\text{Time, sec}} = \frac{\text{Volume of Chamber, gal}}{\text{Flow, gal/sec}}}$$

For Flocculation Basins:

$$\boxed{\frac{\text{Detention}}{\text{Time, min}} = \frac{\text{Volume of Basin, gal}}{\text{Flow, gal/min}}}$$

SUMMARY—Cont'd

**3. Determining Chemical Feeder Setting—
Dry Chemical Feeder (lbs/day)**

$$\text{(mg/}L\text{ Chem.) (MGD flow) (8.34 lbs/gal)} = \begin{array}{l}\text{lbs/day}\\\text{Chemical}\end{array}$$

**4. Determining Chemical Feeder Setting—
Solution Chemical Feeder, gpd**

<u>If the solution strength is expressed as lbs/gal</u>:
(lbs chemical/gal solution)

$$\frac{\text{(mg/}L\text{ Chem.) (MGD flow) (8.34 lbs/gal)}}{\text{lbs Chem/gal Sol'n}} = \begin{array}{l}\text{gpd}\\\text{Chemical}\end{array}$$

<u>If the solution strength is expressed as a percent</u>:
(convert the percent to mg/L)

Simplified Equation:

$$\text{Desired Dose, lbs/day} = \text{Actual Dose, lbs/day}$$

Expanded Equation:

$$\begin{array}{ccc}\text{(mg/}L\text{) (MGD) (8.34)} & & \text{(mg/}L\text{) (MGD) (8.34)}\\\text{Chem. Flow lbs/gal} & = & \text{Sol'n Sol'n lbs/gal}\\\text{Treated} & & \text{Flow}\end{array}$$

**5. Determining Chemical Feeder Setting—
Solution Chemical Feeder, mL/min**

First calculate the gpd setting using either equation given in Section 9.4:

Then, using gpd flow, determine the mL/min setting:

$$\frac{\text{(gpd flow) (3785 mL/gal)}}{\text{1440 min/day}} = \begin{array}{l}\text{Chemical,}\\\text{mL/min}\end{array}$$

6. Percent Strength of Solutions

Percent strength using dry chemicals:

$$\% \text{ Strength} = \frac{\text{Chemical, lbs}}{\text{Solution, lbs}} \times 100$$

Or

$$\% \text{ Strength} = \frac{\text{Chemical, lbs}}{\text{Water, lbs} + \text{Chemical, lbs}} \times 100$$

Percent strength using liquid chemicals:

$$\begin{array}{c}(\text{Liquid Polymer}) \, (\% \text{ Strength}) \\ \text{lbs} \quad\quad \dfrac{\text{of Liq. Poly.}}{100}\end{array} = \begin{array}{c}(\text{Polymer Solution}) \, (\% \text{Strength}) \\ \text{lbs} \quad\quad \dfrac{\text{of Poly. Soln.}}{100}\end{array}$$

7. Mixing Solutions of Different Strength

$$\begin{array}{c}\% \text{ Strength} \\ \text{of Mixture}\end{array} = \frac{\text{Chemical in Mixture, lbs}}{\text{Solution Mixture, lbs}} \times 100$$

<u>Or</u>, if target strength is desired:

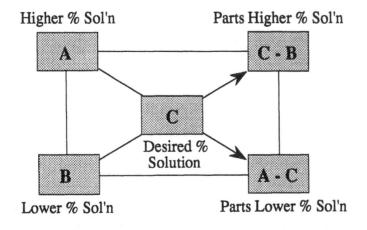

Higher % Sol'n   Parts Higher % Sol'n

A   C - B

C

Desired %
Solution

B   A - C

Lower % Sol'n   Parts Lower % Sol'n

SUMMARY—Cont'd

8. Dry Chemical Feeder Calibration

$$\frac{\text{Chemical Feed}}{\text{Rate, lbs/min}} = \frac{\text{Chemical Used, lbs}}{\text{Length of Application, min}}$$

9. Solution Chemical Feeder Calibration
(Given mL/min or gpd flow)

If chemical feed rate is given as mL/min, first convert mL/min flow to gpd flow:

$$\frac{(\text{mL/min}) \ (1440 \ \text{min/day})}{3875 \ \text{mL/gal}} = \text{gpd}$$

Then calculate chemical dosage, lbs/day:

$$(\text{mg/}L \ \text{Chem.}) \ (\text{MGD flow}) \ (8.34 \ \text{lbs/gal}) = \frac{\text{Chemical,}}{\text{lbs/day}}$$

10. Solution Chemical Feeder Calibration
(Given Drop in Solution Tank Level)

Diameter

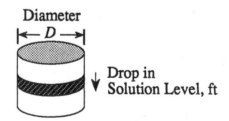

Drop in Solution Level, ft

$$\frac{\text{Flow}}{\text{gpm}} = \frac{\text{Volume Pumped, gal}}{\text{Duration of Test, min}}$$

Or

$$\frac{\text{Flow}}{\text{gpm}} = \frac{(0.785) \ (D^2) \ (\text{Drop, ft}) \ (7.48 \ \text{gal/cu ft})}{\text{Duration of Test, min}}$$

SUMMARY—Cont'd

11. Chemical Use Calculations

First determine the average chemical use:

$$\frac{\text{Average Use}}{\text{lbs/day}} = \frac{\text{Total Chem. Used, lbs}}{\text{Number of Days}}$$

Or

$$\frac{\text{Average Use}}{\text{gpd}} = \frac{\text{Total Chem. Used, gal}}{\text{Number of Days}}$$

Then calculate days' supply in inventory

$$\frac{\text{Days' Supply}}{\text{in Inventory}} = \frac{\text{Total Chem. in Inventory, lbs}}{\text{Average Use, lbs/day}}$$

Or

$$\frac{\text{Days' Supply}}{\text{in Inventory}} = \frac{\text{Total Chem. in Inventory, gal}}{\text{Average Use, gpd}}$$

9.1 CHAMBER OR BASIN VOLUME

Detention time and some chemical feed calibration calculations include a chamber, basin or tank volume calculation. For additional explanation of tank volume calculations, refer to Chapter 1 in this text or Chapter 11 in *Basic Math Concepts*.

VOLUME OF A SQUARE OR RECTANGULAR TANK

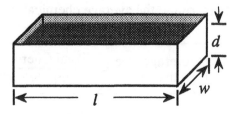

Vol, cu ft = (length, ft) (width, ft) (depth, ft)

Or

Vol, gal = (length, ft) (width, ft) (depth, ft) (7.48 gal/cu ft)

Example 1: (Chamber or Basin Volume)
❑ A flash mix chamber is 3 ft square with water to a depth of 3.5 ft. What is the volume of water (in gallons) in the chamber?

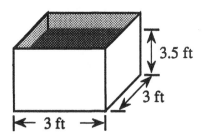

Vol, gal = (length, ft) (width, ft) (depth, ft) (7.48 gal/cu ft)

= (3 ft) (3 ft) (3.5 ft) (7.48 gal/cu ft)

= $\boxed{235.6 \text{ gal}}$

Example 2: (Chamber or Basin Volume)
❏ A flocculation basin is 50 ft long, 15 ft wide, with water to a depth of 8 ft. What is the volume of water (in gallons) in the basin?

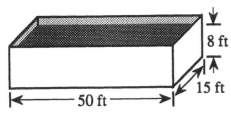

Vol, gal = (length, ft) (width, ft) (depth, ft) (7.48 gal/cu ft)

= (50 ft) (15 ft) (8 ft) (7.48 gal/cu ft)

= | 44,880 gal |

Example 3: (Chamber or Basin Volume)
❏ A flocculation basin is 40 ft long, 20 ft wide, and contains water to a depth of 10 ft 5 in. How many gallons of water are in the tank?

All dimensions must be expressed in terms of <u>feet</u>. Therefore, the 5 inches portion of the depth measurement must be converted to feet:*

$$\frac{5 \text{ in.}}{12 \text{ in./ft}} = 0.4 \text{ ft}$$

Then calculate basin volume:

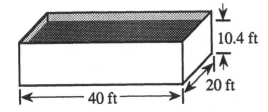

Then calculate volume:

Vol, gal = (length, ft) (width, ft) (depth, ft) (7.48 gal/cu ft)

= (40 ft) (20 ft) (10.4 ft) (7.48 gal/cu ft)

= | 62,234 gal |

CONVERTING INCHES TO FEET

When making the volume calculation, all dimensions must be expressed in the same terms. Since most dimensions are expressed in feet, all dimensions should be converted to feet. To convert from inches to feet, simply divide the inches by 12 in./ft:*

$$\boxed{\frac{\text{Inches}}{12 \text{ in./ft}} = \text{ft}}$$

* For a review of inches and feet conversions, refer to Chapter 8 in *Basic Math Concepts*.

9.2 DETENTION TIME

The detention time for flash mix chambers is measured in seconds, whereas the detention for flocculation basins is generally between 5 and 30 minutes. The equations used to calculate detention time are shown to the right.

MATCHING UNITS

There are many possible ways of calculating detention time, depending on the time unit desired (seconds, minutes, hours, days) and the expression of volume and flow rate.

When calculating detention time, therefore, it is essential that the time and volume units used in the equation are consistent, as illustrated to the right.

The flow rate to the flocculation basin is normally expressed as MGD or gpd. However, since the detention time is desired in minutes, it is important to express the flow rate in the same time frame—gal/min (or gpm):**

$$\frac{\text{Flow, gpd}}{1440 \text{ min/day}} = \text{Flow, gpm}$$

DETENTION TIME IS FLOW-THROUGH TIME

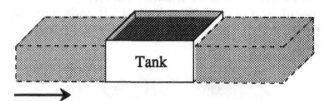

Simplified Equation:*

$$\frac{\text{D.T.}}{\text{min}} = \frac{\text{Volume of Tank, gal}}{\text{Flow Rate, gpm}}$$

Expanded Equation:*

$$\frac{\text{D.T.}}{\text{min}} = \frac{(\text{length})\ (\text{width})\ (\text{depth})\ (7.48)}{\text{ft}\quad\text{ft}\quad\text{ft}\quad\text{gal/cu ft}}{\text{Flow Rate, gpm}}$$

BE SURE YOUR TIME AND VOLUME UNITS MATCH

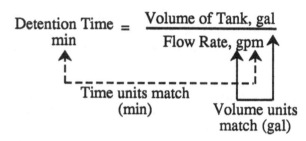

$$\frac{\text{Detention Time}}{\text{min}} = \frac{\text{Volume of Tank, gal}}{\text{Flow Rate, gpm}}$$

Time units match (min)

Volume units match (gal)

Example 1: (Detention Time)
❏ The flow to a flocculation basin 40 ft long, 15 ft wide, and 8 ft deep is 1930 gpm. What is the detention time in the tank, in minutes?

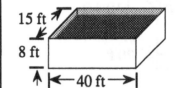

15 ft
8 ft
40 ft

Tank Volume:

(40 ft) (15 ft) (8 ft) (7.48 gal/cu ft)

= 35,904 gal

$$\frac{\text{Detention Time}}{\text{min}} = \frac{\text{Volume of Tank, gal}}{\text{Flow Rate, gpm}}$$

$$= \frac{35,904 \text{ gal}}{1930 \text{ gpm}}$$

$$= \boxed{18.6 \text{ min}}$$

---

* The equation shown is used for flocculation basin detention time calculations. When calculating flash mix chamber detention time (measured in seconds), the flow rate should be expressed in gallons per <u>second</u>, gps.

** For a review of flow conversions, refer to Chapter 2 in this text or Chapter 8 in *Basic Math Concepts*.

Example 2: (Detention Time)
❏ A flocculation basin receives a flow of 2,830,000 gpd. If the basin is 45 ft long, 20 ft wide and has water to a depth of 9 ft, what is the flocculation basin detention time in minutes? (Assume the flow is steady and continuous.)

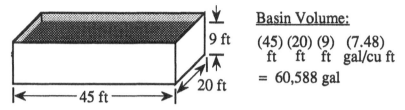

Basin Volume:

$$\underset{\text{ft} \quad \text{ft} \quad \text{ft} \quad \text{gal/cu ft}}{(45)\ (20)\ (9)\ (7.48)}$$

$$= 60,588 \text{ gal}$$

First, convert the flow rate from gpd to gpm so that time units will match. (2,830,000 gpd ÷ 1440 min/day = 1965 gpm). Then calculate detention time:

$$\frac{\text{Detention Time}}{\text{min}} = \frac{\text{Volume of Tank, gal}}{\text{Flow Rate, gpm}}$$

$$= \frac{60,588 \text{ gal}}{1965 \text{ gpm}}$$

$$= \boxed{31 \text{ min}}$$

Example 3: (Detention Time)
❏ A flash mix chamber is 5 ft long, 3.5 ft wide, with water to a depth of 3 ft. If the flow to the flash mix chamber is 5 MGD, what is the chamber detention time in seconds? (Assume the flow is steady and continuous.)

First, convert the flow rate from gpd to gps so that time units will match:

$$\frac{5,000,000 \text{ gpd}}{(1440 \text{ min/day}) (60 \text{ sec/min})} = 58 \text{ gps}$$

Then calculate detention time:

$$\frac{\text{Detention Time}}{\text{sec}} = \frac{\text{Volume of Tank, gal}}{\text{Flow Rate, gps}}$$

$$= \frac{(5 \text{ ft}) (3.5 \text{ ft}) (3 \text{ ft}) (7.48 \text{ gal/cu ft})}{58 \text{ gps}}$$

$$= \boxed{6.8 \text{ sec}}$$

INCLUDING THE VOLUME CALCULATION IN THE NUMERATOR

For calculations pertaining to a particular treatment plant, you would know the normal operating volume of the flash mix chamber or flocculation basin. However, in some situations, such as a change in water level or for a calculation on a certification exam, you will not know the volume of the tank. When this is the case, you can either calculate the volume separately and then use that number in the detention time calculation, as shown in Example 2, or it can be included directly in the numerator of the detention time equation, as shown in Example 3.

CONVERTING FLOW RATES

Detention times for flash mix basins are generally measured in seconds. Because of this, when calculating detention time, the flow rate should be expressed as gallons per second (gps)*. To convert from gpd to gps, the following equation would be used:

$$\frac{\text{Flow Rate, gpd}}{\underset{\text{min/day} \quad \text{sec/min}}{(1440)\ (60)}} = \frac{\text{Flow Rate,}}{\text{gps}}$$

A detailed explanation of flow conversions, including a "box diagram" method of determining conversion factors, is given in Chapter 2 of this text and Chapter 8 of *Basic Math Concepts*.

---

* As an alternative, the detention time in minutes can be calculated and then minutes converted to seconds. In this case, the flow rate would be expressed as gpm (to match the detention time calculated in minutes). Because the detention time conversion from minutes to seconds is often forgotten, it is recommended that the desired detention time units and the flow rate time frame match.

9.3 DETERMINING CHEMICAL FEEDER SETTING—(Dry Chemical Feeder, lbs/day

In chemical dosing, a measured amount of chemical is added to the water. The amount of chemical required depends on such factors as the type of chemical used, the reason for dosing, and the flow rate being treated.

The two expressions most often used to describe the amount of chemical added or required are:

• milligrams per liter (mg/*L*)*, and

• pounds per day (lbs/day)

The equation shown to the right is used to convert mg/*L* dosage to lbs/day dosage. Note that the flow rate must always be expressed in terms of million gallons per day (MGD).

MILLIGRAMS PER LITER IS A MEASURE OF CONCENTRATION

Assume each liter below is divided into 1 million parts:

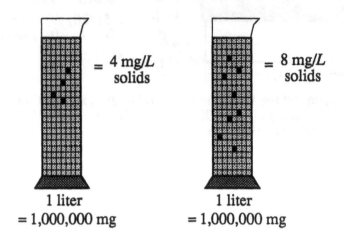

1 liter
= 1,000,000 mg

1 liter
= 1,000,000 mg

The mg/*L* concentration expresses a ratio of the milligrams chemical in each liter of water. For example, if a concentration of 4 mg/*L* is desired, then a total of 12 mg chemical would be required to treat 3 liters:

$$\frac{4\,mg}{L} \times \frac{3}{3} = \frac{12\,mg}{3\,L}$$

The amount of chemical required therefore depends on two factors:

• The desired concentration (mg/*L*), and

• The amount of water to be treated (normally expressed as MGD).

To convert from mg/*L* to lbs/day, the following equation is used:

$$\frac{(mg/L)\ (MGD)\ (8.34)}{Chem.\quad flow\quad lbs/gal} = lbs/day$$

---

* For most water and wastewater calculations, mg/*L* cloncentration and ppm (parts per million) concentration may be used interchangeably. That is, 1 mg/*L* = 1 ppm. Of the two expressions, mg/*L* is preferred.

Example 1: (Chemical Feed Rate)
❑ Jar tests indicate that the best alum dose for a water is 10 mg/*L*. If the flow to be treated is 1,960,000 gpd, what should the lbs/day setting be on the dry alum feeder?

(mg/*L* Chem.) (MGD flow) (8.34 lbs/gal) = lbs/day

(10 mg/*L*) (1.96 MGD) (8.34 lbs/gal) = $\boxed{\begin{array}{c} \text{163 lbs/day} \\ \text{Alum} \end{array}}$

Example 2: (Chemical Feed Rate)
❑ The desired dry polymer dosage is 12 mg/*L*. If the flow to be treated is 3,260,000 gpd, how many lbs/day polymer will be required?

(mg/*L* Polymer) (MGD flow) (8.34 lbs/day) = lbs/day Polymer

(12 mg/*L* Polymer) (3.26 MGD) (8.34 lbs/day) = $\boxed{\begin{array}{c} \text{326 lbs/day} \\ \text{Polymer} \end{array}}$

Example 3: (Chemical Feet Rate)
❑ Determine the desired lbs/day setting on a dry chemical feeder if jar tests indicate an optimum polymer dose of 14 mg/*L* and the flow to be treated is 4.22 MGD.

(mg/*L* Polymer) (MGD flow) (8.34 lbs/gal) = lbs/day Polymer

(14 mg/*L* Polymer) (4.22 MGD) (8.34 lbs/gal) = $\boxed{\begin{array}{c} \text{493 lbs/day} \\ \text{Polymer} \end{array}}$

9.4 DETERMINING CHEMICAL FEEDER SETTING—Solution Chemical Feeder, gpd

**WHEN SOLUTION
CONCENTRATION IS
EXPRESSED AS
LBS CHEM/GAL SOL'N**

When solution concentration is expressed as lbs chemical/gal solution, the required gpd feed rate can be determined as follows:

1. Calculate the lbs/day dry chemical required.

$$\frac{(mg/L)\ (MGD)\ (8.34)}{\text{Chem. flow\quad lbs/gal}} = \frac{\text{Chem.}}{\text{lbs/day}}$$

2. Convert the lbs/day dry chemical to gpd solution (using lbs chemical/gal solution information).

$$\frac{\text{Chem. lbs/day}}{\text{lbs Chem./gal Sol'n}} = \frac{\text{Sol'n}}{\text{gpd}}$$

These two steps can be combined into one equation.

$$\frac{\dfrac{(mg/L)\ (MGD)\ (8.34)}{\text{Chem. Flow\quad lbs/gal}}}{\text{lbs Chem/gal Sol'n}} = \frac{\text{Sol'n}}{\text{gpd}}$$

Example 1 illustrates this type of calculation.

CALCULATING GPD FEEDER SETTING DEPENDS ON HOW SOLUTION CONCENTRATION IS EXPRESSED: (LBS/GAL OR PERCENT)

If the solution strength is expressed as lbs/gal:
(lbs chemical/gal solution)

$$\frac{\dfrac{(mg/L)\ (MGD)\ (8.34)}{\text{Chem.\quad Flow\quad lbs/gal}}}{\text{lbs Chem/gal Sol'n}} = \text{gpd Sol'n}$$

If the solution strength is expressed as a percent:
(Convert percent to mg/L)

Simplified Equation:

$$\text{Desired Dose, lbs/day} = \text{Actual Dose, lbs/day}$$

Expanded Equation:

$$\frac{(mg/L)\ (MGD)\ (8.34)}{\substack{\text{Chem.\quad Flow\quad lbs/gal} \\ \text{Treated}}} = \frac{(mg/L)\ (MGD)\ (8.34)^*}{\substack{\text{Sol'n\quad Sol'n\quad lbs/gal} \\ \text{Flow}}}$$

Example 1: (Feeder Setting, gpd)
❏ Jar tests indicate that the best alum dose for a water is 9 mg/L. The flow to be treated is 1.94 MGD. Determine the gpd setting for the alum solution feeder if the liquid alum contains 5.36 lbs of alum per gallon of solution.

First calculate the lbs/day of dry alum required, using the mg/L to lbs/day equation:

$$\frac{(mg/L)\ (MGD)\ (8.34)}{\text{flow\quad lbs/gal}} = \text{lbs/day}$$

$$(9\ mg/L)\ (1.94\ MGD)\ (8.34\ lbs/gal) = \boxed{\begin{array}{c}146\ lbs/day \\ \text{Dry Alum}\end{array}}$$

Then calculate gpd solution required. (Each gallon of solution contains 5.36 lbs of dry alum. To find how many gallons are required, therefore, you need to determine how many units of 5.36 lbs are needed.)

$$\frac{146\ lbs/day\ alum}{5.36\ lbs\ alum/gal\ solution} = \boxed{\begin{array}{c}27.2\ gpd\ Alum \\ \text{Solution}\end{array}}$$

---

* This assumes the solution weighs 8.34 lbs/gal. In fact, it may weigh as much as 10 lbs/gal. To get an accurate dosage, the actual lbs/gal figure should be used.

Example 2: (Feeder Setting, gpd)
❑ Jar tests indicate that the best alum dose for a water is 9 mg/L. The flow to be treated is 1.94 MGD. Determine the gpd setting for the alum solution feeder if the liquid alum is a 64.3% solution. (Assume the alum solution weighs 8.34 lbs/gal.)

First write the equation, then fill in given information. The solution concentration is 64.3%. This can be re-expressed as 643,000 mg/L for use in the equation.*

Desired Dose, lbs/day = Actual Dose, lbs/day

$$\frac{(mg/L)\ (MGD)\ (8.34)}{Chem.\ Flow\ lbs/gal} = \frac{(mg/L)\ (MGD)\ (8.34)}{Sol'n\ Sol'n\ lbs/gal}$$
Treated   Flow

$$(9\ mg/L)\ (1.94)\ (8.34) = (643,000\ mg/L)\ (x\ MGD)\ (8.34)$$
MGD lbs/gal   lbs/gal

$$\frac{(9)\ (1.94)\ (8.34)}{(643,000)\ (8.34)} = x$$

$$\boxed{0.0000272\ MGD} = x$$

Now convert MGD flow to gpd flow:**

$$0.0000272\ MGD = \boxed{27.2\ gpd\ flow}$$

Example 3: (Feeder Setting, gpd)
❑ The flow to a plant is 3.46 MGD. Jar testing indicates that the optimum alum dose is 12 mg/L. What should the gpd setting be for the solution feeder if the alum solution is a 55% solution? (Assume the alum solution weighs 8.34 lbs/gal.)

A solution concentration of 55% is equivalent to 550,000 mg/L:

Desired Dose, lbs/day = Actual Dose, lbs/day

$$\frac{(mg/L)\ (MGD)\ (8.34)}{Chem.\ Flow\ lbs/gal} = \frac{(mg/L)\ (MGD)\ (8.34)}{Sol'n\ Sol'n\ lbs/gal}$$
Treated   Flow

$$(12\ mg/L)\ (3.46)\ (8.34) = (550,000\ mg/L)\ (x\ MGD)\ (8.34)$$
MGD lbs/gal   lbs/gal

$$\frac{(12)\ (3.46)\ (8.34)}{(550,000)\ (8.34)} = x$$

$$\boxed{0.0000755\ MGD} = x$$

This can be expressed as gpd flow:**

$$0.0000755\ MGD = \boxed{75.5\ gpd\ flow}$$

WHEN SOLUTION CONCENTRATION IS EXPRESSED AS A PERCENT

When the solution concentration is expressed as a percent, it may be converted to mg/L and a different equation may be used, as shown on the facing page. The basis of this equation stated in simple terms is: **the desired dosage rate (lbs/day) must be equal to the actual dosage rate (lbs/day).** In expanded form, the desired dosage rate (lbs/day) is calculated using mg/L desired dosage, flow rate to be treated (in MGD) and 8.34 lbs/gal. The actual dosage rate is calculated using mg/L solution concentration, MGD solution flow, and density (often 8.34 lbs/gal).

Note that Examples 1 and 2 have the same answers. In Example 1, the solution strength was expressed as lbs chemical/gal solution; whereas in Example 2, the same solution strength was expressed as a percent then converted to mg/L.

To convert from a percent to mg/L, use the following equation:

$$\boxed{1\ \% = 10,000\ mg/L}$$

Note there is a decimal move of four places to the right when converting from percent to mg/L:

$$1\ \% = 10,000.\ mg/L$$

WHEN SOLUTION CONCENTRATION IS EXPRESSED AS MILLIGRAMS PER MILLITER SOLUTION

Occasionally solution concentration will be expressed as milligrams chemical per milliliter solution (mg/mL). In this case, simply convert to mg/L. For example:

$$\frac{5\ mg\ x\ 1000}{mL\ x\ 1000} = \frac{5000\ mg}{1000\ mL} = \frac{5000\ mg}{L}$$

* To review the conversion from mg/L to %, and vice versa, refer to Chapter 8 in *Basic Math Concepts*.
** Refer to Chapter 8 in *Basic Math Concepts* for a discussion of flow conversions.

9.5 DETERMINING CHEMICAL FEEDER SETTING—Solution Chemical Feeder, mL/min

Some solution chemical feeders dispense chemical as milliliters per minute (mL/min). To calculate the mL/min solution required, first calculate the gpd feed rate, as described in the previous section. Then convert gpd flow rate to mL/min flow rate. The process, as shown in the box to the right, involves the following conversions:**

$$\frac{gal}{day} \rightarrow \frac{gal}{min} \rightarrow \frac{mL}{min}$$

FIRST DETERMINE GPD FLOW THEN CALCULATE ML/MIN FLOW

Calculate gpd solution flow required: (See Section 9.4)

If the solution strength is expressed as lbs chemical/gal solution, use this equation:

$$\frac{(mg/L\ Chem.)\ (MGD\ Flow)\ (8.34\ lbs/gal)}{lbs\ Chem./gal\ Sol'n} = gpd\ Sol'n$$

If the solution strength is expressed as a percent, use this equation:

$$\underset{\substack{Dose}}{(mg/L)}\ \underset{\substack{Flow \\ Treated}}{(MGD)}\ \underset{\substack{lbs/gal}}{(8.34)} = \underset{\substack{Sol'n}}{(mg/L)}\ \underset{\substack{Sol'n}}{(MGD)}\ \underset{\substack{lbs/gal \\ Flow}}{(8.34)}$$

Then convert gpd to mL/min solution flow required:*

$$\frac{(gal)}{day}\ \frac{(1\ day)}{1440\ min}\ \frac{(3785\ mL)}{1\ gal} = \frac{mL}{min}$$

Or, simplified as:

$$\frac{(gpd)\ (3785\ mL/gal)}{1440\ min/day} = mL/min$$

Example 1: (Feeder Setting, mL/min)
❑ The desired solution feed rate was calculated to be 8 gpd. What is this feed rate expressed as mL/min?

Since the gpd flow has already been determined, the mL/min flow rate can be calculated directly:

$$\frac{(gpd)\ (3785\ mL/gal)}{1440\ min/day} = mL/min$$

$$\frac{(8\ gpd)\ (3785\ mL/gal)}{1440\ min/day} = \boxed{\begin{array}{c} 21\ mL/min \\ Feed\ Rate \end{array}}$$

---

* The equation is written in a form so that dimensional analysis may be used to check the units of the answer. Refer to Chapter 15 in *Basic Math Concepts* for a review of dimensional analysis.
** Refer to Chapter 8 in *Basic Math Concepts* for flow rate and metric conversions.

Example 2: (Feeder Setting, mL/min)
❑ The desired solution feed rate has been calculated to be 30 gpd. What is this feed rate expressed as mL/min?

Since the gpd solution feed rate has been determined, the mL/min may be calculated directly:

$$\frac{(\text{gpd}) \ (3785 \ \text{m}L/\text{gal})}{1440 \ \text{min/day}} \ = \ \text{m}L/\text{min}$$

$$\frac{(30 \ \text{gpd}) \ (3785 \ \text{m}L/\text{gal})}{1440 \ \text{min/day}} \ = \ \boxed{\begin{array}{c} 78.9 \ \text{m}L/\text{min} \\ \text{Feed Rate} \end{array}}$$

Example 3: (Feeder Setting, mL/min)
❑ The optimum alum dose has been determined to be 12 mg/L. The flow to be treated is 980,000 gpd. If the solution to be used contains 60% active polymer, what should the solution chemical feeder setting be, in mL/min? (Assume the alum solution weighs 8.34 lbs/gal.)

First calculate the gpd feed rate required:

$$\underset{\substack{\text{Dose} \quad \text{Flow} \quad \text{lbs/gal} \\ \text{Treated}}}{(\text{mg}/L) \ (\text{MGD}) \ (8.34)} = \underset{\substack{\text{Sol'n} \quad \text{Sol'n} \quad \text{lbs/gal} \\ \text{Flow}}}{(\text{mg}/L) \ (\text{MGD}) \ (8.34)}$$

$$\underset{\substack{\text{Alum} \qquad\qquad \text{lbs/gal}}}{(12 \ \text{mg}/L) \ (0.98 \ \text{MGD}) \ (8.34)} = \underset{\substack{\text{mg}/L \qquad\qquad \text{lbs/gal}}}{(600{,}000) \ (x \ \text{MGD}) \ (8.34)}$$

$$\frac{(12) \ (0.98) \ \cancel{(8.34)}}{(600{,}000) \ \cancel{(8.34)}} = x \ \text{MGD}$$

$$0.0000196 \ \text{MGD} = x$$

$$20 \ \text{gpd} = x$$

Then convert gpd flow rate to mL/min flow rate:

$$\frac{(\text{gpd}) \ (3785 \ \text{m}L/\text{gal})}{1440 \ \text{min/day}} \ = \ \text{m}L/\text{min}$$

$$\frac{(20 \ \text{gpd}) \ (3785 \ \text{m}L/\text{gal})}{1440 \ \text{min/day}} \ = \ \boxed{\begin{array}{c} 52.6 \ \text{m}L/\text{min} \\ \text{Feed Rate} \end{array}}$$

A TWO-STEP CALCULATION

Sometimes you will want to know mL/min solution feed rate but you will not know the gpd solution feed rate. In such cases, calculate the gpd solution feed rate first, using one of the equations shown in the previous section, then convert gpd flow rate to mL/min flow rate. Example 3 illustrates one such calculation.

9.6 PERCENT STRENGTH OF SOLUTIONS

PERCENT STRENGTH USING DRY CHEMICALS

The strength of a solution is a measure of the amount of chemical (solute) dissolved in the solution. Since percent is calculated as "part over whole,"

$$\% = \frac{Part}{Whole} \times 100$$

percent strength is calculated as **part chemical**, in lbs, divided by the **whole solution**, in lbs:

$$\% \text{ Strength} = \frac{\text{Chemical, lbs}}{\text{Solution, lbs}} \times 100$$

The denominator of the equation (lbs solution) includes both chemical (lbs) and water (lbs). Therefore the equation can be written in expanded form as:

$$\frac{\%}{\text{Strength}} = \frac{\text{Chemical, lbs}}{\text{Water} + \text{Chemical} \atop \text{lbs} \quad \text{lbs}} \times 100$$

As the two equations above illustrate, **the chemical added must be expressed in pounds.** If the chemical weight is expressed in ounces (as in Example 1) or grams (as in Example 2), it must first be converted to pounds (to correspond with the other units in the problem) before percent strength is calculated.

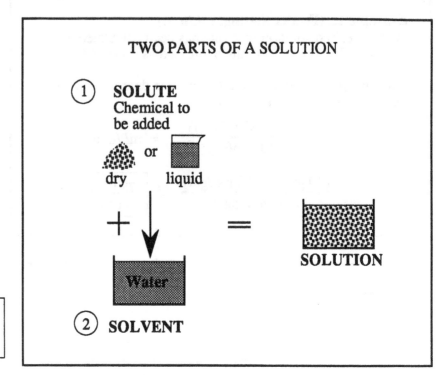

TWO PARTS OF A SOLUTION

① **SOLUTE** Chemical to be added

dry or liquid

+ ↓ = SOLUTION

Water

② **SOLVENT**

Example 1: (Percent Strength)
❏ If a total of 12 ounces of dry polymer are added to 20 gallons of water, what is the percent strength (by weight) of the polymer solution?

Before calculating percent strength, the ounces chemical must be converted to lbs chemical:*

$$\frac{12 \text{ ounces}}{16 \text{ ounces/pound}} = 0.75 \text{ lbs chemical}$$

Now calculate percent strength:

$$\% \text{ Strength} = \frac{\text{Chemical, lbs}}{\text{Water, lbs} + \text{Chemical, lbs}} \times 100$$

$$= \frac{0.75 \text{ lbs Chemical}}{(20 \text{ gal}) (8.34 \text{ lbs/gal}) + 0.75 \text{ lbs}} \times 100$$

$$= \frac{0.75 \text{ lbs Chemical}}{167.55 \text{ lbs Solution}} \times 100$$

$$= \boxed{0.4\%}$$

---

* To review ounces to pounds conversions refer to Chapter 8 in *Basic Math Concepts.*

Example 2: (Percent Strength)
❏ If 80 grams of dry polymer are dissolved in 5 gallons of water, what percent strength is the solution? (1 gram = 0.0022 lbs)

First, convert grams chemical to pounds chemical. Since 1 gram equals 0.0022 lbs, 80 grams is 80 times 0.0022 lbs:

$$(80 \text{ grams}) (0.0022 \text{ lbs/gram}) = 0.18 \text{ lbs}$$
Polymer   Polymer

Now calculate percent strength of the solution:

$$\% \text{ Strength} = \frac{\text{lbs Polymer}}{\text{lbs Water} + \text{lbs Polymer}} \times 100$$

$$= \frac{0.18 \text{ lbs Polymer}}{(5 \text{ gal}) (8.34 \text{ lbs/gal}) + 0.18 \text{ lbs}} \times 100$$

$$= \frac{0.18 \text{ lbs}}{41.88 \text{ lbs}} \times 100$$

$$= \boxed{0.4\%}$$

Example 3: (Percent Strength)
❏ How many pounds of dry polymer must be added to 25 gallons of water to make a 1.2% polymer solution?

First, write the equation as usual and fill in the known information. Then solve for the unknown value.*

$$\% \text{ Strength} = \frac{\text{lbs Chemical}}{\text{lbs Water} + \text{lbs Chemical}} \times 100$$

$$1.2 = \frac{x \text{ lbs Chemical}}{(25 \text{ gal}) (8.34 \text{ lbs/gal}) + x \text{ lbs Chem.}} \times 100$$

$$1.2 = \frac{100 x}{208.5 + x}$$

$$1.2 (208.5 + x) = 100 x$$

$$250.2 + 1.2x = 100 x$$

$$250.2 = 98.8 x$$

$$\boxed{2.5 \text{ lbs Chemical}} = x$$

WHEN GRAMS CHEMICAL ARE USED

The chemical (solute) to be used in making a solution may be measured in grams rather than pounds or ounces. When this is the case, convert grams of chemical to pounds of chemical before calculating percent strength. The following relationship is used for the conversion:

$$\boxed{1 \text{ gram} = 0.0022 \text{ lbs}}$$

SOLVING FOR OTHER UNKNOWN VARIABLES

In the percent strength equation there are three variables:

• % Strength
• lbs Chemical
• lbs Water

In Examples 1 and 2, the unknown value was percent strength. However, the same equation can be used to determine either one of the other two variables. Example 3 illustrates this type of calculation.

Note that gallons water can also be the unknown variable in percent strength calculations. For this type of calculation, you will want to expand the equation shown on the previous page even farther:

$$\% \text{ Strength} = \frac{\text{Chemical, lbs}}{(\text{Water})(8.34) + \text{Chem.}} \times 100$$
gal lbs/gal lbs

* To review "solving for the unknown value", refer to Chapter 2 in *Basic Math Concepts*.

PERCENT STRENGTH USING LIQUID CHEMICALS

When using a liquid chemical to make up a solution, such as liquid polymer, a different calculation is required.

The liquid chemical is shipped from the supplier at a certain percent strength—such as 10 to 12%. Then additional water is added to this liquid chemical to obtain a desired solution of lower percent polymer—such as 1% or 0.5%.

In percent strength calculations, lbs chemical are set equal to lbs chemical, as illustrated in the diagram to the right. The expanded equations are also shown.

THE KEY TO THESE CALCULATIONS— POUNDS CHEMICAL REMAIN CONSTANT
(Liquid polymer is used to illustrate the concept.)

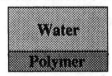

LIQUID POLYMER
(10% Polymer)

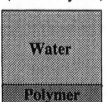

POLYMER SOLUTION
(0.5% Polymer)

Simplified Equation:

$$\text{lbs Polymer in Liquid Polymer} = \text{lbs Polymer in Polymer Solution}$$

Expanded Equation:

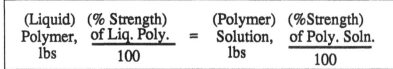

$$\frac{(\text{Liquid}) \ \ (\%\ \text{Strength})}{\text{Polymer, of Liq. Poly.}}{100} = \frac{(\text{Polymer}) \ \ (\%\text{Strength})}{\text{Solution, of Poly. Soln.}}{100}$$

Or

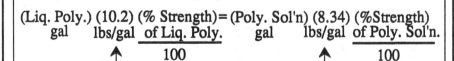

$$\frac{(\text{Liq. Poly.})\ (10.2)\ (\%\ \text{Strength})}{\text{gal} \quad \text{lbs/gal of Liq. Poly.}}{100} = \frac{(\text{Poly. Sol'n})\ (8.34)\ (\%\text{Strength})}{\text{gal} \quad \text{lbs/gal of Poly. Sol'n.}}{100}$$

Use __actual density__ factor here. Liquid polymer often weighs more than 8.34 lbs/gal. Other liquid chemicals may have the same density as water.

Use __actual density__ factor here. Polymer solutions may have a density closer to or equal to 8.34 lbs/gal since the heavy polymer has been diluted.

Example 4: (Percent Strength)
❑ A 10% liquid polymer is to be used in making up a polymer solution. How many lbs of liquid polymer should be mixed with water to produce 130 lbs of a 0.6% polymer solution?

$$\frac{(\text{Liquid Polymer})\ (\%\ \text{Strength})}{\text{lbs} \quad \text{of Liq. Poly.}}{100} = \frac{(\text{Poly. Soln.})\ (\%\text{Strength})}{\text{lbs} \quad \text{of Poly. Soln.}}{100}$$

$$(x\ \text{lbs}) \left(\frac{10}{100}\right) = (130\ \text{lbs}) \frac{(0.6)}{100}$$

$$x = \frac{(130)\,(0.006)}{0.1}$$

$$x = \boxed{7.8\ \text{lbs}}$$

Example 5: (Percent Strength)
❑ How many gallons of 9% liquid polymer should be mixed with water to produce 50 gallons of a 0.5% polymer solution? The density of the polymer liquid is 10.4 lbs/gal. Assume the density of the polymer solution is 8.34 lbs/gal.

Use the expanded form of the equation, filling in known information:

$$\frac{\text{(Liq. Poly.)}}{\text{gal}} \frac{(10.4)}{\text{lbs/gal}} \frac{\text{(\% Strength)}}{\frac{\text{of Liq. Poly.}}{100}} = \frac{\text{(Poly. Soln)}}{\text{gal}} \frac{(8.34)}{\text{lbs/gal}} \frac{\text{(\%Strength)}}{\frac{\text{of Poly. Soln.}}{100}}$$

$$(x \text{ gal}) (10.4) \frac{(9)}{100} = (50 \text{ gal}) (8.34) \frac{(0.5)}{100}$$

$$x = \frac{(50)(8.34)(0.005)}{(10.4)(0.09)}$$

$$x = \boxed{2.2 \text{ gallons}}$$

Example 6: (Percent Strength)
❑ A 10% liquid polymer will be used in making up a solution. How many gallons of liquid polymer should be added to the water to make up 40 gallons of 0.35% polymer solution? The liquid polymer has a specific gravity of 1.1 Assume the polymer solution has a specific gravity of 1.0.

First, convert specific gravity information to density information. The density of the liquid polymer is (8.34 lbs/gal) (1.1) = 9.17 lbs/gal. The density of the polymer solution is 8.34 lbs/gal, the same as water.

$$\frac{\text{(Liq. Poly.)}}{\text{gal}} \frac{(9.17)}{\text{lbs/gal}} \frac{\text{(\% Strength)}}{\frac{\text{of Liq. Poly.}}{100}} = \frac{\text{(Poly. Sol'n)}}{\text{gal}} \frac{(8.34)}{\text{lbs/gal}} \frac{\text{(\%Strength)}}{\frac{\text{of Poly. Sol'n.}}{100}}$$

$$(x \text{ gal}) (9.17 \text{ lbs/gal}) \frac{(10)}{100} = (40 \text{ gal}) (8.34 \text{ lbs/gal}) \frac{(0.35)}{100}$$

$$x = \frac{(40)(8.34)(0.0035)}{(9.17)(0.1)}$$

$$x = \boxed{1.3 \text{ gallons}}$$

* Refer to Chapter 7, Section 7.1, "Density and Specific Gravity".

DENSITY AND SPECIFIC GRAVITY CONSIDERATIONS

As shown in the second expanded equation on the opposite page, **the density of the solution must be included**. Density is the mass per unit volume.* In water and wastewater calculations, 8.34 lbs/gal is used as the density of water. However, the weight of a polymer solution can be as much as 10 or 11 lbs/gal or more. To obtain accurate results using the percent strength equation, it is important to use the appropriate density factor—one for the solute (such as liquid polymer) and another for the solution. When the solution strength is very low, such as 0.5% or 0.1%, the density of the solution is normally much closer to that of water—8.34 lbs/gal.

Occasionally **specific gravity** data may be given for a liquid chemical rather than density information. In fact, density and specific gravity are closely related terms. Density is a measure of the mass per unit volume, and is measured in such terms as lbs/gal. Specific gravity is a comparison of the density of a substance to a standard density. (For liquids, the standard is water. All other densities are compared to the density of water.) So, **a specific gravity of 1.0 means the liquid has the same density as water** (8.34 lbs/gal). A specific gravity of 0.5 means the liquid has a density half that of water, or 4.17 lbs/gal. A specific gravity of 1.5 means the liquid has a density 1.5 times that of water, or 12.51 lbs/gal. Example 6 illustrates a calculation including specific gravity data.

9.7 MIXING SOLUTIONS OF DIFFERENT STRENGTH

There are two types of solution mixture calculations. In one type of calculation, two solutions of different strengths are mixed with no particular target solution strength. The calculation involves determining the percent strength of the solution mixture.

The second type of solution mixture calculation includes a desired or target strength. This type of problem is described in the next section.

WHEN DIFFERENT PERCENT STRENGTH SOLUTIONS ARE MIXED

 +  =

10% Strength Solution

1% Strength Solution

Solution Mixture (% Strength somewhere between 10% and 1% depending on the quantity contributed by each.)

Simplified Equation:

$$\text{\% Strength of Mixture} = \frac{\text{Chemical in Mixture, lbs}}{\text{Solution Mixture, lbs}} \times 100$$

Expanded Equations:

$$\text{\% Strength of Mixture} = \frac{\text{lbs Chem. from Solution 1} + \text{lbs Chem. from Solution 2}}{\text{lbs Solution 1} + \text{lbs Solution 2}} \times 100$$

$$\text{\% Strength of Mixture} = \frac{(\text{Sol'n 1}) \dfrac{(\text{\% Strength})}{\text{lbs} \quad \text{of Sol'n 1}}{100} + (\text{Sol'n 2}) \dfrac{(\text{\% Strength})}{\text{lbs} \quad \text{of Sol'n 2}}{100}}{\text{lbs Solution 1} + \text{lbs Solution 2}} \times 100$$

Example 1: (Solution Mixtures)

❏ If 15 lbs of a 10% strength solution are mixed with 45 lbs of 1% strength solution, what is the percent strength of the solution mixture?

$$\text{\% Strength of Mixture} = \frac{(\text{Sol'n 1}) \dfrac{(\text{\% Strength})}{\text{lbs} \quad \text{of Sol'n 1}}{100} + (\text{Sol'n 2}) \dfrac{(\text{\% Strength})}{\text{lbs} \quad \text{of Sol'n 2}}{100}}{\text{lbs Solution 1} + \text{lbs Solution 2}} \times 100$$

$$= \frac{(15 \text{ lbs}) (0.1) + (45 \text{ lbs}) (0.01)}{15 \text{ lbs} + 45 \text{ lbs}} \times 100$$

$$= \frac{1.5 \text{ lbs} + 0.45 \text{ lbs}}{70 \text{ lbs}} \times 100$$

$$= \boxed{2.8\%}$$

Example 2: (Solution Mixtures)
❑ If 6 gallons of a 9% strength solution are mixed with 60 gallons of a 0.5% strength solution, what is the percent strength of the solution mixture? (Assume the 9% solution weighs 9.58 lbs/gal and the 0.5% solution weighs 8.34 lbs/gal.)

$$\% \text{ Strength of Mixture} = \frac{(\text{Sol'n 1}) \text{ lbs} \frac{(\% \text{ Strength of Sol'n 1})}{100} + (\text{Sol'n 2}) \text{ lbs} \frac{(\% \text{ Strength of Sol'n 2})}{100}}{\text{lbs Solution 1} + \text{lbs Solution 2}} \times 100$$

$$= \frac{(6 \text{ gal}) (9.58 \text{ lbs/gal}) (0.09) + (60 \text{ gal}) (8.34 \text{ lbs/gal}) (0.005)}{(6 \text{ gal}) (9.58 \text{ lbs/gal}) + (60 \text{ gal}) (8.34 \text{ lbs/gal})} \times 100$$

$$= \frac{5.2 \text{ lbs Chem.} + 2.5 \text{ lbs Chem.}}{57.5 \text{ lbs Sol'n 1} + 500.4 \text{ lbs Sol'n 2}} \times 100$$

$$= \frac{7.7 \text{ lbs Chemical}}{557.9 \text{ lbs Solution}} \times 100$$

$$= \boxed{1.4\% \text{ Strength}}$$

Example 3: (Solution Mixtures)
❑ If 12 gallons of a 10% strength solution are added to 48 gallons of 0.6% strength solution, what is the percent strength of the solution mixture? (Assume the 10% strength solution weighs 10.2 lbs/gal and the 0.6% strength solution weighs 8.7 lbs/gal.)

$$\% \text{ Strength of Mixture} = \frac{(\text{Sol'n 1}) \text{ lbs} \frac{(\% \text{ Strength of Sol'n 1})}{100} + (\text{Sol'n 2}) \text{ lbs} \frac{(\% \text{ Strength of Sol'n 2})}{100}}{\text{lbs Solution 1} + \text{lbs Solution 2}} \times 100$$

$$= \frac{(12 \text{ gal}) (10.2 \text{ lbs/gal}) (0.1) + (48 \text{ gal}) (8.7 \text{ lbs/gal}) (0.006)}{(12 \text{ gal}) (10.2 \text{ lbs/gal}) + (48 \text{ gal}) (8.7 \text{ lbs/gal})} \times 100$$

$$= \frac{12.2 \text{ lbs Chem.} + 2.5 \text{ lbs Chem.}}{122.4 \text{ lbs Sol'n 1} + 417.6 \text{ lbs Sol'n 2}} \times 100$$

$$= \frac{14.7 \text{ lbs Chemical}}{540 \text{ lbs Solution}} \times 100$$

$$= \boxed{2.7\% \text{ Strength}}$$

USE DIFFERENT DENSITY FACTORS WHEN APPROPRIATE

Percent strength should be expressed in terms of **pounds chemical per pounds solution.** Therefore, when solutions are given in terms of gallons, the gallons should be expressed as pounds before continuing with the percent strength calculation.

It is important to know what density factor should be used to convert from gallons to pounds. If the solution has a density the same as water, 8.34 lbs/gal is used. If, however, the solution has a higher density, such as some polymer solutions, then the higher density factor should be used. When the density is unknown, it is sometimes possible to weigh the chemical solution to determine the density.

SOLUTION MIXTURES TARGET PERCENT STRENGTH

In the previous section we examined the first type of solution mixture calculation—a calculation where there is no target percent strength. In this type calculation, two solutions are mixed and the percent strength of the mixture is determined.

In the second type of solution mixture calculation, **two different percent strength solutions are mixed in order to obtain a desired quantity of solution at a target percent strength.** These problems may be solved using the same equation shown in Examples 1-3. An illustration of this approach is given in Example 4.

Another approach in solving these problems, and perhaps preferred, is use of the dilution rectangle. Although the initial use of the dilution rectangle can be confusing, the effort to master its use is rewarded—solution mixture problems are quickly calculated. Example 5 uses the dilution rectangle to solve the problem stated in Example 4. Compare the two methods of calculating this type of mixture problem.

Example 4: (Dilution Rectangle)

❑ What weights of a 2% solution and a 7% solution must be mixed to make 850 lbs of a 4% solution?

Use the same equation as shown for Examples 1-3 and fill in given information.* (Note that the lbs of Solution 1 is unknown, x. If lbs of Solution 1 is x, then the lbs of Solution 2 must be the balance of the 850 lbs, or 850–x.)

$$\text{% Strength of Mixture} = \frac{(\text{Sol'n 1}) \; \frac{(\text{% Strength})}{100} + (\text{Sol'n 2}) \; \frac{(\text{% Strength})}{100}}{\text{lbs Solution 1} + \text{lbs Solution 2}} \times 100$$

$$4 = \frac{(x \text{ lbs}) (0.02) + (850 - x \text{ lbs}) (0.07)}{850 \text{ lbs}} \times 100$$

$$\frac{(4)(850)}{100} = 0.02x + 59.5 - 0.07x$$

$$34 = -0.05x + 59.5$$

$$0.05x = 25.5$$

$$x = \boxed{510 \text{ lbs of 2% Solution}}$$

Then $850 - 510 = \boxed{340 \text{ lbs of 7% Solution}}$

THE DILUTION RECTANGLE

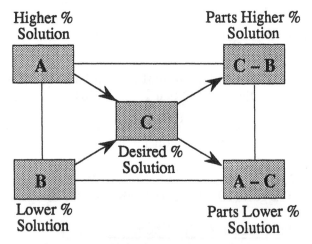

Steps in Using the Dilution Rectangle:

1. Place the % Strength numbers in positions A, B, and C.

2. Calculate parts higher % solution and parts lower % solution, subtracting as indicated.

3. Multiply fractional parts of each solution by the total lbs of solution desired.

---

* Refer to Chapter 2 in *Basic Math Concepts* for a review of solving for the unknown value.

Example 5: (Dilution Rectangle)
❏ What weights of a 2% solution and a 7% solution must be mixed to make 850 lbs of a 4% solution?

Use the Dilution Rectangle to solve this problem. First determine the parts required of each solution:

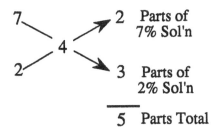

7   2   Parts of 7% Sol'n

    4

2   3   Parts of 2% Sol'n
_____

    5   Parts Total

Thus, 2 parts of the total 5 parts (2/5) come from the 7% solution, and the other three parts (3/5) come from the 2% solution. Now calculate the lbs of 2% and 7% solution, using these fractions:

Amt. of 7% Sol'n: $\dfrac{2}{5}$ (850 lbs) = $\boxed{340 \text{ lbs}}$

Amt. of 2% Sol'n: $\dfrac{3}{5}$ (850 lbs) = $\boxed{510 \text{ lbs}}$

Example 6: (Dilution Rectangle)
❏ How many lbs of a 10% polymer solution and water should be mixed together to form 425 lbs of a 1% polymer solution?

First calculate the parts of each solution required:

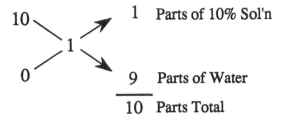

10   1   Parts of 10% Sol'n

    1

0   9   Parts of Water
_____

   10   Parts Total

Then calculate the actual lbs of each solution:

Amt. of 10% Sol'n: $\dfrac{(1)}{10}$ (425 lbs) = $\boxed{\begin{array}{c}42.5 \text{ lbs}\\ 10\% \text{ Sol'n}\end{array}}$

Amt. of Water: $\dfrac{(9)}{10}$ (425 lbs) = $\boxed{\begin{array}{c}382.5 \text{ lbs}\\ \text{Water}\end{array}}$

MIXING A SOLUTION AND WATER

In solution mixing Examples 1-5, two solutions of different strengths are blended. The solution mixing equation and dilution rectangle can also be used when only one solution and water are blended. (Water is considered a 0% strength solution.) Example 6 illustrates such a calculation using the dilution rectangle.

9.8 DRY CHEMICAL FEEDER CALIBRATION

Occasionally you will want to **compare the actual chemical feed rate with the feed rate indicated by the instrumentation.** This is called a calibration calculation.

To calculate the actual chemical feed rate for a dry chemical feeder, place a bucket under the feeder, weigh the bucket when empty, then weigh the bucket again after a specified length of time, such as 30 minutes.

The actual chemical feed rate can then be determined as:

$$\begin{matrix} \text{Chem.} \\ \text{Feed} \\ \text{Rate,} \\ \text{lbs/min} \end{matrix} = \frac{\text{Chem. Applied, lbs}}{\text{Length of Applic, min}}$$

The chemical feed rate can be converted to lbs/day, if desired:

$$\begin{matrix} \text{(Feed Rate)} \\ \text{lbs/min} \end{matrix} \left(1440 \frac{\text{min}}{\text{day}}\right) = \begin{matrix} \text{Feed Rate} \\ \text{lbs/day} \end{matrix}$$

Example 1: (Dry Chemical Feed Calibration)
❑ Calculate the actual chemical feed rate, lbs/day, if a bucket is placed under a chemical feeder and a total of 1.5 lbs is collected during a 30-minute period.

First calculate the lbs/min feed rate:

$$\begin{matrix} \text{Chem. Feed} \\ \text{Rate, lbs/min} \end{matrix} = \frac{\text{Chem. Applied, lbs}}{\text{Length of Application, min}}$$

$$= \frac{1.5 \text{ lbs}}{30 \text{ min}}$$

$$= \boxed{\begin{matrix} 0.05 \text{ lbs/min} \\ \text{Feed Rate} \end{matrix}}$$

Then calculate the lbs/day feed rate:

$$\begin{matrix} \text{Chem. Feed} \\ \text{Rate, lbs/day} \end{matrix} = (0.05 \text{ lbs/min}) (1440 \text{ min/day})$$

$$= \boxed{\begin{matrix} 72 \text{ lbs/day} \\ \text{Feed Rate} \end{matrix}}$$

Example 2: (Dry Chemical Feed Calibration)
❑ Calculate the actual chemical feed rate, lbs/day, if a bucket is placed under a chemical feeder and a total of 1.3 lbs is collected during a 20-minute period.

First calculate the lbs/min feed rate:

$$\begin{matrix} \text{Chem. Feed} \\ \text{Rate, lbs/min} \end{matrix} = \frac{\text{Chem. Applied, lbs}}{\text{Length of Application, min}}$$

$$= \frac{1.3 \text{ lbs}}{20 \text{ min}}$$

$$= \boxed{\begin{matrix} 0.065 \text{ lbs/min} \\ \text{Feed Rate} \end{matrix}}$$

Then calculate the lbs/day feed rate:

$$\begin{matrix} \text{Chem. Feed} \\ \text{Rate, lbs/day} \end{matrix} = (0.065 \text{ lbs/min}) (1440 \text{ min/day})$$

$$= \boxed{\begin{matrix} 94 \text{ lbs/day} \\ \text{Feed Rate} \end{matrix}}$$

Example 3: (Dry Chemical Feed Calibration)
❏ A chemical feeder is to be calibrated. The bucket to be used for collecting chemical is placed under the chemical feeder and weighed (0.25 lbs). After 30 minutes, the weight of the bucket and chemical are found to weigh 2.6 lbs. Based on this test, what is the actual chemical feed rate, in lbs/day?

First calculate the lbs/min feed rate:
(Note that the chemical applied is the wt. of the bucket and chemical minus the wt. of the empty bucket.)

$$\text{Chem. Feed Rate, lbs/min} = \frac{\text{Chem. Applied, lbs}}{\text{Length of Application, min}}$$

$$= \frac{2.6 \text{ lbs} - 0.25 \text{ lbs}}{30 \text{ minutes}}$$

$$= \frac{2.35 \text{ lbs}}{30 \text{ min}}$$

$$= \boxed{\begin{array}{l} 0.078 \text{ lbs/min} \\ \text{Feed Rate} \end{array}}$$

Then calculate the lbs/day feed rate:

$$(0.078 \text{ lbs/min}) (1440 \text{ min/day}) = \boxed{\begin{array}{l} 112 \text{ lbs/day} \\ \text{Feed Rate} \end{array}}$$

Example 4: (Dry Chemical Feed Calibration)
❏ During a 24-hour period, a flow of 1,620,000 gpd water is treated. If a total of 29 lbs of polymer were used for coagulation during that 24-hour period, what is the polymer dosage in mg/L?

Write the mg/L to lbs/day equation, then fill in the given data:

$$(\text{mg/L Chem.}) (\text{MGD Flow}) (8.34 \text{ lbs/gal}) = \text{lbs/day Chem.}$$

$$(x \text{ mg/L}) (1.62 \text{ MGD}) (8.34 \text{ lbs/gal}) = 29 \text{ lbs/day}$$

$$x = \frac{29}{(1.62)(8.34)}$$

$$x = \boxed{2.1 \text{ mg/L}}$$

DETERMINING mg/L DOSAGE

Given the actual chemical feed (lbs) and flow data for a 24-hour period, the mg/L dosage rate can be determined. This simply uses the mg/L to lbs/day equation such as described in Section 9.3:

$$\boxed{\begin{array}{lll} (\text{mg/L}) & (\text{MGD}) & (8.34) = \text{lbs/day} \\ \text{Chem.} & \text{Flow} & \text{lbs/gal} \quad \text{Chem.} \end{array}}$$

Example 4 illustrates this type of calculation.

9.9 SOLUTION CHEMICAL FEEDER CALIBRATION (Given mL/min or gpd Flow)

The calibration calculation for a solution feeder is slightly more difficult than that for a dry chemical feeder if the solution feed is expressed as mL/min rather than gpd.

As with other calibration calculations, the actual chemical feed rate is determined and then compared with the feed rate indicated by the instrumentation.

To calculate the actual chemical feed rate for a solution feeder, first express the solution feed rate in terms of MGD. (The equation for converting from mL/min to gpd is given to the right.) Once the MGD solution flow rate has been calculated, use the mg/L equation to determine chemical dosage in lbs/day.

SOLUTION FEED CALIBRATION

If solution feed is expressed as mL/min, first convert mL/min flow rate to gpd flow rate

$$\frac{(mL/min)\ (1440\ min/day)}{3785\ mL/gal} = gpd$$

Then calculate chemical dosage, lbs/day

$$(mg/L\ Chem.)\ (MGD\ Flow)\ (8.34\ lbs/day) = Chem., lbs/day$$

Example 1: (Solution Chemical Feed Calibration)
❏ A calibration test is conducted for a solution chemical feeder. During the 5-minute test, the pump delivered 960 mL of the 1.25% polymer solution. What is the polymer dosage rate in lbs/day? (Assume the polymer solution weighs 8.34 lbs/gal.)

Normally the mg/L to lbs/day equation* is used to determine the lbs/day feed rate. And in making these calculations, the flow rate must be expressed as MGD. Therefore, the mL/min solution flow rate must first be converted to gpd and then MGD. The mL/min flow rate is calculated as:

$$\frac{960\ mL}{5\ min} = \boxed{192\ mL/min}$$

Next convert the mL/min flow rate to gpd flow rate:

$$\frac{(192\ mL/min)\ (1440\ min/day)}{3785\ mL/gal} = \boxed{73\ gpd \\ flow\ rate}$$

Then calculate the lbs/day polymer feed rate:

$$(12,500\ mg/L)\ (0.000073\ MGD)\ (8.34\ lbs/day) = \boxed{7.6\ lbs/day \\ Polymer}$$

---

* A detailed discussion of the mg/L to lbs/day calculation is given in Chapter 3.

** A solution of 1.2% strength is equivalent to a solution of 12,000 mg/L concentration. Refer to Chapter 8 in *Basic Math Concepts* for a discussion of mg/L to % conversions.

Example 2: (Solution Chemical Feed Calibration)
❏ A calibration test is conducted for a solution chemical feeder. During a 24-hour period, a total of 120 gal of solution is delivered by the solution feeder. The polymer solution is a 1.3% solution. What is the lbs/day feed rate? (Assume the polymer solution weighs 8.34 lbs/gal.)

The solution feed rate is 120 gallons per day or 120 gpd. Expressed as MGD, this is 0.000120 MGD.

Use the mg/L to lbs/day equation to calculate actual feed rate, lbs/day:

$$(\text{mg}/L \text{ Chem.})\ (\text{MGD Flow})\ (8.34\ \text{lbs/day}) = \text{lbs/day Chem.}$$

$$(13{,}000\ \text{mg}/L)\ (0.000120\ \text{MGD})\ (8.34\ \text{lbs/day}) = \boxed{\begin{array}{l}12\ \text{lbs/day}\\ \text{Polymer}\end{array}}$$

Example 3: (Solution Chemical Feed Calibration)
❏ A calibration test is conducted for a solution chemical feeder. During the 5-minute test, the pump delivered 840 mL of the 1.3% polymer solution. The specific gravity of the polymer solution is 1.2. What is the polymer dosage rate in lbs/gal?

First calculate the mL/min flow rate during the 5-minute test:

$$\frac{840\ \text{m}L}{5\ \text{min}} = \boxed{168\ \text{m}L/\text{min}}$$

Then convert the mL/min flow rate to gpd flow rate:

$$\frac{(168\ \text{m}L/\text{min})\ (1440\ \text{min/day})}{3785\ \text{m}L/\text{gal}} = \boxed{\begin{array}{l}63.9\ \text{gpd}\\ \text{flow rate}\end{array}}$$

And calculate the lbs/day polymer feed rate:
(Remember, the specific gravity is 1.2, so the density of the solution is (8.34 lbs/gal) (1.2) = 10 lbs/gal)

$$(13{,}000\ \text{mg}/L)\ (0.0000639\ \text{MGD})\ (10\ \text{lbs/gal}) = \boxed{\begin{array}{l}8.3\ \text{lbs/day}\\ \text{Polymer}\end{array}}$$

TAKING DENSITY AND SPECIFIC GRAVITY INTO CONSIDERATION

In Examples 1 and 2, the polymer solution was assumed to have a density of 8.34 lbs/gal. In many instances, however, polymer solutions have densities different than water—sometimes higher and other times lower.

When the density is different from water, use a factor in the equation other than 8.34 lbs/gal.

Density information is sometimes given as specific gravity*. To determine the density when specific gravity information is given, simply multiply the density of water (8.34 lbs/gal) by the specific gravity:

$$\boxed{\begin{array}{l}(8.34)\ (\text{Specific})\\ \text{lbs/gal}\ \ \text{Gravity}\end{array} = \begin{array}{l}\text{New}\\ \text{Density}\\ \text{lbs/gal}\end{array}}$$

---

* For a review of density and specific gravity, refer to Chapter 7.

9.10 SOLUTION CHEMICAL FEEDER CALIBRATION
(Given Drop in Solution Tank Level)

Actual pumping rates can be determined by calculating the volume pumped during a specified time frame. For example, if 50 gallons are pumped during a 10-minute test, the average pumping rate during the test is 5 gpm.

The gallons pumped can be determined by **measuring the drop in water level during the timed test**, as illustrated in Examples 1 and 2.

The gpm pumping rate can be converted to gpd pumping rate.* The factor used to make the conversion depends on whether the pumping is continuous or intermittent. If the pumping is continuous during the 24 hours, use a factor of 1440 min/day. If, however, the pumping is intermittent, the factor would be the actual number of minutes the pump was in operation during the 24-hour period.

Given the gpd pumping rate and the solution concentration, the lbs/day chemical feed rate can be determined using the mg/L to lbs/day equation:

$$\text{(mg/}L\text{)(MGD)(8.34)} = \frac{\text{lbs/day}}{\text{lbs/gal}} \quad \text{Chem.}$$
Chem. Flow   lbs/gal   Chem.

Such a calculation is illustrated in Example 3.

VOLUME PUMPED IS INDICATED BY DROP IN TANK LEVEL

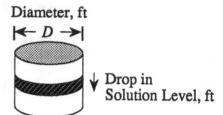

Simplified Equation:

$$\frac{\text{Flow}}{\text{gpm}} = \frac{\text{Volume Pumped, gal}}{\text{Duration of Test, min}}$$

Expanded Equation:

$$\frac{\text{Flow Rate}}{\text{gpm}} = \frac{(0.785)\,(D^2)\,\text{(Drop in Level, ft)}\,(7.48\text{ gal/cu ft})}{\text{Duration of Test, min}}$$

Example 1: (Solution Feeder Calibration)

❑ A pumping rate calibration test is conducted for a 15-minute period. The liquid level in the 3-ft diameter solution tank is measured before and after the test. If the level drops 0.4 ft during the 15-min test, what is the pumping rate in gpm?

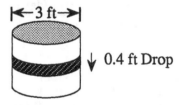

$$\frac{\text{Flow Rate}}{\text{gpm}} = \frac{(0.785)\,(D^2)\,\text{(Drop, ft)}\,(7.48\text{ gal/cu ft})}{\text{Duration of Test, min}}$$

$$= \frac{(0.785)\,(3\text{ ft})\,(3\text{ ft})\,(0.4\text{ ft})\,(7.48\text{ gal/cu ft})}{15\text{ min}}$$

$$= \boxed{\begin{array}{c}\text{1.4 gpm}\\\text{Pumping Rate}\end{array}}$$

---

* For a review of flow conversions, refer to Chapter 8 in *Basic Math Concepts*.

Example 2: (Solution Feeder Calibration)
❑ A pumping rate calibration test is conducted for a
10-minute period. The liquid level in the 4-ft diameter tank is
measured before and after the pumping test. If the level drop
is 3 inches during the test, what is the pumping rate in gpm?

$$\frac{3 \text{ in}}{12 \text{ in/ft}} = 0.25 \text{ ft drop}$$

$$\frac{\text{Flow Rate}}{\text{gpm}} = \frac{(0.785)\,(D^2)\,(\text{Drop, ft})\,(7.48 \text{ gal/cu ft})}{\text{Duration of Test, min}}$$

$$= \frac{(0.785)\,(4 \text{ ft})\,(4 \text{ ft})\,(0.25 \text{ ft})\,(7.48 \text{ gal/cu ft})}{10 \text{ min}}$$

$$= \boxed{\begin{array}{c} 2.3 \text{ gpm} \\ \text{Pumping Rate} \end{array}}$$

Example 3: (Solution Feeder Calibration)
❑ During a 30-minute pumping rate calibration test, the
solution level in the 3 ft diameter chemical tank dropped
1 inch. Assume the pump operates at the same rate for
24 hours. If the polymer solution is a 1.4% solution what is
the lbs/day chemical feed? (Assume the polymer solution
weighs 8.34 lbs/gal.)

First calculate the gpm solution flow:
(1 in. + 12 in./ft = 0.08 ft)

$$\frac{\text{Flow Rate}}{\text{gpm}} = \frac{(0.785)\,(3 \text{ ft})\,(3 \text{ ft})\,(0.08 \text{ ft})\,(7.48 \text{ gal/cu ft})}{30 \text{ min}}$$

$$= 0.14 \text{ gpm}$$

The gpd flow rate is therefore:

(0.14 gpm) (1440 min/day) = 202 gpd

And the chemical feed rate is:

(14,000 mg/*L*) (0.000202 MGD) (8.34 lbs/gal) = 23.6 lbs/day

---

* Refer to Chapter 2 in *Basic Math Concepts* for a review of solving for the unknown value.

9.11 CHEMICAL USE CALCULATIONS

The lbs/day or gpd chemical use should be recorded each day. From this data, you can calculate the average daily use of the chemical or solution.

This information will also enable you to forecast expected chemical use, compare it with chemical in inventory, and determine when additional chemical supplies will be required.

AVERAGE CHEMICAL USE

First determine the average chemical use:

$$\text{Average Use lbs/day} = \frac{\text{Total Chem. Used, lbs}}{\text{Number of Days}}$$

Or

$$\text{Average Use gpd} = \frac{\text{Total Chem. Used, gal}}{\text{Number of Days}}$$

Then calculate days' supply in inventory:*

$$\text{Days' Supply in Inventory} = \frac{\text{Total Chem. in Inventory, lbs}}{\text{Average Use, lbs/day}}$$

Or

$$\text{Days' Supply in Inventory} = \frac{\text{Total Chem. in Inventory, gal}}{\text{Average Use, gpd}}$$

Example 1: (Average Use)
❑ The chemical used for each day during a week is given below. Based on this data, what was the average lbs/day chemical use during the week?

Monday—90 lbs/day   Friday—98 lbs/day
Tuesday—96 lbs/day   Saturday—91 lbs/day
Wednesday—92 lbs/day   Sunday—87 lbs/day
Thursday—89 lbs/day

$$\text{Average Use lbs/day} = \frac{\text{Total Chem. Used, lbs}}{\text{Number of Days}}$$

$$= \frac{643 \text{ lbs}}{7 \text{ days}}$$

$$= \boxed{91.9 \text{ lbs/day} \atop \text{Aver. Use}}$$

---

* Note how similar these equations are to detention time equations. Refer to Chapter 5 for detention time calculations.

Example 2: (Average Use)
❑ The average chemical use at a plant is 78 lbs/day. If the chemical inventory in stock is 2400 lbs, how many days' supply is this?

$$\text{Days' Supply in Inventory} = \frac{\text{Total Chem. in Inventory, lbs}}{\text{Average Use, lbs/day}}$$

$$= \frac{2400 \text{ lbs in Inventory}}{78 \text{ lbs/day Aver. Use}}$$

$$= \boxed{\begin{array}{l}30.8 \text{ days'} \\ \text{Supply in Inventory}\end{array}}$$

Example 3: (Average Use)
❑ The average gallons polymer solution used each day at a treatment plant is 86 gpd. A chemical feed tank has a diameter of 3 ft and contains solution to a depth of 4.1 ft. How many days' supply are represented by the solution in the tank?

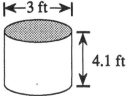

|←— 3 ft —→|

4.1 ft

$$\text{Days' Supply in Tank} = \frac{\text{Total Solution in Tank, gal}}{\text{Average Use, gpd}}$$

$$x \text{ days} = \frac{(0.785)\ (3 \text{ ft})\ (3 \text{ ft})\ (4.1 \text{ ft})\ (7.48 \text{ gal/cu ft})}{86 \text{ gpd}}$$

$$= \boxed{\begin{array}{l}2.5 \text{ days'} \\ \text{Supply in Tank}\end{array}}$$

NOTES:

10 *Sedimentation*

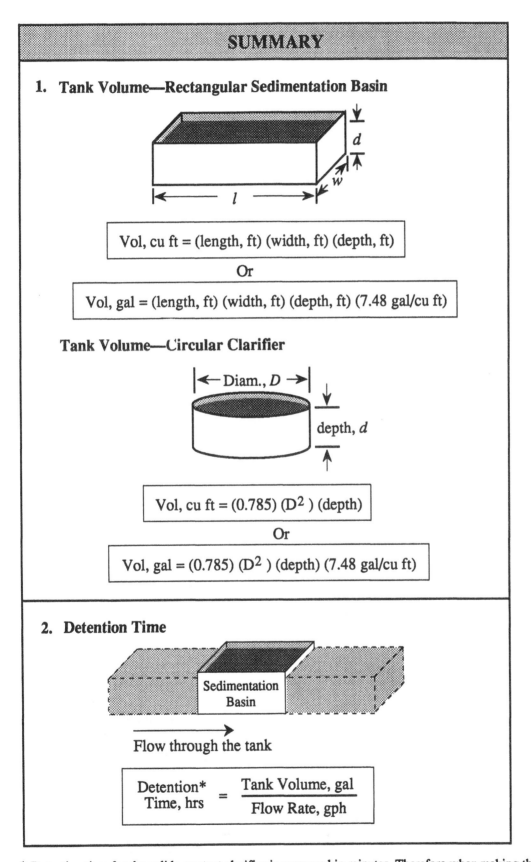

SUMMARY

1. **Tank Volume—Rectangular Sedimentation Basin**

$$\boxed{\text{Vol, cu ft} = (\text{length, ft}) \,(\text{width, ft}) \,(\text{depth, ft})}$$

Or

$$\boxed{\text{Vol, gal} = (\text{length, ft}) \,(\text{width, ft}) \,(\text{depth, ft}) \,(7.48 \text{ gal/cu ft})}$$

Tank Volume—Circular Clarifier

$$\boxed{\text{Vol, cu ft} = (0.785) \,(D^2) \,(\text{depth})}$$

Or

$$\boxed{\text{Vol, gal} = (0.785) \,(D^2) \,(\text{depth}) \,(7.48 \text{ gal/cu ft})}$$

2. **Detention Time**

Flow through the tank

$$\boxed{\frac{\text{Detention*}}{\text{Time, hrs}} = \frac{\text{Tank Volume, gal}}{\text{Flow Rate, gph}}}$$

* Detention time for the solids-contact clarifier is measured in minutes. Therefore when making the detention time calculation for this unit, the flow rate must be expressed as gpm.

SUMMARY—Cont'd

3. Surface Overflow Rate

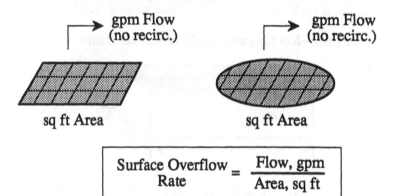

$$\text{Surface Overflow Rate} = \frac{\text{Flow, gpm}}{\text{Area, sq ft}}$$

4. Mean Flow Velocity

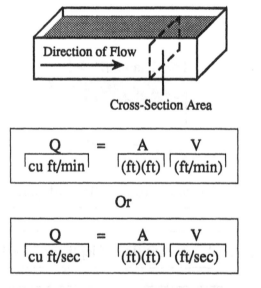

Direction of Flow

Cross-Section Area

$$\frac{Q}{\text{cu ft/min}} = \frac{A}{\text{(ft)(ft)}} \; \frac{V}{\text{(ft/min)}}$$

Or

$$\frac{Q}{\text{cu ft/sec}} = \frac{A}{\text{(ft)(ft)}} \; \frac{V}{\text{(ft/sec)}}$$

5. Weir Loading Rate

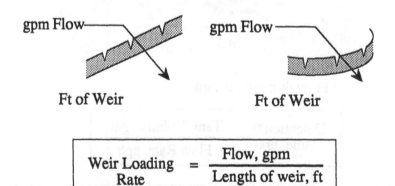

$$\text{Weir Loading Rate} = \frac{\text{Flow, gpm}}{\text{Length of weir, ft}}$$

Calculations associated with the solids contact clarification process include:

- Detention Time,

- Percent Settled Sludge,
  (sometimes called "volume over volume" test)

- Lime Dose Required, mg/L, and

- Lime Dose Required, lbs/day or grams/min.

The detention time calculation has been described above (see calculation #2). Equations for the remaining calculations are given below.

6. Percent Settled Sludge (V/V Test)

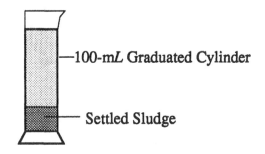

—100-mL Graduated Cylinder

— Settled Sludge

$$\text{\% Settled Sludge} = \frac{\text{Settled Sludge Volume, m}L}{\text{Total Sample Volume, m}L} \times 100$$

SUMMARY—Cont'd

7. Lime Dose Required, mg/L

The lime dose requirement (in mg/L) for the solids contact clarification process is calculated in three steps:

- **Step 1**—Determine the <u>total alkalinity</u> (HCO_3^-) required to react with the alum to be added and provide proper precipitation: (Use 1 mg/L alum reacts with 0.45 mg/L alkalinity, HCO_3^-.)

$$\begin{array}{ccc} \text{Total Alk.} & & \text{Alk. Required to} & & \text{Alk. Required to} \\ \text{Required,} & = & \text{React with} & + & \text{Assure Proper} \\ \text{mg/L} & & \text{the Alum,} & & \text{Precipitation of} \\ & & \text{mg/L} & & \text{Alum,} \\ & & & & \text{mg/L} \end{array}$$

- **Step 2**—Determine alkalinity (HCO_3^-) to be added to the water:

$$\begin{array}{ccc} \text{Total Alkalinity} & & \text{Alkalinity} & & \text{Alkalinity} \\ \text{Required, mg/L} & - & \text{Present in Water,} & = & \text{to be Added} \\ & & \text{mg/L} & & \text{to the Water, mg/L} \end{array}$$

- **Step 3**—Determine the lime required to meet this alkalinity need: (Using the relationships 1 mg/L alum reacts with 0.45 mg/L alkalinity, HCO_3^-, and 1 mg/L alum reacts with 0.35 mg/L lime, the following proportion can be constructed.)

From Step 2 above

$$\frac{0.45 \text{ mg/L } HCO_3^-}{0.35 \text{ mg/L } Ca(OH)_2} = \frac{\square \text{ mg/L } HCO_3^-}{x \text{ mg/L } Ca(OH)_2}$$

8. Lime Dose Required, lbs/day

$$\frac{(\text{mg/L}) \ (\text{Flow}) \ (8.34)}{\text{Lime} \quad \text{MGD} \quad \text{lbs/gal}} = \frac{\text{lbs/day}}{\text{Lime}}$$

9. Lime Dose Required, g/min
(Using 1 lb = 453.6 g/lb)

$$\frac{(\text{Lime, lbs/day}) \ (453.6 \text{ g/lb})}{1440 \text{ min/day}} = \frac{\text{Lime,}}{\text{g/min}}$$

NOTES:

10.1 TANK VOLUME CALCULATIONS

The two common tank shapes of sedimentation tanks are rectangular and cylindrical. The equations for calculating the volume for each type tank are shown to the right. For further review of volume calculations, refer to Chapter 1.

CALCULATING TANK VOLUME

For Rectangular Sedimentation Basins:

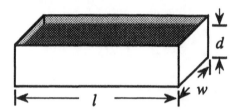

Vol., gal = (length, ft) (width, ft) (depth, ft) (7.48 gal/cu ft)

For Circular Clarifiers:

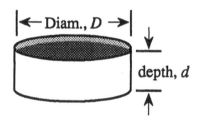

Vol, gal = (0.785) (D^2) (depth, ft) (7.48 gal/cu ft)

Example 1: (Tank Volume)
❏ A sedimentation basin is 20 ft wide, 70 ft long, and contains water to a depth of 12 ft. What is the volume of water in the basin, in gallons?

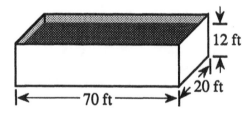

Vol., gal = (length, ft) (width, ft) (depth, ft) (7.48 gal/cu ft)

= (70 ft) (20 ft) (12 ft) (7.48 gal/cu ft)

= ⟨ 125,664 gal ⟩

Example 2: (Tank Volume)

❑ A sedimentation basin is 25 ft wide and 80 ft long. When the basin contains 150,000 gallons, what would the water depth be?

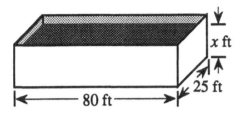

Vol., gal = (length, ft) (width, ft) (depth, ft) (7.48 gal/cu ft)

150,000 gal = (80 ft) (25 ft) (x ft) (7.48 gal/cu ft)

$$\frac{150,000}{(80)(25)(7.48)} = x \text{ ft}$$

$$\boxed{10 \text{ ft}} = x$$

SOLVING FOR OTHER VARIABLES

In most volume calculations, volume is the unknown variable. However, the same equations can be used to solve for another unknown variable, such as depth. In this type problem, write the equation as usual, fill in given data, then solve for the unknown value.* Example 2 illustrates a calculation of this type.

Example 3: (Tank Volume)

❑ The diameter of a circular clarifier is 60 ft. When the water depth is 13 ft, what is the volume of water in the circular clarifier, in gallons?

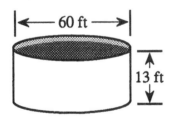

Vol., gal = (0.785) (D^2) (d) (7.48 gal/cu ft)

= (0.785) (60 ft) (60 ft) (13 ft) (7.48 gal/cu ft)

= $\boxed{274,800 \text{ gal}}$

---

* For a review of solving for the unknown value, refer to Chapter 2 in *Basic Math Concepts*.

10.2 DETENTION TIME

The detention time for clarifiers may vary from one to three hours. The equations used to calculate detention time are shown to the right.

MATCHING UNITS

When calculating detention time, it is essential that the time and volume units used in the equation are consistent, as illustrated in the diagram.

The flow rate to the clarifier is normally expressed as MGD or gpd. However, since the detention time is desired in hours, it is important to express the flow rate in the same time frame—gal/hr (or gph):*

$$\frac{\text{Flow, gpd}}{24 \text{ hrs/day}} = \text{Flow, gph}$$

DETENTION TIME IS FLOW-THROUGH TIME

Simplified Equation:

$$\text{Detention Time, hrs} = \frac{\text{Volume of Tank, gal}}{\text{Flow Rate, gph}}$$

Expanded Equations:

For rectangular sedimentation basins

$$\text{Detention Time, hrs} = \frac{(\text{Length, ft})(\text{Width, ft})(\text{Depth, ft})(7.48 \text{ gal/cu ft})}{\text{Flow Rate, gph}}$$

For circular clarifiers

$$\text{Detention Time, hrs} = \frac{(0.785)(D^2)(\text{Depth, ft})(7.48 \text{ gal/cu ft})}{\text{Flow Rate, gph}}$$

BE SURE YOUR TIME AND VOLUME UNITS MATCH

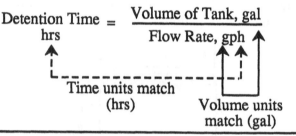

Example 1: (Detention Time)
❑ A sedimentation tank has a volume of 131,000 gallons. If the flow to the tank is 116,000 gph, what is the detention time in the tank, in hours?

$$\text{Detention Time, hrs} = \frac{\text{Volume of Tank, gal}}{\text{Flow Rate, gph}}$$

$$= \frac{131,000 \text{ gal}}{116,000 \text{ gph}}$$

$$= \boxed{1.1 \text{ hours}}$$

---

* For a review of flow rate conversions, refer to Chapter 8 in *Basic Math Concepts.*

Example 2: (Detention Time)

❑ A sedimentation basin is 65 ft long, 20 ft wide, and has water to a depth of 12 ft. If the flow to the basin is 1,600,000 gpd, what is the sedimentation basin detention time? (Assume the flow is steady and continuous.)

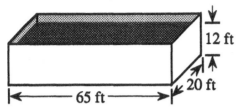

12 ft

20 ft

|← 65 ft →|

First, convert the flow rate from gpd to gph so that time units will match. (1,600,000 gpd ÷ 24 hrs/day = 66,667 gph). Then calculate detention time:

$$\frac{\text{Detention Time}}{\text{hrs}} = \frac{\text{Volume of Tank, gal}}{\text{Flow Rate, gph}}$$

$$= \frac{(65 \text{ ft})(20 \text{ ft})(12 \text{ ft})(7.48 \text{ gal/cu ft})}{66,667 \text{ gph}}$$

$$= \boxed{1.8 \text{ hours}}$$

USING THE EXPANDED EQUATION

The flow rate conversion from gpd to gph is often completed separately, before detention time is calculated.* The volume calculation, however, is often incorporated in the numerator of the detention time calculation, as shown in the expanded equation on the previous page. Examples 2 and 3 illustrate the use of the expanded equations.

Example 3: (Detention Time)

❑ A circular clarifier is 50 ft in diameter and 12 ft deep. At what flow rate (MGD) would the detention time be 2 hours? (Assume the flow is steady and continuous.)

Use the detention time equation, filling in known data. Since the equation normally used includes flow rate as gph, first calculate gph flow rate:**

$$\frac{\text{Detention Time}}{\text{hrs}} = \frac{\text{Volume of Tank, gal}}{\text{Flow Rate, gph}}$$

$$2 \text{ hrs} = \frac{(0.785)(50 \text{ ft})(50 \text{ ft})(12 \text{ ft})(7.48 \text{ gal/cu ft})}{x \text{ gph}}$$

$$x \text{ gph} = \frac{(0.785)(50 \text{ ft})(50 \text{ ft})(12 \text{ ft})(7.48 \text{ gal/cu ft})}{2 \text{ hrs}}$$

$$x = 88,077 \text{ gph}$$

Then convert gph to gpd and MGD flow rate:

$$(88,077 \text{ gph}) (24 \text{ hrs/day}) = 2,113,848 \text{ gpd}$$

$$\text{or } = \boxed{2.1 \text{ MGD}}$$

SOLVING FOR OTHER VARIABLES

In most detention time calculations, detention time is the unknown variable. However, the same equation can be used to solve for other unknown values. Since tank dimensions are not variable (except during design), the other variables that could be calculated using the detention time equation are water depth and flow rate. Example 3 illustrates a calculation when detention time is known and flow rate is the unknown variable.

---

* If the flow rate (denominator of the detention time calculation) conversion were incorporated as part of the expanded equation, it would necessitate division by a fraction. This is a more complex type of math problem and more easily results in errors.

** Refer to Chapter 8 in *Basic Math Concepts* for a review of flow conversions.

10.3 SURFACE OVERFLOW RATE

Surface overflow rate is used to determine loading on sedimentation basins and circular clarifiers. It is similar to hydraulic loading rate—flow per unit area. However, hydraulic loading rate measures the total water entering the process (plant flow plus any recirculation) whereas **surface overflow rate measures only the water overflowing the process (plant flow only).**

As indicated in the diagram to the right, **surface overflow rate calculations do not include recirculated flows.** This is because recirculated flows are taken from the bottom of the clarifier and hence do not flow up and out of the clarifier (overflow).

Since surface overflow rate is a measure of flow (Q) divided by area (A), surface overflow is an indirect measure* of the **upward velocity** of water as it overflows the clarifier:**

$$V = \frac{Q}{A}$$

This calculation is important in maintaining proper sedimentation basin or clarifier operation since settling solids will be drawn upward and out of the clarifier if surface overflow rates are too high.

Other terms used synonymously with surface overflow rate are:

• Surface Loading Rate, and

• Surface Settling Rate

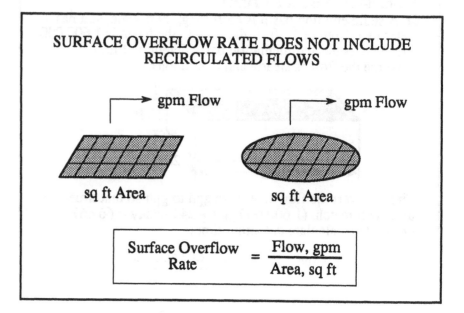

SURFACE OVERFLOW RATE DOES NOT INCLUDE RECIRCULATED FLOWS

→ gpm Flow   → gpm Flow

sq ft Area   sq ft Area

$$\text{Surface Overflow Rate} = \frac{\text{Flow, gpm}}{\text{Area, sq ft}}$$

Example 1: (Surface Overflow Rate)
❑ A circular clarifier has a diameter of 60 ft. If the flow to the clarifier is 1600 gpm, what is the surface overflow rate in gpm/sq ft?

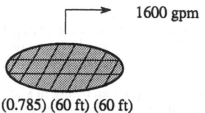

1600 gpm

(0.785) (60 ft) (60 ft)

$$\text{Surface Overflow Rate} = \frac{\text{Flow, gpm}}{\text{Area, sq ft}}$$

$$= \frac{1600 \text{ gpm}}{(0.785) (60 \text{ ft}) (60 \text{ ft})}$$

$$= \boxed{0.57 \text{ gpm/sq ft}}$$

---

* It is not a <u>direct</u> measure sof velocity since dividing out the units in the numerator and denominator does not result in ft/min.

** Refer to Chapter 2 for a review of $Q = AV$ problems.

Example 2: (Surface Overflow Rate)
❏ A sedimentation basin 75 ft by 20 ft receives a flow of 900 gpm. What is the surface overflow rate in gpm/sq ft?

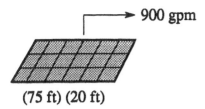

→ 900 gpm

(75 ft) (20 ft)

$$\text{Surface Overflow Rate} = \frac{\text{Flow, gpm}}{\text{Area, sq ft}}$$

$$= \frac{900 \text{ gpm}}{(75 \text{ ft}) (20 \text{ ft})}$$

$$= \boxed{0.6 \text{ gpm/sq ft}}$$

Example 3: (Surface Overflow Rate)
❏ The flow to a sedimentation basin is 3.2 MGD. If the length of the basin is 90 ft and the width is 45 ft, what is the surface overflow rate in gpm/sq ft?

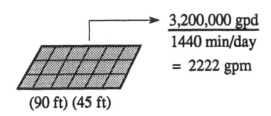

→ $\frac{3,200,000 \text{ gpd}}{1440 \text{ min/day}}$

= 2222 gpm

(90 ft) (45 ft)

$$\text{Surface Overflow Rate} = \frac{\text{Flow, gpm}}{\text{Area, sq ft}}$$

$$= \frac{2222 \text{ gpm}}{(90 \text{ ft}) (45 \text{ ft})}$$

$$= \boxed{0.55 \text{ gpm/sq ft}}$$

CONVERTING GPD FLOW RATE TO GPM FLOW RATE

Surface overflow rate is expressed as gpm/sq ft. Therefore, if the flow rate is expressed in terms other than gpm (such as gpd or MGD), that flow rate should be converted to gpm before calculating surface overflow rate.* To convert gpd flow rate to gpm, use the following equation:

$$\frac{\text{Flow Rate, gpd}}{1440 \text{ min/day}} = \text{Flow Rate, gpm}$$

Example 3 illustrates a calculation in which flow rate must be converted from gpd to gpm.

---

* For a review of flow rate conversions, refer to Chapter 8 in *Basic Math Concepts*.

10.4 MEAN FLOW VELOCITY

Mean flow velocity is a measure of the average velocity of the water as it travels through a rectangular sedimentation basin. This calculation is important in maintaining proper operation of the sedimentation process since settling solids will be drawn out of the sedimentation basin if velocities are too high.

Mean flow velocity is calculated using the Q = AV equation,* where Q represents flow rate, A represents cross-sectional area of the channel or pipe, and V represents the water velocity through the sedimentation basin.

Note that the cross-sectional area is measured perpendicular to the direction of flow. Since the direction of flow is down the length of the basin, the cross-sectional area is measured using the width of the tank and the water depth.

One essential aspect of the Q = AV calculation is how the units are expressed. As shown in the equations given at the top of the page, the <u>time frame</u> for the flow rate and velocity must match.

In the first equation, a flow rate expressed in cu ft/min (or cfm) results when the velocity is expressed as ft/min (or fpm). And in the second equation , a flow rate expressed in cu ft/sec (or cfs) results when the velocity is expressed as ft/sec (or fps). If a flow rate expressed in gal/min (gpm) were used for Q, **an additional factor of 7.48 gal/cu ft would be required** on the

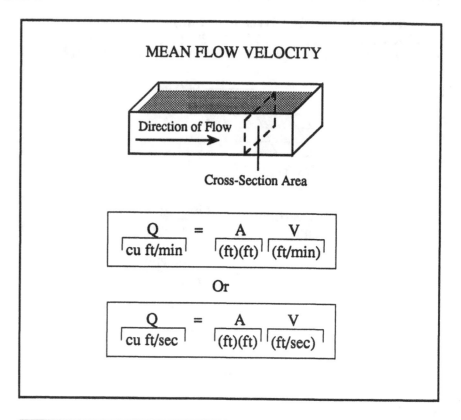

MEAN FLOW VELOCITY

Direction of Flow

Cross-Section Area

$$\underset{\text{cu ft/min}}{Q} = \underset{\text{(ft)(ft)}}{A} \underset{\text{(ft/min)}}{V}$$

Or

$$\underset{\text{cu ft/sec}}{Q} = \underset{\text{(ft)(ft)}}{A} \underset{\text{(ft/sec)}}{V}$$

Example 1: (Mean Flow Velocity)
❏ A sedimentation basin 50 ft long and 15 ft wide has water to a depth of 10 ft. When the flow through the basin is 850,000 gpd, what is the mean flow velocity in the basin, in ft/min?

Since velocity is desired in ft/min, the flow rate in the Q = AV equation must be expressed in cu ft/min (or cfm):**

$$\frac{850,000 \text{ gpd}}{(1440 \text{ min/day}) (7.48 \text{ gal/cu ft})} = 79 \text{ cfm}$$

Then use the Q = AV equation to calculate velocity:

$$Q_{cfm} = A V_{fpm}$$

$$79 \text{ cfm} = (15 \text{ ft}) (10 \text{ ft}) (x \text{ fpm})$$

$$\frac{79}{(15)(10)} = x$$

$$\boxed{0.5 \text{ fpm}} = x$$

---

* Refer to Chapter 2 for a review of *Q = AV* problems.

** Refer to Chapter 8 in *Basic Math Concepts* for a review of flow conversions.

Example 2: (Mean Flow Velocity)
❏ A rectangular sedimentation basin 60 ft long and 25 ft wide has a water depth of 11 ft. If the flow to the basin is 1,980,000 gpd, what is the mean flow velocity in ft/min?

Since velocity is desired in ft/min, the flow rate in the Q = AV equation must be expressed in cu ft/min (or cfm):

$$\frac{1,980,000 \text{ gpd}}{(1440 \text{ min/day}) (7.48 \text{ gal/cu ft})} = 184 \text{ cfm}$$

Then use the Q = AV equation to calculate velocity:

$$Q_{cfm} = A \, V_{fpm}$$

$$184 \text{ cfm} = (25 \text{ ft}) (11 \text{ ft}) (x \text{ fpm})$$

$$\frac{184}{(25)(11)} = x$$

$$\boxed{0.7 \text{ fpm}} = x$$

Example 3: (Mean Flow Velocity)
❏ A sedimentation basin 70 ft long and 20 ft wide operates at a depth of 12 ft. If the treatment plant treats a total of 1,470,000 gpd, what is the mean flow velocity in the basin, in ft/min?

First convert the flow rate to cfm:

$$\frac{1,470,000 \text{ gpd}}{(1440 \text{ min/day}) (7.48 \text{ gal/cu ft})} = 136 \text{ cfm}$$

Then calculate velocity, using the Q = AV equation:

$$Q_{cfm} = A \, V_{fpm}$$

$$136 \text{ cfm} = (20) (12 \text{ ft}) (60 \text{ ft}) (x \text{ fpm})$$

$$\frac{136}{(20)(12)} = x$$

$$\boxed{0.6 \text{ fpm}} = x$$

right side of the equation to convert cubic feet to gallons:

$$\underset{\text{gpm}}{Q} = \underset{(ft)(ft)}{A} \quad \underset{(ft/min)}{V} \quad \underset{\text{cu ft}}{(7.48 \text{ gal})}$$

To avoid errors in making Q = AV calculations, it is recommended that you:

- Always write the equation in the same form—Q = AV. Use this form of the equation *regardless* of which variable is the unknown factor. Once the data has been placed in the equation, then terms can be moved to solve for the unknown factor.

- Always express the flow rate in terms of cubic feet (cfm, cfs, etc.). This eliminates the need for the 7.48 gal/cu ft factor on the right side of the equation.

Examples 1-3 illustrate the calculation of mean flow velocity (ft/min). Note that in each example, the Q = AV equation is written in the same form and the flow rate is expressed in terms of cubic feet per minute.

---

* Refer to Chapter 2 in *Basic Math Concepts* for a review of solving for the unknown value.

10.5 WEIR LOADING RATE

The calculation of weir loading rate (also referred to as weir overflow rate) is important in detecting high velocities near the weir, which would affect the efficiency of the sedimentation process. With excessively high velocities near the weir, the settling solids are pulled over the weirs and into the effluent troughs, thus preventing settling as desired.

Weir loading rate is a measure of the **gallons per minute** flowing over each **foot of weir**.

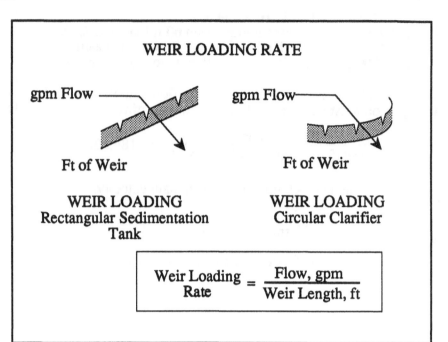

WEIR LOADING RATE

gpm Flow — Ft of Weir

gpm Flow — Ft of Weir

WEIR LOADING
Rectangular Sedimentation Tank

WEIR LOADING
Circular Clarifier

$$\text{Weir Loading Rate} = \frac{\text{Flow, gpm}}{\text{Weir Length, ft}}$$

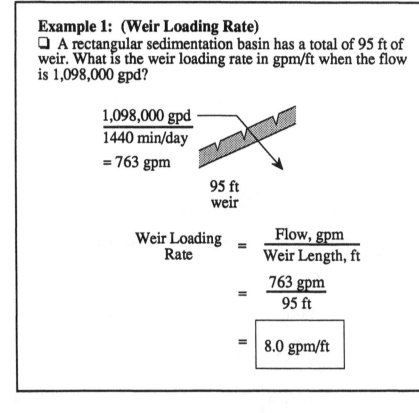

Example 1: (Weir Loading Rate)
❑ A rectangular sedimentation basin has a total of 95 ft of weir. What is the weir loading rate in gpm/ft when the flow is 1,098,000 gpd?

$$\frac{1{,}098{,}000 \text{ gpd}}{1440 \text{ min/day}} = 763 \text{ gpm}$$

95 ft weir

$$\text{Weir Loading Rate} = \frac{\text{Flow, gpm}}{\text{Weir Length, ft}}$$

$$= \frac{763 \text{ gpm}}{95 \text{ ft}}$$

$$= \boxed{8.0 \text{ gpm/ft}}$$

Example 2: (Weir Loading Rate)
❏ A circular clarifier receives a flow of 3.45 MGD. If the diameter of the weir is 80 ft, what is the weir loading rate in gpm/ft?

The total ft of weir is not given directly in this problem. However, weir diameter is given (80 ft) and from that information, the total ft of weir can be determined.

$$\frac{3,450,000 \text{ gpd}}{1440 \text{ min/day}}$$

$$= 2396 \text{ gpm}$$

ft of weir:
$$= (3.14)\ (80 \text{ ft})$$
$$= 251 \text{ ft}$$

$$\text{Weir Loading Rate} = \frac{\text{Flow, gpm}}{\text{Weir Length, ft}}$$

$$= \frac{2396 \text{ gpm}}{251 \text{ ft}}$$

$$= \boxed{9.5 \text{ gpm/ft}}$$

Example 3: (Weir Loading Rate)
❏ A clarifier receives a flow of 1.98 MGD. If the diameter of the weir is 70 ft, what is the weir loading rate in gpm/ft?

First calculate the gpm flow:

$$\frac{1,980,000 \text{ gpd}}{1440 \text{ min/day}} = 1375 \text{ gpm}$$

Then calculate weir loading rate:

$$\text{Weir Loading Rate} = \frac{\text{Flow, gpm}}{\text{Weir Length, ft}}$$

$$= \frac{1375 \text{ gpm}}{(3.14)\ (70 \text{ ft})}$$

$$= \boxed{6.3 \text{ gpm/ft}}$$

CALCULATING WEIR CIRCUMFERENCE

In some calculations of weir loading rate, you will have to calculate the total weir length, given the weir diameter. To calculate the length around any circle, you need to know the relationship between the diameter and circumference of a circle. **The distance around any circle (circumference) is about three times the distance across that circle (diameter);** or more precisely, the circumference is 3.14 times the diameter:*

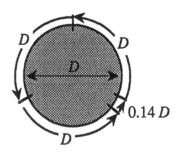

Therefore, when you know the weir diameter, you can calculate the total feet of weir:

$$\boxed{\begin{array}{l}\text{Total Ft} \\ \text{of Weir}\end{array} = (3.14)\ \begin{array}{l}(\text{Weir Diam.}) \\ \text{in ft}\end{array}}$$

---

* For a review of circumference calculations, refer to Chapter 9, "Linear Measurement", in *Basic Math Concepts*.

10.6 PERCENT SETTLED SLUDGE

The percent settled sludge calculation for solids contact clarifiers gives a general indication of the settleability of sludge in the solids contact unit. The percent settled sludge test (sometimes called the "volume over volume" test, or V/V test) is conducted by collecting a 100-mL slurry sample from the solids contact unit and allowing it to settle for ten minutes. After ten minutes, the volume of settled sludge at the bottom of the 100-mL graduated cylinder is measured and recorded. The equation used to calculate percent settled sludge is shown to the right.* Examples 1-3 illustrate this calculation.

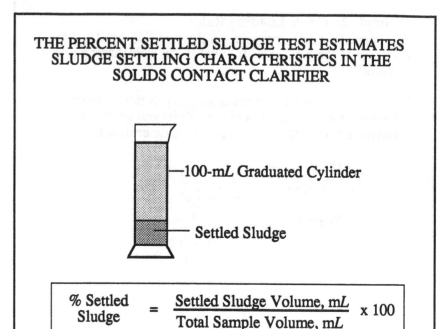

THE PERCENT SETTLED SLUDGE TEST ESTIMATES SLUDGE SETTLING CHARACTERISTICS IN THE SOLIDS CONTACT CLARIFIER

—100-mL Graduated Cylinder

— Settled Sludge

$$\text{\% Settled Sludge} = \frac{\text{Settled Sludge Volume, m}L}{\text{Total Sample Volume, m}L} \times 100$$

Example 1: (% Settled Sludge)
❏ A 100-mL sample of slurry from a solids contact unit is placed in a graduated cylinder and allowed to set for 10 minutes. The settled sludge at the bottom of the graduated cylinder after 10 minutes is 19 mL. What is the percent settled sludge of the sample?

$$\text{\% Settled Sludge} = \frac{\text{Settled Sludge, m}L}{\text{Total Sample, m}L} \times 100$$

$$= \frac{19 \text{ m}L}{100 \text{ m}L} \times 100$$

$$= \boxed{\text{19\% Settled Sludge}}$$

---

* For a review of percent calculations, refer to Chapter 5 of *Basic Math Concepts*.

Example 2: (**% Settled Sludge**)

❏ A 100-mL sample of slurry from a solids contact unit is placed in a graduated cylinder. After 10 minutes, a total of 22 mL of sludge settled to the bottom of the cylinder. What is the percent settled sludge of the sample?

$$\begin{aligned} \text{\% Settled Sludge} &= \frac{\text{Settled Sludge, m}L}{\text{Total Sample, m}L} \times 100 \\[2mm] &= \frac{22 \text{ m}L}{100 \text{ m}L} \times 100 \\[2mm] &= \boxed{22\% \text{ Settled Sludge}} \end{aligned}$$

Example 3: (**% Settled Sludge**)

❏ A percent settled sludge test is run on a 100-mL sample of solids-contact unit slurry. After 10 minutes of settling, a total of 20 mL of sludge settled to the bottom of the cylinder. What is the percent settled sludge of the sample?

$$\begin{aligned} \text{\% Settled Sludge} &= \frac{\text{Settled Sludge, m}L}{\text{Total Sample, m}L} \times 100 \\[2mm] &= \frac{20 \text{ m}L}{100 \text{ m}L} \times 100 \\[2mm] &= \boxed{20\% \text{ Settled Sludge}} \end{aligned}$$

10.7 LIME DOSE REQUIRED, mg/*L*

When alum is used for coagulation in the solids contact clarification process, lime is sometimes added to provide adequate alkalinity (HCO_3^-) for the coagulation and precipitation of the solids.

There are three steps in determining the lime dose required, in mg/*L*, as illustrated to the right. Generally speaking, in the first two steps you are determining how much alkalinity should be added to the water, and in the third step you are determining how much lime is required to provide that level of alkalinity.

In **Step 1**, the total alkalinity required is calculated. This total amount reflects two considerations: (1) the alkalinity that will react with the alum added to the water; and (2) the "residual" alkalinity (usually at least 30 mg/*L*) that must be present in the water to promote complete precipitation. This type problem is similar to a chlorine dosage/demand/residual problem described in Chapter 12.* Examples 1 and 2 illustrate this calculation.

THREE STEPS IN CALCULATING LIME DOSE REQUIRED, mg/*L*

• **Step 1**—Determine the <u>total alkalinity</u> (HCO_3^-) required to react with the alum to be added and provide proper precipitation: (Examples 1 and 2 illustrate this calculation.)

$$\begin{array}{ccccc} \text{Total Alk.} & & \text{Alk. that will} & & \text{Alk. Required to} \\ \text{Required} & = & \text{React with} & + & \text{Assure Proper} \\ \text{mg}/L & & \text{the Alum, mg}/L & & \text{Precipitation of} \\ & & & & \text{Alum, mg}/L \end{array}$$

Use: 1 mg/*L* alum reacts with 0.45 mg/*L* alkalinity, HCO_3^-

• **Step 2**—Determine the alkalinity (HCO_3^-) to be added to the water: (Examples 3 and 4 illustrate this calculation.)

$$\begin{array}{ccccc} \text{Alkalinity} & & \text{Total Alkalinity} & & \text{Alkalinity Present} \\ \text{to be Added} & = & \text{Required, mg}/L & - & \text{in the Water,} \\ \text{to the Water,} & & & & \text{mg}/L \\ \text{mg}/L & & & & \end{array}$$

• **Step 3**—Determine the lime ($Ca(OH)_2$) required to meet this alkalinity need (alkalinity to be added, calculated in Step 2). Using the relationships 1 mg/*L* alum reacts with 0.45 mg/*L* alkalinity, HCO_3^-, and 1 mg/*L* alum reacts with 0.35 mg/*L* lime, the following proportion can be constructed: (Examples 5 and 6 illustrate this calculation.)

From Step 2 above

$$\frac{0.45 \text{ mg}/L \text{ } HCO_3^-}{0.35 \text{ mg}/L \text{ } Ca(OH)_2} = \frac{\Box \text{ mg}/L \text{ } HCO_3^-}{x \text{ mg}/L \text{ } Ca(OH)_2}$$

Examples 7 and 8 illustrate the calculation of all three steps in determining lime dose in mg/*L*.

---

* Chlorine Dosage = Chlorine Demand + Chlorine Residual.

Example 1: (Lime Dose Required, mg/L)
❑ A raw water requires an alum dose of 47 mg/L, as determined by jar testing. If a "residual" 30 mg/L alkalinity (HCO_3^-) must be present in the water to ensure complete precipitation of the alum added, what is the total alkalinity required, in mg/L? (1 mg/L alum reacts with 0.45 mg/L alkalinity, HCO_3^-.)

First, calculate the alkalinity that will react with 45 mg/L alum:*

$$\frac{0.45 \text{ mg/L Alk.}}{1 \text{ mg/L Alum}} = \frac{x \text{ mg/L Alk.}}{47 \text{ mg/L Alum}}$$

$$(0.45)(47) = x$$

$$21.2 \text{ mg/L} = x$$
$$\text{Alk.}$$

Then calculate the total alkalinity required:

$$\begin{array}{c}\text{Total Alk.} \\ \text{Req'd, mg/L}\end{array} = \begin{array}{c}\text{Alk. to React with} \\ \text{Alum, mg/L}\end{array} + \begin{array}{c}\text{"Residual" Alk.,} \\ \text{mg/L}\end{array}$$

$$= 21.2 \text{ mg/L} + 30 \text{ mg/L}$$

$$= \boxed{51.2 \text{ mg/L}}$$

Example 2: (Lime Dose Required, mg/L)
❑ Jar tests indicate that 38 mg/L alum are optimum for a particular raw water. If a "residual" 30 mg/L alkalinity must be present to promote complete precipitation of the alum added, what is the total alkalinity required, in mg/L? (1 mg/L alum reacts with 0.45 mg/L alkalinity, HCO_3^-.)

First, calculate the alkalinity that will react with 38 mg/L alum:

$$\frac{0.45 \text{ mg/L Alk.}}{1 \text{ mg/L Alum}} = \frac{x \text{ mg/L Alk.}}{38 \text{ mg/L Alum}}$$

$$(0.45)(38) = x$$

$$17.1 \text{ mg/L} = x$$

Then calculate the total alkalinity required:

$$\begin{array}{c}\text{Total Alk.} \\ \text{Req'd, mg/L}\end{array} = 17.1 \text{ mg/L} + 30 \text{ mg/L}$$

$$= \boxed{47.1 \text{ mg/L}}$$

* For a review of ratios and proportions, refer to Chapter 7 in *Basic Math Concepts*. This proportion could also be set up as: $\dfrac{0.45 \text{ mg/L Alk.}}{x \text{ mg/L Alk.}} = \dfrac{1 \text{ mg/L Alum}}{45 \text{ mg/L Alum}}$

The total alkalinity <u>required</u> and that actually <u>added</u> to the water are quite often different numbers. That is because some alkalinity is already present in the water. For example, if a total of 45 mg/*L* alkalinity is required for the water (based on Step 1 calculation) and the alkalinity already in the water is 10 mg/*L*, then only 35 mg/*L* alkalinity would have to be added to the water. This comparison between required alkalinity and alkalinity already in the raw water, is the heart of the **Step 2** calculation. Examples 3 and 4 illustrate this calculation.

Example 3: (Lime Dose Required, mg/*L*)

❑ A total of 46 mg/*L* alkalinity is required to react with alum and ensure proper precipitation. If the raw water has an alkalinity of 31 mg/*L* as bicarbonate (HCO_3^-), how many mg/*L* alkalinity (HCO_3^-) should be added to the water?

$$\text{Alk. to be Added to the Water, mg/}L = \text{Total Alk. Req'd, mg/}L - \text{Alk. Present in the Water, mg/}L$$

$$= 46 \text{ mg/}L - 31 \text{ mg/}L$$

$$= \boxed{15 \text{ mg/}L \text{ Alk. to be Added}}$$

Example 4: (Lime Dose Required, mg/*L*)

❑ A total of 53 mg/*L* alkalinity is required to react with alum and ensure complete precipitation. If the raw water has an alkalinity of 25 mg/*L* as bicarbonate (HCO_3^-), how many mg/*L* alkalinity (HCO_3^-) should be added to the water?

$$\text{Alk. to be Added to the Water, mg/}L = \text{Total Alk. Req'd, mg/}L - \text{Alk. Present in the Water, mg/}L$$

$$= 53 \text{ mg/}L - 25 \text{ mg/}L$$

$$= \boxed{28 \text{ mg/}L \text{ Alk. to be Added}}$$

Example 5: (Lime Dose Required, mg/L)
❑ It has been calculated that 15 mg/L alkalinity (HCO_3^-) must be added to a raw water. How many mg/L lime will be required to provide this amount of alkalinity? (1 mg/L alum reacts with 0.45 mg/L and 1 mg/L alum reacts with 0.35 mg/L lime.)

To determine the mg/L lime required, use a proportion that relates bicarbonate alkalinity to lime:*

$$\frac{0.45 \text{ mg/}L \text{ Alk.}}{0.35 \text{ mg/}L \text{ Lime}} = \frac{15 \text{ mg/}L \text{ Alk.}}{x \text{ mg/}L \text{ Lime}}$$

Using cross-multiplication:

$$0.45 \, x = (15)(0.35)$$

$$x = \frac{(15)(0.35)}{0.45}$$

$$x = \boxed{11.7 \text{ mg/}L \text{ Lime}}$$

Once the amount of alkalinity to be added to the water has been calculated, it is necessary to determine how much lime (the source of the alkalinity) should therefore be added. The ratio given in **Step 3** allows you to make the conversion from mg/L alkalinity to be added to mg/L lime to be added. Examples 5 and 6 illustrate this calculation.

Example 6: (Lime Dose Required, mg/L)
❑ A total of 26 mg/L alkalinity (HCO_3^-) must be added to a raw water. How many mg/L lime will be required to provide this amount of alkalinity? (1 mg/L alum reacts with 0.45 mg/L and 1 mg/L alum reacts with 0.35 mg/L lime.)

Use a proportion that relates alkalinity (HCO_3^-) and lime:

$$\frac{0.45 \text{ mg/}L \text{ Alk.}}{0.35 \text{ mg/}L \text{ Lime}} = \frac{26 \text{ mg/}L \text{ Alk.}}{x \text{ mg/}L \text{ Lime}}$$

Then, using cross-multiplication:

$$0.45 \, x = (26)(0.35)$$

$$x = \frac{(26)(0.35)}{0.45}$$

$$x = \boxed{20.2 \text{ mg/}L \text{ Lime}}$$

---

* Refer to Chapter 7 in *Basic Math Concepts* for a review of ratios and proportions.

In Examples 7 and 8, all three steps in determining lime dose (mg/L) are required.

Example 7: (Lime Dose Required, mg/L)
❑ Given the data below, calculate the lime dose required, in mg/L.

- Alum dose required, as determined by jar tests—54 mg/L
- • 1 mg/L alum reacts with 0.45 mg/L alkalinity
- "Residual" alkalinity req'd for precipitation—30 mg/L
- • Raw water alkalinity—38 mg/L
- 1 mg/L alum reacts with 0.35 mg/L lime

To calculate the total alkalinity required, you must first calculate the alkalinity that will react with 54 mg/L alum:

$$\frac{0.45 \text{ mg/}L \text{ Alk.}}{1 \text{ mg/}L \text{ Alum}} = \frac{x \text{ mg/}L \text{ Alk.}}{54 \text{ mg/}L \text{ Alum}}$$

$$(0.45)\,(54) = x$$

$$24.3 \text{ mg/}L = x$$
$$\text{Alk.}$$

The total alkalinity requirement can now be determined:

$$\begin{array}{c}\text{Total Alk.}\\\text{Req'd, mg/}L\end{array} = \begin{array}{c}\text{Alk. to React}\\\text{with Alum, mg/}L\end{array} + \begin{array}{c}\text{"Residual"}\\\text{Alk., mg/}L\end{array}$$

$$= 24.3 \text{ mg/}L + 30 \text{ mg/}L$$

$$= \boxed{\begin{array}{c}54.3 \text{ mg/}L \text{ Total}\\\text{Alk. Required}\end{array}}$$

Now calculate how much alkalinity must be <u>added</u> to the water:

$$\begin{array}{c}\text{Alk. to be Added}\\\text{to the Water, mg/}L\end{array} = \begin{array}{c}\text{Total Alk.}\\\text{Req'd, mg/}L\end{array} - \begin{array}{c}\text{Alk. Present}\\\text{in the Water, mg/}L\end{array}$$

$$= 54.3 \text{ mg/}L - 38 \text{ mg/}L$$

$$= \boxed{\begin{array}{c}16.3 \text{ mg/}L \text{ Alk.}\\\text{to be Added to the Water}\end{array}}$$

And finally, calculate the lime required to provide this additional alkalinity:

$$\frac{0.45 \text{ mg/}L \text{ Alk.}}{0.35 \text{ mg/}L \text{ Lime}} = \frac{16.3 \text{ mg/}L \text{ Alk.}}{x \text{ mg/}L \text{ Lime}}$$

$$0.45\,x = (16.3)\,(0.35)$$

$$x = \frac{(16.3)\,(0.35)}{0.45}$$

$$x = \boxed{12.7 \text{ mg/}L \text{ Lime}}$$

Example 8: (Lime Dose Required, mg/L)
❑ Given the data below, calculate the lime dose required, in mg/L.

- Alum dose required per jar tests—48 mg/L

- 1 mg/L alum reacts with 0.45 mg/L alkalinity

- Raw water alkalinity—31 mg/L

- 1 mg/L alkalinity (HCO_3^-) reacts with 0.35 mg/L lime

- "Residual" alkalinity req'd for precipitation—30 mg/L

To determine total alkalinity required, you must first calculate the alkalinity that will react with 48 mg/L alum:

$$\frac{0.45 \text{ mg/}L \text{ Alk.}}{1 \text{ mg/}L \text{ Alum}} = \frac{x \text{ mg/}L \text{ Alk.}}{48 \text{ mg/}L \text{ Alum}}$$

$$(0.45)(48) = x$$

$$21.6 \text{ mg/}L = x$$
$$\text{Alk.}$$

Now calculate the total alkalinity required:

$$\begin{array}{c} \text{Total Alk.} \\ \text{Req'd, mg/}L \end{array} = \begin{array}{c} \text{Alk. to React} \\ \text{with Alum, mg/}L \end{array} + \begin{array}{c} \text{"Residual"} \\ \text{Alk., mg/}L \end{array}$$

$$= 21.6 \text{ mg/}L + 30 \text{ mg/}L$$

$$= \boxed{\begin{array}{c} 51.6 \text{ mg/}L \text{ Total} \\ \text{Alk. Required} \end{array}}$$

Next calculate how much of this alkalinity must actually be <u>added</u> to the water:

$$\begin{array}{c} \text{Alk. to be Added} \\ \text{to the Water, mg/}L \end{array} = \begin{array}{c} \text{Total Alk.} \\ \text{Req'd, mg/}L \end{array} - \begin{array}{c} \text{Alk. Present} \\ \text{in the Water, mg/}L \end{array}$$

$$= 51.6 \text{ mg/}L - 31 \text{ mg/}L$$

$$= \boxed{\begin{array}{c} 20.6 \text{ mg/}L \text{ Alk.} \\ \text{to be Added to the Water} \end{array}}$$

And then calculate how much lime is required to provide this additional alkalinity:

$$\frac{0.45 \text{ mg/}L \text{ Alk.}}{0.35 \text{ mg/}L \text{ Lime}} = \frac{20.6 \text{ mg/}L \text{ Alk.}}{x \text{ mg/}L \text{ Lime}}$$

$$0.45 x = (20.6)(0.35)$$

$$x = \frac{(20.6)(0.35)}{0.45}$$

$$x = \boxed{16.0 \text{ mg/}L \text{ Lime}}$$

10.8 LIME DOSE REQUIRED, lbs/day

Once the lime dose has been determined in terms of mg/L, it is a fairly simple matter to calculate the lime dose in lbs/day, since it is one of the most common calculations in water and wastewater treatment. To convert from mg/L to lbs/day* lime dose, use the following equation:

$$\frac{(mg/L)}{Lime} \frac{(MGD)}{flow} \frac{(8.34)}{lbs/gal} = \frac{lbs/day}{Lime}$$

Examples 1-4 illustrate this calculation.

Example 1: (Lime Dose Required, lbs/day)
❏ The lime dose for a raw water has been calculated to be 14.1 mg/L. If the flow to be treated is 2.2 MGD, how many lbs/day lime will be required?

$$\frac{(mg/L)}{Lime} \frac{(MGD)}{flow} \frac{(8.34)}{lbs/gal} = \frac{lbs/day}{Lime}$$

$$(14.1 \text{ mg/}L) \text{ (2.2 MGD) (8.34 lbs/gal)} = \boxed{259 \text{ lbs/day} \atop Lime}$$

Example 2: (Lime Dose Required, lbs/day)
❏ The lime dose for a solids contact unit has been calculated to be 13.8 mg/L. If the flow to be treated is 890,000 gpd, how many lbs/day lime will be required?

$$\frac{(mg/L)}{Lime} \frac{(MGD)}{flow} \frac{(8.34)}{lbs/gal} = \frac{lbs/day}{Lime}$$

$$(13.8 \text{ mg/}L) \text{ (0.89 MGD) (8.34 lbs/gal)} = \boxed{102 \text{ lbs/day} \atop Lime}$$

---

* An entire chapter has been devoted to the mg/L to lbs/day calculation. Refer to Chapter 3 for a review of this calculation.

Example 3: (Lime Dose Required, lbs/day)
❑ The flow to a solids contact clarifier is 2,400,000 gpd. If the lime dose required is determined to be 11.9 mg/L, how many lbs/day lime will be required?

$$\underset{\text{Lime}}{(\text{mg}/L)} \; \underset{\text{flow}}{(\text{MGD})} \; \underset{\text{lbs/gal}}{(8.34)} = \underset{\text{Lime}}{\text{lbs/day}}$$

$$(11.9 \text{ mg}/L)\,(2.4 \text{ MGD})\,(8.34 \text{ lbs/gal}) = \boxed{\begin{array}{c} 238 \text{ lbs/day} \\ \text{Lime} \end{array}}$$

Example 4: (Lime Dose Required, lbs/day)
❑ A solids contact clarification unit receives a flow of 1.7 MGD. Alum is to be used for coagulation purposes. If a lime dose of 16 mg/L will also be required, how many lbs/day lime is this?

$$\underset{\text{Lime}}{(\text{mg}/L)} \; \underset{\text{flow}}{(\text{MGD})} \; \underset{\text{lbs/gal}}{(8.34)} = \underset{\text{Lime}}{\text{lbs/day}}$$

$$(16 \text{ mg}/L)\,(1.7 \text{ MGD})\,(8.34 \text{ lbs/gal}) = \boxed{\begin{array}{c} 227 \text{ lbs/day} \\ \text{Lime} \end{array}}$$

10.9 LIME DOSE REQUIRED, g/min

There are at least two methods of converting from mg/L lime to grams/min (g/min) lime. Perhaps the easiest method is to rely on a conversion path that is partially familiar:

mg/L ⟶ lbs/day ⟶ g/day ⟶ g/min

The conversion of lime dose **from mg/L to lbs/day** is described in the previous section. This is the familiar part of the conversion:

$$\frac{(mg/L)\ (MGD)\ (8.34)}{Lime\quad flow\quad lbs/gal} = \frac{lbs/day}{Lime}$$

In order to convert from **lbs/day to g/day** you will need to know the relationship between pounds and grams: (1 lb = 453.6 g)

$$\frac{\cancel{lbs}}{day} \times \frac{453.6\ g}{\cancel{lb}} = \frac{g}{day}$$

Then to convert from g/day to g/min, simply use the 1440 min/day factor as follows.

$$\frac{g/day}{1440\ min/day} = \frac{g}{min}$$

These steps can be consolidated into one equation, as shown at the top of the page.

CONVERTING LIME DOSE
(from lbs/day to g/min*)

The basic conversion:

$$\frac{(\cancel{lbs})}{\cancel{day}}\ \frac{(453.6\ g)}{\cancel{lb}}\ \frac{(1\ \cancel{day})}{1440\ min} = \frac{g}{min}$$

This equation can be rewritten as:

$$\frac{(lbs/day)\ (453.6\ g/lb)}{1440\ min/day} = g/min$$

Example 1: (Lime Dose Required, g/min)

❏ A total of 260 lbs/day lime will be required to raise the alkalinity of the water passing through a solids-contact clarification process. How many g/min lime does this represent? (1 lb = 453.6 g)

The conversion from lbs/day to g/min is a two step process: (1) converting pounds to grams, and (2) converting days to minutes. The equation given below includes both conversions:

$$\frac{(lbs/day)\ (453.6\ g/lb)}{1440\ min/day} = g/min$$

$$\frac{(260\ lbs/day)\ (453.6\ g/lb)}{1440\ min/day} = \boxed{81.9\ g/min \\ Lime}$$

---

*The units in this problem have been canceled to verify resulting units. Refer to Chapter 15 in *Basic Math Concepts* for a review of dimensional analysis.

Example 2: (Lime Dose Required, g/min)
❏ A lime dose of 140 lbs/day is required for a solids-contact clarification process. How many g/min lime does this represent? (1 lb = 453.6 g)

To convert from lbs/day to g/min, use the following equation:

$$\frac{(lbs/day)\ (453.6\ g/lb)}{1440\ min/day} = g/min$$

$$\frac{(140\ lbs/day)\ (453.6\ g/lb)}{1440\ min/day} = \boxed{\begin{array}{c} 44.1\ g/min \\ Lime \end{array}}$$

Example 3: (Lime Dose Required, g/min)
❏ A lime dose of 14 mg/L is required to add the required alkalinity to a raw water. If the flow to be treated is 2,650,000 gpd, what is the g/min lime dose required? (1 lb = 453.6 g)

First calculate the lbs/day lime required:

$$\begin{array}{ccc} (mg/L) & (MGD) & (8.34) \\ Lime & flow & lbs/gal \end{array} = \begin{array}{c} lbs/day \\ Lime \end{array}$$

$$(14\ mg/L)(2.65\ MGD)(8.34\ lbs/gal) = \begin{array}{c} 309\ lbs/day \\ Lime \end{array}$$

Then convert lbs/day to g/min lime dose:

$$\frac{(309\ lbs/day)\ (453.6\ g/lb)}{1440\ min/day} = \boxed{\begin{array}{c} 97.3\ g/min \\ Lime \end{array}}$$

NOTES:

11 *Filtration*

1. Flow Rate Through A Filter

<u>Using Flow Meter</u>

$$\frac{\text{Flow Rate, gpd}}{1440 \text{ min/day}} = \text{Flow Rate, gpm}$$

<u>Using Total Gallons Produced</u>

$$\frac{\text{Flow Rate,}}{\text{gpm}} = \frac{\text{Total Gallons Produced}}{\text{Filter Run, min}}$$

<u>Using Water Drop Data</u>
(This is a Q = AV problem)

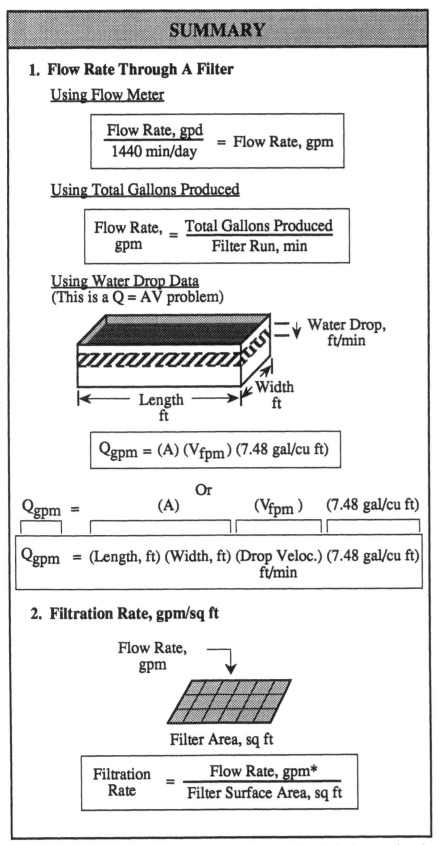

$$Q_{gpm} = (A) \ (V_{fpm}) \ (7.48 \text{ gal/cu ft})$$

Or

$$Q_{gpm} = \underbrace{(A)}_{} \quad \underbrace{(V_{fpm})}_{} \quad (7.48 \text{ gal/cu ft})$$

$$Q_{gpm} = \frac{(\text{Length, ft}) \ (\text{Width, ft}) \ (\text{Drop Veloc.}) \ (7.48 \text{ gal/cu ft})}{\text{ft/min}}$$

2. Filtration Rate, gpm/sq ft

Flow Rate, gpm

Filter Area, sq ft

$$\frac{\text{Filtration}}{\text{Rate}} = \frac{\text{Flow Rate, gpm*}}{\text{Filter Surface Area, sq ft}}$$

---

* Sometimes gpm flow rate is known and can be used directly in the equation shown. Other times, flow rate must first be calculated (using equations described in Section 11.1) before it can be used to calculate filtration rate.

SUMMARY—Cont'd

3. Unit Filter Run Volume (UFRV), gal/sq ft

$$\text{UFRV} = \frac{\text{Total Water Filtered, gal}}{\text{Filter Surface Area, sq ft}}$$

Filtration rate data can also be used to calculate UFRV:

$$\text{UFRV} = (\text{Filtration Rate, gpm/sq ft})(\text{Filter Run Time, min})$$

Or

$$\text{UFRV} = (\text{Filtration Rate, gpm/sq ft})(\text{Filter Run})(60 \text{ min/hr})$$
$$\text{Time, hrs}$$

4. Backwash Rate, gpm/sq ft

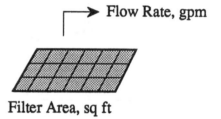

Flow Rate, gpm

Filter Area, sq ft

$$\text{Backwash Rate} = \frac{\text{Flow Rate, gpm}}{\text{Filter Surface Area, sq ft}}$$

Backwash rate is sometimes expressed as in./min. There are two ways to convert from gpm/sq ft to in./min backwash rate:*

$$\frac{(\text{Backwash Rate,})(12 \text{ in./ft})}{7.48 \text{ gal/cu ft}} = \text{Backwash Rate, in./min}$$
$$\text{gpm/sq ft}$$

Or

$$(\text{Backwash Rate,})(1.6) = \text{Backwash Rate, in./min}$$
$$\text{gpm/sq ft}$$

* Note that in the second equation 12 in./ft + 7.48 gal/cu ft has merely been simplified to 1.6.

SUMMARY—Cont'd

5. Volume of Backwash Water Required, gal

$$\begin{array}{c} \text{Backwash Water} \\ \text{Required, gal} \end{array} = \begin{array}{c} \text{(Backwash Flow,)} \\ \text{gpm} \end{array} \begin{array}{c} \text{(Duration of)} \\ \text{Backwash, min} \end{array}$$

6. Required Depth of Backwash Water Tank, ft*

<u>If Tank is Cylindrical</u>

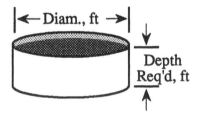

$$\begin{array}{c} \text{Tank Vol.} \\ \text{Req'd, gal} \end{array} = (0.785) (D^2) (\text{Depth, ft}) (7.48 \text{ gal/cu ft})$$

<u>If Tank is Rectangular</u>

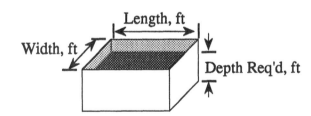

$$\begin{array}{c} \text{Tank Vol.} \\ \text{Req'd, gal} \end{array} = \begin{array}{c} \text{(Length,)} \\ \text{ft} \end{array} \begin{array}{c} \text{(Width,)} \\ \text{ft} \end{array} \begin{array}{c} \text{(Depth,)} \\ \text{ft} \end{array} (7.48 \text{ gal/cu ft})$$

7. Backwash Pumping Rate, gpm

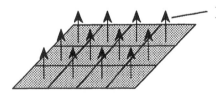

Backwash flow rate through <u>each</u> sq ft of filter area

$$\begin{array}{c} \text{Backwash Pumping} \\ \text{Rate, gpm} \end{array} = \begin{array}{c} \text{(Desired Backwash)} \\ \text{Rate, gpm/sq ft} \end{array} \begin{array}{c} \text{(Filter Area,)} \\ \text{sq ft} \end{array}$$

* This calculation is essentially a tank volume calculation. For a review of tank volume problems, refer to Chapter 1.

SUMMARY—Cont'd

8. Percent of Product Water Used for Backwashing

$$\begin{array}{c}\% \text{ Backwash} \\ \text{Water}\end{array} = \frac{\text{Backwash Water, gal}}{\text{Water Filtered, gal}} \times 100$$

9. Percent Mud Ball Volume

$$\begin{array}{c}\% \text{ Mud Ball} \\ \text{Volume}\end{array} = \frac{\text{Mud Ball Volume, m}L}{\text{Total Sample Volume, m}L} \times 100$$

NOTES:

11.1 FLOW RATE THROUGH A FILTER, gpm

USING FLOW METER

The flow rate (in gpm) through a filter can be determined by simply converting the gpd flow rate, as indicated on the flow meter,* to gpm flow rate:

$$\frac{\text{Flow Rate, gpd}}{1440 \text{ min/day}} = \text{Flow Rate, gpm}$$

Example 1 illustrates this calculation.**

USING TOTAL GALLONS PRODUCED

The gpm flow rate through a filter can also be calculated, given total gallons produced during the filter run and the duration of that filter run:

Simplified Equation:

$$\frac{\text{Flow Rate,}}{\text{gpm}} = \frac{\text{Total Gal. Produced}}{\text{Filter Run, min}}$$

Expanded Equation:

$$\frac{\text{Flow Rate,}}{\text{gpm}} = \frac{\text{Total Gal. Produced}}{(\text{Filter Run,})(60 \frac{\text{min}}{\text{hr}})}$$

Examples 2-4 illustrate the calculation of flow rate using filter run data.

Example 1: (Flow Rate Through A Filter)
❏ The flow rate through a filter is 3.15 MGD. What is this flow rate expressed as gpm?

$$\frac{\text{Flow Rate, gpd}}{1440 \text{ min/day}} = \text{Flow Rate, gpm}$$

$$\frac{3,150,000 \text{ gpd}}{1440 \text{ min/day}} = \boxed{2188 \text{ gpm}}$$

Example 2: (Flow Rate Through A Filter)
❏ During a 75-hour filter run, a total of 21.2 million gallons of water are filtered. What is the average flow rate through the filter in gpm during this filter run?

$$\frac{\text{Flow Rate,}}{\text{gpm}} = \frac{\text{Total Gallons Produced}}{\text{Filter Run, min}}$$

$$= \frac{21,200,000 \text{ gal}}{(75 \text{ hrs}) (60 \text{ min/hr})}$$

$$= \boxed{4711 \text{ gpm}}$$

---

* The flow rate on a meter is generally read as MGD. It can easily be converted to gpd by multiplying by 1,000,000.

** For a review of flow conversions, refer to Chapter 8 in *Basic Math Concepts*.

Example 3: (Flow Rate Through A Filter)
❑ A total of 18,140,000 gallons of water are filtered during a 72-hour filter run. What was the average gpm flow rate through the filter during this filter run?

$$\text{Flow Rate, gpm} = \frac{\text{Total Gallons Produced}}{\text{Filter Run, min}}$$

$$= \frac{18,140,000 \text{ gal}}{(72 \text{ hrs}) (60 \text{ min/hr})}$$

$$= \boxed{4199 \text{ gpm}}$$

CALCULATING OTHER VARIABLES

In Examples 2 and 3, the unknown variable was flow rate. The same equation may be used in calculating total gallons produced or filter run time. Example 4 illustrates one such calculation.

Example 4: (Flow Rate Through A Filter)
❑ At an average flow rate of 4200 gpm, how long a filter run (in hours) would be required to produce 20 MG of filtered water.

Write the equation as usual, filling in known data:

$$\text{Flow Rate, gpm} = \frac{\text{Total Gallons Produced}}{\text{Filter Run, min}}$$

$$4200 \text{ gpm} = \frac{20,000,000 \text{ gal}}{(x \text{ hrs}) (60 \text{ min/hr})}$$

Then solve for x*:

$$x = \frac{20,000,000 \text{ gal}}{(4200) (60)}$$

$$x = \boxed{79.4 \text{ hrs}}$$

---

* For a review of solving for x, refer to Chapter 2 in *Basic Math Concepts*.

USING WATER DROP DATA

The gpm flow rate through a filter can be determined a third way—by measuring the water drop in a filter when the influent valve to the filter has been closed.

Water drop indicates the **velocity of the flow** through the filter. Then by using the $Q = AV$ equation, the flow rate through the filter can be calculated, as shown in the diagram to the right.

Examples 5-7 illustrate the calculation of gpm flow rate using water drop information.

FLOW RATE USING WATER DROP IS A Q = AV CALCULATION

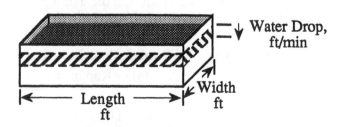

Simplified Equation:*

$$Q_{cfm} = (A)(V_{fpm})$$

Expanded Equation:

$$Q_{cfm} = \underset{ft}{(Length)} \; \underset{ft}{(Width)} \; \underset{ft/min}{(Drop\ Veloc.)}$$

Example 5: (Flow Rate Through A Filter)
❏ The influent valve to a filter is closed for 5 minutes. During this time the water level in the filter drops 10 inches (0.8 ft). If the filter is 40 ft long and 20 ft wide, what is the gpm flow rate through the filter?

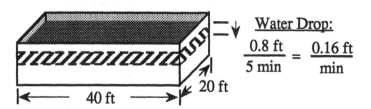

Water Drop:
$$\frac{0.8\ ft}{5\ min} = \frac{0.16\ ft}{min}$$

First calculate cfm flow rate using the $Q = AV$ equation:

$$Q_{cfm} = (Length,\ ft)\ (Width,\ ft)\ (Drop\ Veloc.,\ ft/min)$$

$$= (40\ ft)\ (20\ ft)\ (0.16\ ft/min)$$

$$= 128\ cfm$$

Then convert cfm flow rate to gpm flow rate:

$$(128\ cfm)\ (7.48\ gal/cu\ ft) = \boxed{957\ gpm}$$

---

* The flow rate on a meter is generally read as MGD. It can easily be converted to gpd by multiplying by 1,000,000.
** For a review of flow conversions, refer to Chapter 8 in *Basic Math Concepts*.

Example 6: (Flow Rate Through A Filter)
❏ A filter is 30 ft long and 20 ft wide. To verify the flow rate through the filter, the influent valve to the filter is closed for a 5-minute period. During that time the water level drops 14 inches. What is the gpm flow rate through the filter?

Water drop must be expressed in terms of ft
(14 in. ÷ 12 in./ft = 1.2 ft)

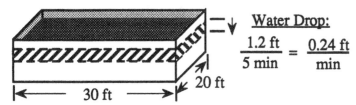

Water Drop:
$$\frac{1.2 \text{ ft}}{5 \text{ min}} = \frac{0.24 \text{ ft}}{\text{min}}$$

First calculate cfm flow rate using the Q = AV equation:

Q_{cfm} = (Length, ft) (Width, ft) (Drop Veloc., ft/min)

     = (30 ft) (20 ft) (0.24 ft/min)

     = 144 cfm

Then convert cfm flow rate to gpm flow rate:

(144 cfm) (7.48 gal/cu ft) = $\boxed{1077 \text{ gpm}}$

Example 7: (Flow Rate Through A Filter)
❏ The influent valve to a filter is closed for a period of 7 minutes. The water drop during that time was 18 inches. If the filter is 35 ft long and 25 ft wide, what is the gpm flow rate through the filter?

Water drop data must be expressed in terms of ft:
$$18 \text{ in.} \div 12 \text{ in./ft} = 1.5 \text{ ft}$$

Now use the expanded Q = AV equation to calculate gpm flow rate directly:

$$Q_{gpm} = \underset{\text{ft}}{(\text{Length})} \; \underset{\text{ft}}{(\text{Width})} \; \underset{\text{ft/min}}{(\text{Drop Veloc.})} \; \underset{\text{gal/cu ft}}{(7.48)}$$

$$= (35 \text{ ft}) (25 \text{ ft}) \left(\frac{1.5 \text{ ft}}{7 \text{ min}}\right) \left(\frac{7.48 \text{ gal}}{\text{cu ft}}\right)$$

$$= \frac{9817.5}{7}$$

$$= \boxed{1403 \text{ gpm}}$$

CALCULATING GPM FLOW RATE DIRECTLY

It has been recommended elsewhere in this text that flow rate be calculated in terms of cubic feet (cfm or cfs) when using the Q = AV equation, as illustrated in Examples 5 and 6. The cfm flow rate is then converted to gpm flow rate. Some people, however, prefer to calculate gpm flow rate <u>directly</u> using the Q = AV equation. In this case, a factor of 7.48 gal/cu ft must be added to the right side of the equation, as follows:

$$Q_{gpm} = (A) (V_{fpm}) (7.48 \text{ gal/cu ft})$$

Example 7 illustrates this calculation.

11.2 FILTRATION RATE

Filtration rate is one measure of filter production. Along with filter run time, it provides valuable information for the operation of filters. It is the gallons per minute of water filtered through each square foot of filter area.

$$\text{Filtration Rate} = \frac{\text{Flow Rate, gpm}}{\text{Filter Area, sq ft}}$$

Filtration rates generally range from 2-10 gpm/sq ft. Examples 1-8 illustrate the calculation of filtration rate.

Example 1: (Filtration Rate)

❑ A filter 20 ft by 25 ft receives a flow of 1810 gpm. What is the filtration rate in gpm/sq ft?

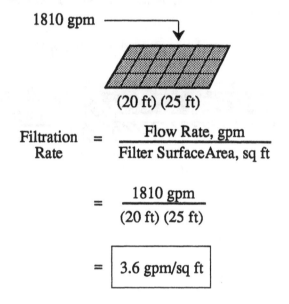

1810 gpm

(20 ft) (25 ft)

$$\text{Filtration Rate} = \frac{\text{Flow Rate, gpm}}{\text{Filter Surface Area, sq ft}}$$

$$= \frac{1810 \text{ gpm}}{(20 \text{ ft}) (25 \text{ ft})}$$

$$= \boxed{3.6 \text{ gpm/sq ft}}$$

Example 2: (Filtration Rate)

❑ A filter is 40 ft long and 20 ft wide treats a flow of 1900 gpm. What is the filtration rate in gpm/sq ft?

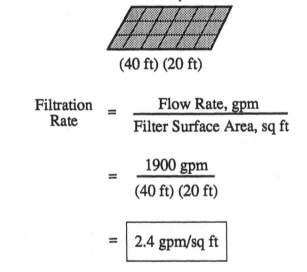

1900 gpm

(40 ft) (20 ft)

$$\text{Filtration Rate} = \frac{\text{Flow Rate, gpm}}{\text{Filter Surface Area, sq ft}}$$

$$= \frac{1900 \text{ gpm}}{(40 \text{ ft}) (20 \text{ ft})}$$

$$= \boxed{2.4 \text{ gpm/sq ft}}$$

Example 3: (Filtration Rate)
❑ A filter 30 ft long and 20 ft wide treats a flow of 3.8 MGD. What is the filtration rate in gpm/sq ft?

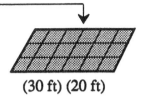

$$\frac{3,800,000 \text{ gpd}}{1440 \text{ min/day}}$$

$$= 2639 \text{ gpm}$$

(30 ft) (20 ft)

$$\text{Filtration Rate} = \frac{\text{Flow Rate, gpm}}{\text{Filter Surface Area, sq ft}}$$

$$= \frac{2639 \text{ gpm}}{(30 \text{ ft}) (20 \text{ ft})}$$

$$= \boxed{4.4 \text{ gpm/sq ft}}$$

Example 4: (Filtration Rate)
❑ A filter has a surface area of 35 ft by 25 ft. If the filter receives a flow of 3,140,000 gpd, what is the filtration rate in gpm/sq ft?

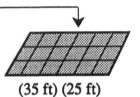

$$\frac{3,140,000 \text{ gpd}}{1440 \text{ min/day}}$$

$$= 2181 \text{ gpm}$$

(35 ft) (25 ft)

$$\text{Filtration Rate} = \frac{\text{Flow Rate, gpm}}{\text{Filter Surface Area, sq ft}}$$

$$= \frac{2181 \text{ gpm}}{(35 \text{ ft}) (25 \text{ ft})}$$

$$= \boxed{2.5 \text{ gpm/sq ft}}$$

In Examples 1-4, the flow rate was provided directly (MGD or gpd). In some calculations of filtration rate, however, **the flow rate data is provided more indirectly and must first be calculated before the filtration rate can be determined.*** Examples 5-8 illustrate this type of calculation.

Example 5: (Filtration Rate)

❑ A filter 45 ft long and 25 ft wide produces a total of 20 MG during a 74.5-hour filter run. What is the average filtration rate in gpm/sq ft for this filter run?

First calculate the gpm flow rate through the filter:*

$$\text{Flow Rate gpm} = \frac{\text{Total Gallons Produced}}{\text{Filter Run, min}}$$

$$= \frac{20,000,000 \text{ gal}}{(74.5 \text{ hrs}) (60 \text{ min/hr})}$$

$$= 4474 \text{ gpm}$$

Then calculate filtration rate:

$$\text{Filtration Rate} = \frac{\text{Flow Rate, gpm}}{\text{Filter Area, sq ft}}$$

$$= \frac{4474 \text{ gpm}}{(45 \text{ ft}) (25 \text{ ft})}$$

$$= \boxed{4.0 \text{ gpm/sq ft}}$$

Example 6: (Filtration Rate)

❑ A filter is 35 ft long and 20 ft wide produces a total of 17.1 MG during a 71-hour filter run. What is the average filtration rate for this filter run?

First calculate gpm flow rate through the filter:

$$\text{Flow Rate gpm} = \frac{\text{Total Gallons Produced}}{\text{Filter Run, min}}$$

$$= \frac{17,100,000 \text{ gal}}{(71 \text{ hrs}) (60 \text{ min/hr})}$$

$$= 4014 \text{ gpm}$$

Then determine filtration rate:

$$\text{Filtration Rate} = \frac{\text{Flow Rate, gpm}}{\text{Filter Area, sq ft}}$$

$$= \frac{4014 \text{ gpm}}{(35 \text{ ft}) (20 \text{ ft})}$$

$$= \boxed{5.7 \text{ gpm/sq ft}}$$

---

* For a review of calculating these flow rates, refer to Section 11.1 of this chapter.

Example 7: (Filtration Rate)
❑ A filter is 40 ft long and 20 ft wide. During a test of filter flow rate, the influent valve to the filter is closed for 5 minutes. The water level drop during this period is 20 inches. What is the filtration rate for the filter in gpm/sq ft?

First calculate gpm flow rate, using the Q = AV equation: (20 in. + 12 in./ft = 1.7 ft)

$$Q_{gpm} = \underset{ft}{(Length)}\ \underset{ft}{(Width)}\ \underset{ft/min}{(Drop\ Veloc.)}\ \underset{gal/cu\ ft}{(7.48)}$$

$$= (40\ ft)\ (20\ ft)\ \frac{(1.7\ ft)}{5\ min}\ \frac{(7.48\ gal)}{cu\ ft}$$

$$= 2035\ gpm$$

Then calculate filtration rate:

$$\begin{matrix} Filtration \\ Rate \end{matrix} = \frac{Flow\ Rate,\ gpm}{Filter\ Area,\ sq\ ft}$$

$$= \frac{2035\ gpm}{(40\ ft)\ (20\ ft)}$$

$$= \boxed{2.5\ gpm/sq\ ft}$$

Example 8: (Filtration Rate)
❑ A filter is 45 ft long and 25 ft wide. During a test of flow rate, the influent valve to the filter is closed for six minutes. The water level drop during this period is 17 inches. What is the filtration rate for the filter in gpm/sq ft?

First calculate gpm flow rate, using the Q = AV equation: (17 in. + 12 in./ft = 1.4 ft)

$$Q_{gpm} = \underset{ft}{(Length)}\ \underset{ft}{(Width)}\ \underset{ft/min}{(Drop\ Veloc.)}\ \underset{gal/cu\ ft}{(7.48)}$$

$$= (45\ ft)\ (25\ ft)\ \frac{(1.4\ ft)}{6\ min}\ \frac{(7.48\ gal)}{cu\ ft}$$

$$= 1964\ gpm$$

Then calculate filtration rate:

$$\begin{matrix} Filtration \\ Rate \end{matrix} = \frac{Flow\ Rate,\ gpm}{Filter\ Area,\ sq\ ft}$$

$$= \frac{1964\ gpm}{(45\ ft)\ (25\ ft)}$$

$$= \boxed{1.7\ gpm/sq\ ft}$$

11.3 UNIT FILTER RUN VOLUME (UFRV)

The unit filter run volume (UFRV) calculation indicates the total gallons passing through each square foot of filter surface area during an entire filter run. This calculation is used to compare and evaluate filter runs. The equation to be used in these calculations is shown to the right.

UFRV's are usually at least 5000 gal/sq ft and generally in the range of 10,000 gpd/sq ft. As the performance of the filter begins to deteriorate, the UFRV value will begin to decline as well.

UNIT FILTER RUN VOLUME

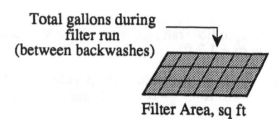

Total gallons during filter run (between backwashes)

Filter Area, sq ft

$$\text{UFRV} = \frac{\text{Total Gallons Filtered}}{\text{Filter Surface Area, sq ft}}$$

Example 1: (UFRV)

❏ The total water filtered during a filter run (between backwashes) is 2,810,000 gal. If the filter is 20 ft by 20 ft, what is the unit filter run volume (UFRV) in gal/sq ft?

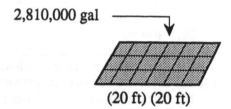

2,810,000 gal

(20 ft) (20 ft)

$$\text{UFRV} = \frac{\text{Total Gallons Filtered}}{\text{Filter Surface Area, sq ft}}$$

$$= \frac{2,810,000 \text{ gal}}{(20 \text{ ft}) (20 \text{ ft})}$$

$$= \boxed{7025 \text{ gal/sq ft}}$$

Example 2: (UFRV)

❑ The total water filtered during a filter run is 4,450,000 gallons. If the filter is 25 ft by 20 ft what is the UFRV in gal/sq ft?

4,450,000 gal

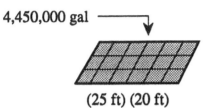

(25 ft) (20 ft)

$$\text{UFRV} \ = \ \frac{\text{Total Gallons Filtered}}{\text{Filter Surface Area, sq ft}}$$

$$= \ \frac{4{,}450{,}000 \text{ gal}}{(25 \text{ ft}) (20 \text{ ft})}$$

$$= \ \boxed{8900 \text{ gal/sq ft}}$$

Example 3: (UFRV)

❑ The total water filtered during a filter run is 4,960,000 gallons. If the filter is 30 ft by 20 ft, what is the unit filter run volume in gal/sq ft?

4,960,000 gal

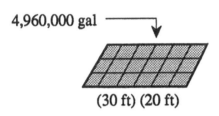

(30 ft) (20 ft)

$$\text{UFRV} \ = \ \frac{\text{Total Gallons Filtered}}{\text{Filter Surface Area, sq ft}}$$

$$= \ \frac{4{,}960{,}000 \text{ gal}}{(30 \text{ ft}) (20 \text{ ft})}$$

$$= \ \boxed{8267 \text{ gal/sq ft}}$$

CALCULATING UFRV USING FILTRATION RATE DATA

The unit filter run volume can also be calculated given filtration rate and filter run data. To do this use either of the following equations:

Simplified Equation:

$$UFRV = \text{(Filtration Rate, gpm/sq ft)} \text{(Filter Run, min)}$$

Expanded Equation:

$$UFRV = \text{(Filtration Rate, gpm/sq ft)} \text{(Filter Run, hrs)} \left(\frac{60 \text{ min}}{hr}\right)$$

minutes

Once you know the number of gallons passing through each square foot of filter area **each minute,** multiplying by the total number of minutes of filter run will give you the **total number of gallons** entering each square foot of filter area during that filter run.

An analysis of the units of the problem using dimensional analysis,* illustrates that the resulting units are gal/sq ft, as desired:

$$UFRV = \text{(gpm/sq ft)(min)}$$

$$= \frac{\frac{gal}{min}}{\frac{sq\ ft}{1}} \cdot \frac{min}{1}$$

$$= \frac{gal}{\cancel{min}} \cdot \frac{1}{sq\ ft} \cdot \frac{\cancel{min}}{1}$$

$$= \frac{gal}{sq\ ft}$$

Example 4: (UFRV)

❏ The average filtration rate for a filter was determined to be 2.2 gpm/sq ft. If the filter run time was 4320 minutes, what was the unit filter run volume in gal/sq ft?

$$UFRV = \text{(Filtration Rate, gpm/sq ft)} \text{(Filter Run Time, min)}$$

$$= \text{(2.2 gpm/sq ft) (4320 min)}$$

$$= \boxed{9504 \frac{gal}{sq\ ft}}$$

This problem indicates that at an average filtration rate of 2.2 gallons entering each square foot of filter <u>each minute,</u> the total gallons entering during the total filter run is 4320 times that amount.

Example 5: (UFRV)

❏ The average filtration rate for a filter was determined to be 3.6 gpm/sq ft. If the filter run time was 3180 minutes, what was the unit filter run volume in gal/sq ft?

$$UFRV = \text{(Filtration Rate, gpm/sq ft)} \text{(Filter Run Time, min)}$$

$$= \text{(3.6 gpm/sq ft) (3180 min)}$$

$$= \boxed{11,448 \text{ gal/sq ft}}$$

---

* For a review of dimensional analysis, refer to Chapter 15 in *Basic Math Concepts.*

Example 6: (UFRV)
❑ A filter operated 60.7 hours between backwashes. If the average filtration rate was determined to be 2.9 gpm/sq ft, what was the unit filter run volume in gal/sq ft?

$$UFRV = \text{(Filtration Rate,)} \text{(Filter Run,)} \left(60 \frac{min}{hr}\right)$$
$$ \text{gpm/sq ft} \qquad \text{hrs}$$

$$= (2.9 \text{ gpm/sq ft}) (60.7 \text{ hrs}) \left(60 \frac{min}{hr}\right)$$

$$= \boxed{10,562 \text{ gal/sq ft}}$$

Example 7: (UFRV)
❑ The average filtration rate during a particular filter run was determined to be 3.4 gpm/sq ft. If the filter run time was 58.8 hours, what was the UFRV in gal/sq ft for the filter run?

$$UFRV = \text{(Filtration Rate,)} \text{(Filter Run,)} \left(60 \frac{min}{hr}\right)$$
$$ \text{gpm/sq ft} \qquad \text{hrs}$$

$$= (3.4 \text{ gpm/sq ft}) (58.8 \text{ hrs}) \left(60 \frac{min}{hr}\right)$$

$$= \boxed{11,995 \text{ gal/sq ft}}$$

11.4 BACKWASH RATE

Filter backwash rate is a measure of the gpm flowing upward through each sq ft of filter surface area. The calculation of backwash rate is similar to filtration rate.

$$\text{Backwash Rate} = \frac{\text{Flow Rate, gpm}}{\text{Filter Area, sq ft}}$$

Backwash rates will range from 10-25 gpm/sq ft.

Example 1: (Backwash Rate)

❏ A filter with a surface area of 300 sq ft has a backwash flow rate of 2980 gpm. What is the filter backwash rate in gpm/sq ft?

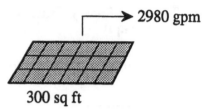

2980 gpm

300 sq ft

$$\text{Backwash Rate} = \frac{\text{Flow Rate, gpm}}{\text{Filter Area, sq ft}}$$

$$= \frac{2980 \text{ gpm}}{150 \text{ sq ft}}$$

$$= \boxed{19.9 \text{ gpm/sq ft}}$$

Example 2: (Backwash Rate)

❏ A filter 25 ft by 10 ft has a backwash rate of 3500 gpm. What is the backwash rate in gpm/sq ft?

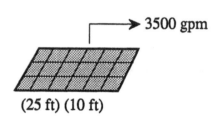

3500 gpm

(25 ft) (10 ft)

$$\text{Backwash Rate} = \frac{\text{Flow Rate, gpm}}{\text{Filter Area, sq ft}}$$

$$= \frac{3500 \text{ gpm}}{(25 \text{ ft}) (10 \text{ ft})}$$

$$= \boxed{14 \text{ gpm/sq ft}}$$

Example 3: (Backwash Rate)
❑ A filter 15 ft by 20 ft has a backwash flow rate of 3620 gpm. What is the filter backwash rate in gpm/sq ft?

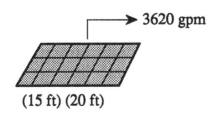

3620 gpm

(15 ft) (20 ft)

$$\text{Backwash Rate} = \frac{\text{Flow Rate, gpm}}{\text{Filter Area, sq ft}}$$

$$= \frac{3620 \text{ gpm}}{(15 \text{ ft}) (20 \text{ ft})}$$

$$= \boxed{12.1 \text{ gpm/sq ft}}$$

Example 4: (Backwash Rate)
❑ A filter 15 ft long and 15 ft wide has a backwash flow rate of 4.74 MGD. What is the filter backwash rate in gpm/sq ft?

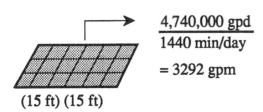

$$\frac{4{,}740{,}000 \text{ gpd}}{1440 \text{ min/day}}$$

$$= 3292 \text{ gpm}$$

(15 ft) (15 ft)

$$\text{Backwash Rate} = \frac{\text{Flow Rate, gpm}}{\text{Filter Area, sq ft}}$$

$$= \frac{3292 \text{ gpm}}{(15 \text{ ft}) (15 \text{ ft})}$$

$$= \boxed{14.6 \text{ gpm/sq ft}}$$

WHEN THE FLOW RATE IS EXPRESSED AS GPD

Normally the backwash flow rate is expressed as gpm. If it is expressed in any other flow rate terms, simply convert the given flow rate to gpm.* For example, if gpd flow rate is given, convert the gpd flow rate as follows:

$$\boxed{\frac{\text{Flow Rate, gpd}}{1440 \text{ min/day}} = \text{Flow Rate, gpm}}$$

* For a review of flow rate conversions, refer to Chapter 8 in *Basic Math Concepts*.

BACKWASH RATE EXPRESSED AS IN./MIN "RISE RATE"

Backwash rate is occasionally expressed as in./min rise. This is a measure of the upward velocity of the water during backwashing.

To convert from gpm/sq ft backwash rate to in./min rise rate, use either of the equations shown to the right.

Examining this conversion more closely, note the progression of units during the process of conversion:

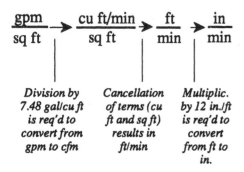

$$\frac{gpm}{sq\ ft} \rightarrow \frac{cu\ ft/min}{sq\ ft} \rightarrow \frac{ft}{min} \rightarrow \frac{in}{min}$$

Division by 7.48 gal/cu ft is req'd to convert from gpm to cfm

Cancellation of terms (cu ft and sq ft) results in ft/min

Multiplic. by 12 in./ft is req'd to convert from ft to in.

In making these conversions, remember that **in./min rise rate will always be a larger number than gpm/sq ft backwash rate.** Examples 5-7 illustrate the calculation of in./min rise rate.

BACKWASH RATE AS "RISE RATE"

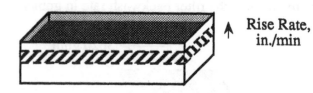

Rise Rate, in./min

To express backwash rate as in./min "rise rate," use one of these equations:

$$\frac{(Backwash\ Rate,)\ (12\ in./ft)}{gpm/sq\ ft} = \text{Backwash Rate,} \atop 7.48\ gal/cu\ ft} \quad \text{in./min}$$

Or

$$(Backwash\ Rate,)\ (1.6) = \text{Backwash Rate,} \atop gpm/sq\ ft \qquad \text{in./min}$$

Example 5: (Backwash Rate)
❏ A filter has a backwash rate of 17 gpm/sq ft. What is this backwash rate expressed as in./min rise rate?

$$\frac{(Backwash\ Rate,)\ (12\ in./ft)}{gpm/sq\ ft} \over 7.48\ gal/cu\ ft} = \text{Backwash Rate,} \atop \text{in./min}$$

$$\frac{(17\ gpm/sq\ ft)\ (12\ in./ft)}{7.48\ gal/cu\ ft} = \boxed{27.3\ in./min}$$

Example 6: (Backwash Rate)
❏ A filter 25 ft long and 10 ft wide has a backwash rate of 3450 gpm. What is this backwash rate expressed as in./min rise?

First calculate the backwash rate as gpm/sq ft:

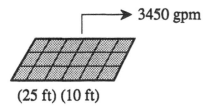

→ 3450 gpm

(25 ft) (10 ft)

$$\frac{\text{Backwash}}{\text{Rate}} = \frac{\text{Flow Rate, gpm}}{\text{Filter Area, sq ft}}$$

$$= \frac{3450 \text{ gpm}}{(25 \text{ ft}) (10 \text{ ft})}$$

$$= 13.8 \text{ gpm/sq ft}$$

Then convert gpm/sq ft to in./min rise rate:

$$\frac{(13.8 \text{ gpm/sq ft}) (12 \text{ in./ft})}{7.48 \text{ gal/cu ft}} = \boxed{22.1 \text{ in./min}}$$

USING THE "BOX METHOD" OF CONVERSION

The "box method" of conversions* can be used to convert from gpm/ft to in./min, using the following equation:

$$\boxed{1 \text{ gpm/sq ft} = 1.6 \text{ in./min}}$$

Written in "box method" terms, this would be:

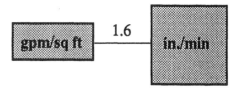

When moving from the smaller box (gpm/sq ft) to the larger box (in./min), **multiplication** is indicated. Moving from the larger to the smaller box, **division** is indicated.

Example 7: (Backwash Rate)
❏ Convert a backwash rate of 12 gpm/sq ft to in./min rise rate using the box method of conversion.
(1 gpm/sq ft = 1.6 in./min)

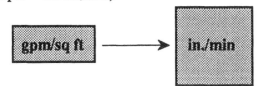

Converting from gpm/sq ft to in./min is a move from a smaller box to a larger box. Therefore multiplication by 1.6 is required:

$$(12 \text{ gpm/sq ft}) (1.6) = \boxed{19.2 \text{ in./min}}$$

---

* For a review of the "box method" of conversion, refer to Chapter 8 in *Basic Math Concepts*.

11.5 VOLUME OF BACKWASH WATER REQUIRED, gal

To determine the volume of water required for backwashing, you must know both the desired backwash flow rate (gpm) and the duration of backwash (min):

$$\begin{array}{c}\text{Backwash} \\ \text{Water Vol.,} \\ \text{gal}\end{array} = \begin{array}{c}\text{(Backwash)} \\ \text{Flow Rate,} \\ \text{gpm}\end{array} \begin{array}{c}\text{(Duration of)} \\ \text{Backwash,} \\ \text{min}\end{array}$$

For example, a backwash flow rate of 5000 gpm for a duration of 5 minutes would require a total volume of backwash water of 25,000 gallons (i.e. 5000 gpm x 5 min).

Example 1: (Backwash Water Req'd, gal)
❑ For a backwash flow rate of 8,000 gpm, and a total backwash time of 7 minutes, how many gallons of water will be required for backwashing?

$$\begin{array}{c}\text{Backwash} \\ \text{Water Vol.,} \\ \text{gal}\end{array} = \begin{array}{c}\text{(Backwash)} \\ \text{Flow Rate,} \\ \text{gpm}\end{array} \begin{array}{c}\text{(Duration of)} \\ \text{Backwash,} \\ \text{min}\end{array}$$

$$= (8,000 \text{ gpm}) (7 \text{ min})$$

$$= \boxed{56,000 \text{ gal}}$$

Example 2: (Backwash Water Req'd, gal)
❑ A backwash flow rate of 9,400 gpm for a total backwashing period of 6 minutes would require how many gallons of water for backwashing?

$$\begin{array}{c}\text{Backwash} \\ \text{Water Vol.,} \\ \text{gal}\end{array} = \begin{array}{c}\text{(Backwash)} \\ \text{Flow Rate,} \\ \text{gpm}\end{array} \begin{array}{c}\text{(Duration of)} \\ \text{Backwash,} \\ \text{min}\end{array}$$

$$= (9,400 \text{ gpm}) (6 \text{ min})$$

$$= \boxed{56,400 \text{ gal}}$$

Example 3: (Backwash Water Req'd, gal)
❏ A backwash flow rate of 5,120 gpm for a total of 8 minutes would require how many gallons of water?

$$\begin{array}{ll} \text{Backwash} \\ \text{Water Vol.,} \\ \text{gal} \end{array} = \begin{array}{ll} \text{(Backwash)} & \text{(Duration of)} \\ \text{Flow Rate,} & \text{Backwash,} \\ \text{gpm} & \text{min} \end{array}$$

$$= (5{,}120 \text{ gpm}) (8 \text{ min})$$

$$= \boxed{40{,}960 \text{ gal}}$$

Example 4: (Backwash Water Req'd, gal)
❏ How many gallons of water would be required to provide a backwash flow rate of 4,950 gpm for a total of 6 minutes?

$$\begin{array}{ll} \text{Backwash} \\ \text{Water Vol.,} \\ \text{gal} \end{array} = \begin{array}{ll} \text{(Backwash)} & \text{(Duration of)} \\ \text{Flow Rate,} & \text{Backwash,} \\ \text{gpm} & \text{min} \end{array}$$

$$= (4{,}950 \text{ gpm}) (6 \text{ min})$$

$$= \boxed{29{,}700 \text{ gal}}$$

11.6 REQUIRED DEPTH OF BACKWASH WATER TANK, ft

Once the volume of water required for backwashing has been calculated (as described in the previous section), the required depth of water in the backwash water tank can be determined.

To make this calculation, simply use the volume equation shown to the right, fill in the known data, then solve for the depth.**

Examples 1-3 illustrate this calculation.

CALCULATING THE REQUIRED DEPTH OF THE BACKWASH WATER TANK IS A <u>VOLUME</u> CALCULATION

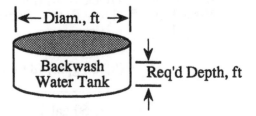

Use the equation for the volume of a cylinder:*

$$\text{Vol., gal} = (0.785)(D^2)(\text{Depth, ft})(7.48 \text{ gal/cu ft})$$

Example 1: (Depth of Backwash Water Tank)

❑ The volume of water required for backwashing has been calculated to be 75,000 gallons. What is the required depth of water in the backwash water tank to provide this amount of water if the diameter of the tank is 50 ft?

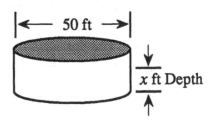

Use the volume equation for a cylindrical tank, filling in known data; then solve for x:

$$\text{Vol., gal} = (0.785)(D^2)(\text{Depth, ft})(7.48 \text{ gal/cu ft})$$

$$75,000 \text{ gal} = (0.785)(50 \text{ ft})(50 \text{ ft})(x \text{ ft})(7.48 \text{ gal/cu ft})$$

$$\frac{75,000}{(0.785)(50)(50)(7.48)} = x$$

$$\boxed{5.1 \text{ ft}} = x$$

---

* For a review of volume calculations, refer to Chapter 1 in this text or Chapter 11 in *Basic Math Concepts*.
** For a review of solving for the unknown value, refer to Chapter 2 in *Basic Math Concepts*.

Example 2: (Depth of Backwash Water Tank)
❏ A total of 56,000 gallons of water will be required for backwashing a filter at a rate of 7000 gpm for an 8-minute period. What depth of water is required in the backwash water tank to provide this backwashing capability? The tank has a diameter of 40 ft.

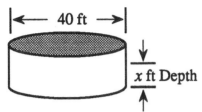

Use the volume equation for cylindrical tanks:

$$\text{Vol., gal} = (0.785)(D^2)(\text{Depth, ft})(7.48 \text{ gal/cu ft})$$

$$56,000 \text{ gal} = (0.785)(40 \text{ ft})(40 \text{ ft})(x \text{ ft})(7.48 \text{ gal/cu ft})$$

$$\frac{56,000}{(0.785)(40)(40)(7.48)} = x$$

$$\boxed{6.0 \text{ ft}} = x$$

Example 3: (Depth of Backwash Water Tank)
❏ A backwash rate of 6130 gpm is desired for a total backwash time of 7 minutes. What depth of water is required in the backwash water tank to provide this much water? The diameter of the tank is 45 ft.

First determine the volume of backwash water required:*

$$\begin{array}{ll} \text{Backwash} \\ \text{Water Vol.,} = & \begin{array}{cc} \text{(Backwash)} & \text{(Duration of)} \\ \text{Flow Rate,} & \text{Backwash,} \\ \text{gpm} & \text{min} \end{array} \\ \text{gal} \end{array}$$

$$= (6130 \text{ gpm})(7 \text{ min})$$

$$= 42,910 \text{ gal}$$

Then calculate the depth of water required in the backwash water tank:

$$\text{Vol., gal} = (0.785)(D^2)(\text{Depth, ft})(7.48 \text{ gal/cu ft})$$

$$42,910 \text{ gal} = (0.785)(45 \text{ ft})(45 \text{ ft})(x \text{ ft})(7.48 \text{ gal/cu ft})$$

$$\frac{42,910}{(0.785)(45)(45)(7.48)} = x$$

$$\boxed{3.6 \text{ ft}} = x$$

* For a review of this type calculation, refer to Section 11.5.

11.7 BACKWASH PUMPING RATE, gpm

The desired backwash pumping rate (gpm) for a filter depends on two factors:

- The desired backwash rate in gpm/sq ft, and

- The sq ft area of the filter.

As illustrated in the diagram, once you know the desired gpm backwash flow through <u>one sq ft of filter area</u>, the total gpm backwash flow through the filter can be determined by multiplying the gpm/sq ft flow by the total sq ft of filter area. Examining the units of this problem confirms the equation given:*

$$gpm = \frac{(gpm)}{\cancel{sq\,ft}} \frac{(\cancel{sq\,ft})}{1}$$

Examples 1-3 illustrate this calculation.

BACKWASH PUMPING RATE, GPM

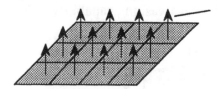

Backwash flow rate through <u>each</u> sq ft of filter area

To calculate the total backwash flow rate through the filter (the backwash pumping rate), **multiply** the backwash rate for one sq ft by the entire filter area:

$$\begin{matrix} \text{Backwash Pumping} \\ \text{Rate, gpm} \end{matrix} = \begin{matrix} \text{(Desired Backwash)} \\ \text{Rate, gpm/sq ft} \end{matrix} \begin{matrix} \text{(Filter Area,)} \\ \text{sq ft} \end{matrix}$$

Example 1: (Backwash Pumping Rate)

❑ A filter is 30 ft long and 25 ft wide. If the desired backwash rate is 22 gpm/sq ft, what backwash pumping rate (gpm) will be required?

The desired backwash flow through each square foot of filter area is 22 gpm. The total gpm flow through the filter is therefore 22 gpm times the entire square foot area of the filter:

$$\begin{matrix} \text{Backwash Pumping} \\ \text{Rate, gpm} \end{matrix} = \begin{matrix} \text{(Desired Backwash)} \\ \text{Rate, gpm/sq ft} \end{matrix} \begin{matrix} \text{(Filter Area,)} \\ \text{sq ft} \end{matrix}$$

$$= (22 \text{ gpm/sq ft}) (30 \text{ ft}) (25 \text{ ft})$$

$$= \boxed{16,500 \text{ gpm}}$$

---

*For a review of dimensional analysis, refer to Chapter 15 in *Basic Math Concepts*.

Example 2: (Backwash Pumping Rate)
❑ A filter is 40 ft long and 20 ft wide. If the desired backwash rate is 20 gpm/sq ft, what backwash pumping rate (gpm) will be required?

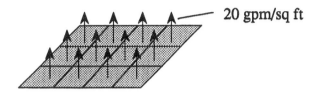

20 gpm/sq ft

$$\begin{array}{ll} \text{Backwash Pumping} \\ \text{Rate, gpm} \end{array} = \begin{array}{l} \text{(Desired Backwash)} \\ \text{Rate, gpm/sq ft} \end{array} \begin{array}{l} \text{(Filter Area,)} \\ \text{sq ft} \end{array}$$

$$= (20 \text{ gpm/sq ft}) (40 \text{ ft}) (20 \text{ ft})$$

$$= \boxed{16,000 \text{ gpm}}$$

Example 3: (Backwash Pumping Rate)
❑ The desired backwash pumping rate for a filter is 15 gpm/sq ft. If the filter is 30 ft long and 20 ft wide, what backwash pumping rate (gpm) will be required?

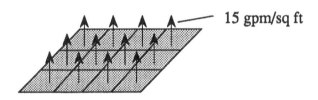

15 gpm/sq ft

$$\begin{array}{ll} \text{Backwash Pumping} \\ \text{Rate, gpm} \end{array} = \begin{array}{l} \text{(Desired Backwash)} \\ \text{Rate, gpm/sq ft} \end{array} \begin{array}{l} \text{(Filter Area,)} \\ \text{sq ft} \end{array}$$

$$= (15 \text{ gpm/sq ft}) (30 \text{ ft}) (20 \text{ ft})$$

$$= \boxed{9,000 \text{ gpm}}$$

11.8 PERCENT OF PRODUCT WATER USED FOR BACKWASHING

Filter performance is measured in several ways, including filtration rate and filter run time. Another aspect of filter operation that is monitored is the percent of product water used for backwashing.

The typical equation for percent calculations to be used is:*

$$\% = \frac{Part}{Whole} \times 100$$

To calculate the percent product water used for backwashing, the equation is:

$$\frac{Backwash}{Water, \%} = \frac{Bkwash\ Water,\ gal}{Water\ Filtered,\ gal} \times 100$$

Examples 1-4 illustrate this calculation.

Example 1: (% of Product Water for Backwashing)
❏ A total of 17,800,000 gallons of water were filtered during a filter run. If 72,000 gallons of this product water were used for backwashing, what percent of the product water was used for backwashing?

$$\frac{Backwash}{Water, \%} = \frac{Backwash\ Water,\ gal}{Water\ Filtered,\ gal} \times 100$$

$$= \frac{72,000\ gal}{17,800,000\ gal} \times 100$$

$$= \boxed{0.4\%\ Backwash\ Water}$$

Example 2: (% of Product Water for Backwashing)
❏ A total of 9,943,000 gallons of water are filtered between backwashes. If 56,700 gallons of this product water are used for backwashing, what percent of the product water is used for backwashing?

$$\frac{Backwash}{Water, \%} = \frac{Backwash\ Water,\ gal}{Water\ Filtered,\ gal} \times 100$$

$$= \frac{56,700\ gal}{9,943,000\ gal} \times 100$$

$$= \boxed{0.57\%\ Backwash\ Water}$$

---

* For a review of percent calculations, refer to Chapter 5 in *Basic Math Concepts.*

Example 3: (% of Product Water for Backwashing)
❑ 59,100 gallons of product water are used for filter backwashing at the end of a filter run. If a total of 12,573,000 gallons are filtered during the filter run, what percent of the product water is used for backwashing?

$$\text{Backwash Water, \%} = \frac{\text{Backwash Water, gal}}{\text{Water Filtered, gal}} \times 100$$

$$= \frac{59,100 \text{ gal}}{12,573,000 \text{ gal}} \times 100$$

$$= \boxed{0.47\% \text{ Backwash Water}}$$

Example 4: (% of Product Water for Backwashing)
❑ A total of 11,480,000 gallons of water are filtered during a filter run. If 49,500 gallons of product water are used for backwashing, what percent of the product water is used for backwashing?

$$\text{Backwash Water, \%} = \frac{\text{Backwash Water, gal}}{\text{Water Filtered, gal}} \times 100$$

$$= \frac{49,500 \text{ gal}}{11,480,000 \text{ gal}} \times 100$$

$$= \boxed{0.43\% \text{ Backwash Water}}$$

11.9 PERCENT MUD BALL VOLUME

Mud balls are floc particles and sand that form together in "balls" as a result of insufficient or improper backwashing of a filter. To detect the presence of mud balls, the filter media is checked periodically using a mud ball sampler, such as that shown in the diagram to the right.

After the filter is backwashed, the filter is drained until the water level is at least 1 ft below the surface of the media. Five samples, taken at various locations on the filter, are placed in a bucket so that the media and mud balls can be separated using a sieve.

Once the mud balls have been separated from the rest of the filter media, the volume of mud balls is determined. Theoretically, the volume of mud balls could be determined <u>directly</u> by measuring the diameter of the mud ball. This calculated volume would only be approximate since the mud ball is not a perfect sphere. Another very simple method is used to determine mud ball volume. This method uses the concept of **positive displacement**. The mud balls are placed in a graduated cylinder with a known volume of water (such as 500 mL or 1000 mL). The water level in the graduated cylinder rises as a result of placing the mud balls in the cylinder.

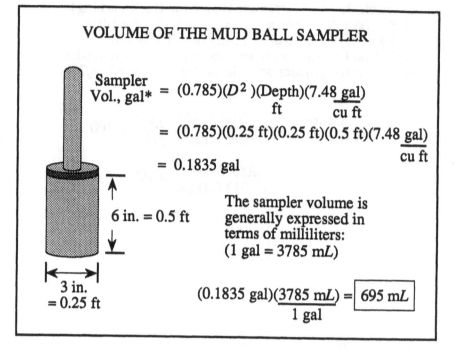

VOLUME OF THE MUD BALL SAMPLER

$$\text{Sampler Vol., gal*} = (0.785)(D^2)(\text{Depth})\frac{(7.48 \text{ gal})}{\text{cu ft}}$$

$$= (0.785)(0.25 \text{ ft})(0.25 \text{ ft})(0.5 \text{ ft})\frac{(7.48 \text{ gal})}{\text{cu ft}}$$

$$= 0.1835 \text{ gal}$$

6 in. = 0.5 ft

3 in. = 0.25 ft

The sampler volume is generally expressed in terms of milliliters:
(1 gal = 3785 mL)

$$(0.1835 \text{ gal})\frac{(3785 \text{ mL})}{1 \text{ gal}} = \boxed{695 \text{ mL}}$$

Example 1: (% Mud Ball Volume)
❑ A 3475-mL sample of filter media was taken for mud ball evaluation. The volume of water in the graduated cylinder rose from 500 mL to 528 mL when mud balls were placed in the cylinder. What is the percent mud ball volume of the sample?

First, determine the volume of mud balls in the sample:

$$528 \text{ mL} - 500 \text{ mL} = 28 \text{ mL}$$

Then calculate the percent mud ball volume:

$$\frac{\text{\% Mud Ball}}{\text{Volume}} = \frac{\text{Mud Ball Vol., mL}}{\text{Total Sample Vol., mL}} \times 100$$

$$= \frac{28 \text{ mL}}{3475 \text{ mL}} \times 100$$

$$= \boxed{0.8\%}$$

---

* For a review of volume calculations, refer to Chapter 2 in this text or Chapter 11 in *Basic Math Concepts*.

Example 2: (% Mud Ball Volume)
❑ A filter is tested for the presence of mud balls. The mud ball sampler has a total sample volume of 695 mL. Five samples were taken from the filter. When the mud balls were placed in 500-mL of water, the water level rose to 563 mL. What is the percent mud ball volume of the sample?

$$\text{\% Mud Ball Volume} = \frac{\text{Mud Ball Vol., m}L}{\text{Total Sample Vol., m}L} \times 100$$

The mud ball volume is the volume the water rose:
563 mL − 500 mL = 63 mL.

Since 5 samples of media were taken, the total sample volume is 5 times the sampler volume:
(5)(695 mL) = 3475 mL

$$\text{\% Mud Ball Volume} = \frac{63\ \text{m}L}{3475\ \text{m}L} \times 100$$

$$= \boxed{1.8\%}$$

In fact, **the rise in water level corresponds precisely to the volume of mud balls.** Therefore, the volume of mud balls can be determined by calculating the **change in volume** in the cylinder:

$$\text{Mud Ball Vol., m}L = \begin{matrix}\text{Vol. of}\\ \text{Water With}\\ \text{Mud Balls}\end{matrix} - \begin{matrix}\text{Vol. of}\\ \text{Water}\\ \text{Without}\\ \text{Mud Balls}\end{matrix}$$

Once the volume of mud balls in the sample has been determined, the percent mud ball volume of the sample can be calculated:

$$\text{\% Mud Ball Vol.} = \frac{\text{Mud Ball Vol., m}L}{\text{Sample Vol., m}L} \times 100$$

Examples 1-3 illustrate the calculation of percent mud ball volume.

Example 3: (% Mud Ball Volume)
❑ Five samples of filter media are taken for mud ball evaluation. The volume of water in the graduated cylinder rises from 500 mL to 542 mL when the mud balls are placed in the water. What is the percent mud ball volume of the sample? (The mud ball sampler has a sample volume of 695 mL.)

First, calculate the volume of mud balls in the sample:

$$542\ \text{m}L - 500\ \text{m}L = 42\ \text{m}L$$
rise in volume

Then calculate the percent mud ball volume:

$$\text{\% Mud Ball Volume} = \frac{\text{Mud Ball Vol., m}L}{\text{Total Sample Vol., m}L} \times 100$$

$$= \frac{42\ \text{m}L}{(5)(695\ \text{m}L)} \times 100$$

$$= \frac{42\ \text{m}L}{3475\ \text{m}L} \times 100$$

$$= \boxed{1.2\%}$$

NOTES:

12 *Chlorination*

SUMMARY

1. Chlorine Feed Rate

$$(\text{mg}/L\ Cl_2)\ (\text{MGD flow})\ (8.34\ \text{lbs/gal}) = \begin{array}{l}\text{lbs/day} \\ \text{Chlorine}\end{array}$$

2. Chlorine Dose, Demand and Residual

$$Cl_2\ \text{Dose} =\ Cl_2\ \text{Demand} + Cl_2\ \text{Residual}$$

To determine if chlorination is above the **breakpoint**, compare the expected increase in residual with the actual increase in residual:

Expected Increase in Residual

$$\begin{array}{l}(\text{mg}/L)\ (\text{MGD flow})\ (8.34\ \text{lbs/gal}) \\ \text{Expected} \\ \text{Increase}\end{array} = \begin{array}{l}\text{lbs/day} \\ \text{Increase in} \\ Cl_2\ \text{Dose}\end{array}$$

Actual Increase in Residual

$$\begin{array}{l}\text{Actual} \\ \text{Increase} \\ \text{in Resid.,} \\ \text{mg}/L\end{array} = \begin{array}{l}\text{New} \\ \text{Residual,} \\ \text{mg}/L\end{array} - \begin{array}{l}\text{Old} \\ \text{Residual,} \\ \text{mg}/L\end{array}$$

3. Dry Hypochlorite Feed Rate

Simplified Equation:

$$\begin{array}{l}\text{Hypochlorite,} \\ \text{lbs/day}\end{array} = \dfrac{\text{lbs/day } Cl_2}{\dfrac{\text{\% Available } Cl_2}{100}}$$

Expanded Equations:

$$\begin{array}{l}\text{Hypochlorite,} \\ \text{lbs/day}\end{array} = \dfrac{(\text{mg}/L\ Cl_2)\ (\text{MGD flow})\ (8.34\ \text{lbs/gal})}{\dfrac{\text{\% Strength of Hypochl.}}{100}}$$

Or

$$\begin{array}{l}\text{Hypochlorite,} \\ \text{lbs}\end{array} = \dfrac{(\text{mg}/L\ Cl_2)\ (\text{MG Tank Vol.})\ (8.34\ \text{lbs/gal})}{\dfrac{\text{\% Strength of Hypochl.}}{100}}$$

SUMMARY—Cont'd

4. Hypochlorite Solution Feed Rate

Simplified Equation:

$$\begin{array}{ccc} \text{Actual Dose} & = & \text{Solution Feeder Dose} \\ \text{lbs/day} & & \text{lbs/day} \end{array}$$

Expanded Equation:

$$\begin{array}{c} \underset{\text{Dose \quad Treated \quad lbs/gal}}{(\text{mg}/L)\ (\text{MGD Flow})\ (8.34)} = \underset{\text{Sol'n \quad Sol'n \quad lbs/gal}}{(\text{mg}/L)\ (\text{MGD})\ (8.34)} \end{array}$$

5. Percent Strength of Solutions

Percent strength using dry chlorine:

$$\begin{array}{c} \% \ Cl_2 \\ \text{Strength} \end{array} = \frac{\text{Chlorine, lbs}}{\text{Solution, lbs}} \times 100$$

Or

$$\begin{array}{c} \% \ Cl_2 \\ \text{Strength} \end{array} = \frac{\text{Chlorine, lbs}}{\text{Water, lbs} + \text{Chlorine, lbs}} \times 100$$

Or

$$\begin{array}{c} \% \ Cl_2 \\ \text{Strength} \end{array} = \frac{\dfrac{(\text{Hypochl., lbs})(\% \ \text{Avail. } Cl_2)}{100}}{\text{Water, lbs} + \dfrac{(\text{Hypochl., lbs})(\% \ \text{Avail. } Cl_2)}{100}} \times 100$$

Percent strength using liquid chlorine:

$$\begin{array}{ccc} \text{lbs Chlorine} & = & \text{lbs Chlorine} \\ \text{in Liquid Hypochlorite} & & \text{in Hypochlorite Solution} \end{array}$$

Or

$$\underset{\text{lbs}}{(\text{Liquid Hypochl.})} \ \dfrac{\underset{\text{of Liq. Hypo.}}{(\% \ \text{Strength})}}{100} = \underset{\text{lbs}}{(\text{Hypochl. Solution})} \ \dfrac{\underset{\text{of Hypo. Sol'n.}}{(\% \text{Strength})}}{100}$$

Or

$$\underset{\text{gal \quad lbs/gal}}{(\text{Liq. Hypo.})\ (8.34)} \ \dfrac{\underset{\text{of Liq. Hypo.}}{(\% \ \text{Strength})}}{100} = \underset{\text{gal \quad lbs/gal}}{(\text{Hypo. Sol'n})\ (8.34)} \ \dfrac{\underset{\text{of Liq. Hypo.}}{(\% \ \text{Strength})}}{100}$$

SUMMARY—Cont'd

6. **Mixing Hypochlorite Solutions**

Simplified Equation:

$$\begin{array}{c}\% \text{ Cl}_2 \text{ Strength} \\ \text{of Mixture}\end{array} = \frac{\text{Chlorine in Mixture, lbs}}{\text{Solution Mixture, lbs}} \times 100$$

Expanded Equations:

$$\begin{array}{c}\% \text{ Cl}_2 \\ \text{Strength} \\ \text{of Mixture}\end{array} = \frac{\begin{array}{c}\text{lbs Cl}_2 \text{ from} \\ \text{Solution 1}\end{array} + \begin{array}{c}\text{lbs Cl}_2 \text{ from} \\ \text{Solution 2}\end{array}}{\text{lbs Solution 1} + \text{lbs Solution 2}} \times 100$$

Or

$$\begin{array}{c}\% \text{ Cl}_2 \\ \text{Strength} \\ \text{of Mixture}\end{array} = \frac{\begin{array}{c}(\text{Sol'n 1}) \dfrac{(\% \text{ Avail. Cl}_2)}{\text{of Sol'n 1}} \\ \text{lbs} \quad\quad\quad 100\end{array} + \begin{array}{c}(\text{Sol'n 2}) \dfrac{(\% \text{ Avail. Cl}_2)}{\text{of Sol'n 2}} \\ \text{lbs} \quad\quad\quad 100\end{array}}{\text{lbs Solution 1} + \text{lbs Solution 2}} \times 100$$

<u>Or</u>, if target strength is desired:

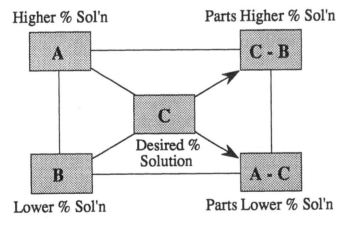

SUMMARY—Cont'd

7. Chemical Use Calculations

Average Chemical Use:

$$\text{Average Use} \atop \text{lbs/day} = \frac{\text{Total Chem. Used, lbs}}{\text{Number of Days}}$$

Or

$$\text{Average Use} \atop \text{gpd} = \frac{\text{Total Chem. Used, gal}}{\text{Number of Days}}$$

Days' Supply In Inventory:

$$\text{Days' Supply} \atop \text{in Inventory} = \frac{\text{Total Chem. in Inventory, lbs}}{\text{Average Use, lbs/day}}$$

Or

$$\text{Days' Supply} \atop \text{in Inventory} = \frac{\text{Total Chem. in Inventory, gal}}{\text{Average Use, gpd}}$$

Total Chemical Use, lbs:

$$\text{Total Chemical Used, lbs} = \frac{(\text{Chem. Use,}) (\text{Days Use})}{\text{lbs/day}}$$

Or

$$\text{Total Chemical Used, lbs} = \frac{(\text{Chem. Use,}) (\text{Hrs Use})}{\text{lbs/hr}}$$

Chlorine Cylinders Required:

$$\text{Required Cylinders/Week} = \frac{\text{Chem. Use, lbs/week}}{\text{lbs Chem./Cylinder}}$$

Or

$$\text{Required Cylinders/Month} = \frac{\text{Chem. Use, lbs/month}}{\text{lbs Chem./Cylinder}}$$

NOTES:

12.1 CHLORINE FEED RATE

Chlorine is added to water to kill any disease-causing organisms which may be present in the water or may enter the water as it travels through the distribution system.

The two expressions most often used to describe the amount of chlorine added or required are:

- milligrams per liter (mg/L)*, and

- pounds per day (lbs/day)

The equation shown to the right can be used to calculate <u>either</u> mg/L or lbs/day chlorine dosage. The equation has three variables: mg/L chlorine, MGD flow rate, and lbs/day chlorine. When <u>any</u> <u>two</u> of the variables are known, the third can be calculated. Examples 1-3 illustrate lbs/day dosage calculations; and Examples 4-7 illustrate mg/L dosage calculations.

MILLIGRAMS PER LITER IS A MEASURE OF CONCENTRATION

Assume each liter below is divided into 1 million parts:

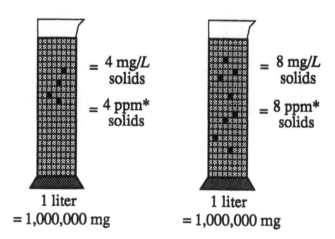

1 liter
= 1,000,000 mg

1 liter
= 1,000,000 mg

The mg/L concentration expresses a ratio of the milligrams chemical in each liter of water. For example, if a concentration of 4 mg/L is desired, then a total of 12 mg chemical would be required to treat 3 liters:

$$\frac{4 \text{ mg}}{L} \times \frac{\times 3}{\times 3} = \frac{12 \text{ mg}}{3 L}$$

The amount of chlorine required therefore depends on two factors:

- The desired concentration (mg/L), and

- The amount of water to be treated (normally expressed as MGD).

To convert from mg/L to lbs/day, or vice versa, the following equation is used:

$$\begin{array}{ccc} (\text{mg/L}) & (\text{MGD}) & (8.34) \\ \text{Cl}_2 & \text{flow} & \text{lbs/gal} \end{array} = \begin{array}{c} \text{lbs/day} \\ \text{Cl}_2 \end{array}$$

---

* For most water and wastewater calculations, mg/L concentration and ppm (parts per million) concentration may be used interchangeably. That is, 1 mg/L = 1 ppm. Of the two expressions, mg/L is preferred.

Example 1: (Chlorine Feed Rate)
❑ Determine the chlorinator setting (lbs/day) needed to treat a flow of 3 MGD with a chlorine dose of 4 mg/L.

First write the equation. Then fill in the information given:

(mg/L Cl_2) (MGD flow) (8.34 lbs/gal) = lbs/day Cl_2

(4 mg/L Cl_2) (3 MGD) (8.34 lbs/gal) = $\boxed{100 \text{ lbs/day } Cl_2}$

Example 2: (Chlorine Feed Rate)
❑ A flow of 875,000 gpd is to receive a chlorine dose of 2.7 mg/L. What should the chlorinator setting be (in lbs/day)?

(mg/L Cl_2) (MGD flow) (8.34 lbs/gal) = lbs/dayCl_2

(2.7 mg/L) (0.875 MGD) (8.34 lbs/gal) = $\boxed{19.7 \text{ lbs/day}}$

Example 3: (Chlorine Feed Rate)
❑ A pipeline 12-inches in diameter and 1,200 ft long is to be treated with a chlorine dose of 50 mg/L. How many lbs of chlorine will this require?

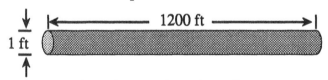

To use the mg/L to lbs equation, the gallon volume of the pipeline must first be determined:*

Vol. gal = (0.785) (D^2) (Length, ft) (7.48 gal/cu ft)

= (0.785) (1 ft) (1 ft) (1200 ft) (7.48 gal/cu ft)

= 7046 gal

The lbs chlorine required can now be calculated:

(mg/L Cl_2) (MG Vol.) (8.34 lbs/gal) = lbs Cl_2

(50 mg/L) (0.007046 MG) (8.34 lbs/gal) = $\boxed{2.9 \text{ lbs}}$

DOSAGE IN A TANK

To calculate chlorine dose for tanks or pipelines, a slightly modified equation must be used. Instead of MGD flow rate, MG volume* is used:

$$\boxed{\begin{array}{ccccc} (\text{mg/}L) & (\text{MG}) & (8.34) = & \text{lbs} \\ Cl_2 & \text{Tank} & \text{lbs/gal} & Cl_2 \\ & \text{Vol.} \end{array}}$$

---

* For a review of volume calculations, refer to Chapter 1 in this text or Chapter 11 in *Basic Math Concepts*.

Whether the unknown variable is lbs/day or mg/*L* chlorine, use the **same equation** to calculate the answer. Once the equation has been filled in with known data, then move the factors as needed to solve for the unknown value.*

Example 4: (Chlorine Feed Rate)
❑ A chlorinator setting is 35 lbs per 24 hrs. If the flow being chlorinated is 1.15 MGD, what is the chlorine dosage expressed as mg/*L*?

$$(\text{mg}/L\ \text{Cl}_2)\ (\text{MGD flow})\ (8.34\ \text{lbs/gal}) = \text{lbs/day Cl}_2$$

$$(x\ \text{mg}/L)\ (1.15\ \text{MGD})\ (8.34\ \text{lbs/gal}) = 35\ \text{lbs/day}$$

Now solve for *x*:*

$$x = \frac{35}{(1.15)\ (8.34)}$$

$$x = \boxed{3.6\ \text{mg}/L}$$

Example 5: (Chlorine Feed Rate)
❑ The current setting on a chlorinator is 42 lbs per 24 hrs. For a flow of 2,050,000 gpd, what is this chlorine dosage expressed in mg/*L*?

$$(\text{mg}/L\ \text{Cl}_2)\ (\text{MGD flow})\ (8.34\ \text{lbs/gal}) = \text{lbs/day Cl}_2$$

$$(x\ \text{mg}/L)\ (2.05\ \text{MGD})\ (8.34\ \text{lbs/gal}) = 42\ \text{lbs/day}$$

Then solve for *x*:

$$x = \frac{42}{(2.05)\ (8.34)}$$

$$x = \boxed{2.5\ \text{mg}/L}$$

---

* For a review of solving for the unknown value, refer to Chapter 2 in *Basic Math Concepts*.

Example 6: (Chlorine Feed Rate)
❏ A flow of 1880 gpm is to be chlorinated. At a chlorinator setting of 51 lbs per 24 hrs, what would be the chlorine dosage in mg/*L*?

First convert the gpm flow rate to MGD flow rate:*

$$(1880 \text{ gpm}) (1440 \text{ min/day}) = 2,707,200 \text{ gpd}$$
$$= 2.71 \text{ MGD}$$

Then the chlorine dosage in mg/L can be calculated:

$$(\text{mg/}L \text{ Cl}_2) (\text{MGD flow}) (8.34 \text{ lbs/gal}) = \text{lbs/day Cl}_2$$

$$(x \text{ mg/}L) (2.71 \text{ MGD}) (8.34 \text{ lbs/gal}) = 51 \text{ lbs/day}$$

$$x = \frac{51}{(2.71)(8.34)}$$

$$x = \boxed{2.3 \text{ mg/}L}$$

Example 7: (Chlorine Feed Rate)
❏ A chlorinator setting is 51 lbs per 24 hours. If the flow being treated is 2.71 MGD, what is the mg/*L* concentration of the chlorine dose? (Note this is the same problem as Example 6.)

$$\frac{\text{mg}}{L} = \frac{\text{parts}}{1,000,000 \text{ parts}} = \frac{\text{lbs Chlorine}}{1,000,000 \text{ lbs Water}}$$

So, lbs chlorine must be in the numerator and lbs water treated (expressed in <u>million lbs</u>) must be in the denominator:**

$$\frac{51 \text{ lbs Cl}_2}{(2.71 \text{ MGD}) (8.34 \text{ lbs/gal})} = \frac{51 \text{ lbs Cl}_2}{22.6 \text{ million lbs/day flow}}$$

$$= \frac{2.3 \text{ lbs Chem.}}{1 \text{ million lbs/day flow}}$$

$$\text{or} = \boxed{2.3 \text{ mg/}L}$$

WHEN FLOW RATE IS EXPRESSED IN OTHER TERMS

The mg/*L* to lbs/day equation requires that the **flow rate be expressed as MGD flow**. If the flow rate is expressed in any other terms (gpd, gpm, cfm, etc.), it must first be converted to MGD flow rate before continuing with the mg/*L* to lbs/day calculation.

USING THE "PARTS PER MILLION" APPROACH

Some people use another approach in solving these mg/*L* dosage problems. This approach is based on the concept that mg/*L* concentration may also be expressed as "parts per million" (ppm) concentration.

As illustrated in the diagram on the previous page, 4 mg/*L* is a concentration of 4 mg/1,000,000 mg or 4 parts/1,000,000 parts, or 4 lbs/1,000,000 lbs:

$$\frac{4 \text{ mg}}{L} = \frac{4 \text{ mg}}{1,000,000 \text{ mg}} = \frac{4 \text{ lbs}}{1,000,000 \text{ lbs}}$$

It is the last ratio, pounds per million pounds, which is the heart of the alternative approach. Once you have calculated the **pounds** of chlorine used **per million pounds** of water treated, you have in fact calculated parts per million or mg/*L* concentration as well. Example 7 illustrates this calculation.

---

* For a review of flow conversions, refer to Chapter 8 in *Basic Math Concepts.*

** Note the similarity of the equation $51 \text{ lbs} \div \left[(2.71 \text{ MGD}) (8.34 \text{ lbs/gal}) \right]$ to the <u>second to last</u> equation in Example 6.

12.2 CHLORINE DOSE, DEMAND AND RESIDUAL

The chlorine dose required depends on two considerations—the chlorine demand and the desired chlorine residual:

Dose = Demand + Residual
mg/*L* mg/*L* mg/*L*

The **chlorine demand** is the amount of chlorine used in reacting with various components of the water such as harmful organisms and other organic and inorganic substances. When the chlorine demand has been satisfied, these reactions stop.

In some cases, such as perhaps during the initial phase of treatment, chlorinating just to meet the chlorine demand is sufficient. In other cases, however, such as at the end of the treatment process, it is desirable to have an additional amount of chlorine in the water available for disinfection as it travels through the distribution system. This additional chlorine is called the **chlorine residual**.

Examples 1-6 illustrate the calculation of chlorine dose, demand, and residual.

Example 1: (Chlorine Dose, Demand, Residual)
❑ A water is tested and found to have a chlorine demand of 1.9 mg/*L*. If the desired chlorine residual is 0.8 mg/*L*, what is the desired chlorine dose in mg/*L*?

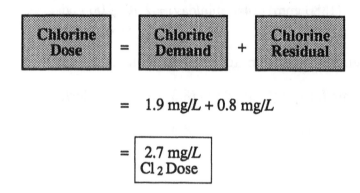

$$= \quad 1.9 \text{ mg/}L + 0.8 \text{ mg/}L$$

$$= \boxed{\begin{array}{l} 2.7 \text{ mg/}L \\ Cl_2 \text{ Dose} \end{array}}$$

Example 2: (Chlorine Dose, Demand, Residual)
❑ The chlorine demand of a water is 2.1 mg/*L*. If a chlorine residual of 0.6 mg/*L* is desired, what is the desired chlorine dosage in mg/*L*?

$$\boxed{\text{Chlorine Dose}} = \boxed{\text{Chlorine Demand}} + \boxed{\text{Chlorine Residual}}$$

$$= \quad 2.1 \text{ mg/}L + 0.6 \text{ mg/}L$$

$$= \boxed{\begin{array}{l} 2.7 \text{ mg/}L \\ Cl_2 \text{ Dose} \end{array}}$$

Example 3: (Chlorine Dose, Demand, Residual)
❑ The chlorine dosage for a water is 2.9 mg/*L*. If the chlorine residual after 30 minutes contact time is found to be 0.6 mg/*L*, what is the chlorine demand expressed in mg/*L*?

$$\boxed{\text{Chlorine Dose}} = \boxed{\text{Chlorine Demand}} + \boxed{\text{Chlorine Residual}}$$

$$2.9 \text{ mg/}L = x \text{ mg/}L + 0.6 \text{ mg/}L$$

$$2.9 \text{ mg/}L - 0.6 \text{ mg/}L = x \text{ mg/}L$$

$$2.3 \text{ mg/}L = x$$
$$\text{Cl}_2 \text{ Demand}$$

CALCULATING OTHER UNKNOWN VARIABLES

In Examples 1 and 2, the unknown variable was chlorine dosage, mg/*L*. However, the same equation may be used when chlorine demand or chlorine residual is unknown. Example 3 illustrates this calculation.

Example 4: (Chlorine Dose, Demand, Residual)
❑ What should the chlorinator setting be (lbs/day) to treat a flow of 2.46 MGD if the chlorine demand is 3.1 mg/*L* and a chlorine residual of 0.8 mg/*L* is desired?

First calculate the chlorine dosage in mg/*L*:

$$\boxed{\text{Chlorine Dose}} = \boxed{\text{Chlorine Demand}} + \boxed{\text{Chlorine Residual}}$$

$$= 3.1 \text{ mg/}L + 0.8 \text{ mg/}L$$

$$= 3.9 \text{ mg/}L$$

Then calculate the chlorine dosage (feed rate) in lbs/day:

$$(\text{mg/}L \text{ Cl}_2)(\text{MGD flow})(8.34 \text{ lbs/gal}) = \text{lbs/day Cl}_2$$

$$(3.9 \text{ mg/}L)(2.46 \text{ MGD})(8.34 \text{ lbs/gal}) = \boxed{80 \text{ lbs/day Chlorine}}$$

COMBINING WITH FEED RATE CALCULATIONS

Once the chlorine dosage (mg/*L*) has been calculated using the dose/demand/residual equation, the chlorine dosage in lbs/day can be calculated. Example 4 illustrates this type of problem.

BREAKPOINT CHLORINATION

Various chemical reactions occur when chlorine is added to water. First, chlorine reacts with various minerals in the water. At this stage no disinfection of the water occurs (Point 1 in the illustration to the right.)

As more chlorine is added, it begins to react with organics and ammonia in the water, forming chlororganics and chloramines (Point 2). These compounds are only weak disinfectants and their presence in water can result in chlorine taste and odor problems.

With the addition of more chlorine, some of the chlororganics and chloramines are destroyed (Point 3). And adding slightly more chlorine, the **breakpoint** is reached. At this point, there is the formation of **free available chlorine residual** (Point 4). The addition of any chlorine beyond this point simply results in a higher level of free available chlorine residual. Many treatment plants chlorinate just beyond the breakpoint since free available chlorine residual is a very effective disinfectant of the water.

As mentioned above, once the breakpoint has been reached, **the addition of any chlorine will result in a corresponding increase in residual chlorine.** Therefore, to determine if the chlorinator setting is above the breakpoint, compare an increased chlorine dose and the expected increase in chlorine residual with the actual increase in chlorine residual, as shown in the box to the right. If the expected increase and actual increase are approximately the same, it is assumed that the water is in fact being chlorinated above the breakpoint. Examples 5 and 6 illustrate this calculation.

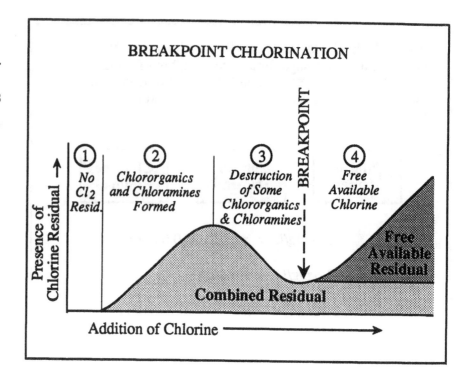

BREAKPOINT CHLORINATION

① No Cl₂ Resid. ② Chlororganics and Chloramines Formed ③ Destruction of Some Chlororganics & Chloramines BREAKPOINT ④ Free Available Chlorine

Free Available Residual

Combined Residual

Presence of Chlorine Residual →

Addition of Chlorine ⟶

COMPARE <u>EXPECTED</u> INCREASE IN RESIDUAL WITH <u>ACTUAL</u> INCREASE IN RESIDUAL

Expected increase in residual is reflected directly by the increase in Cl_2 dose (lbs/day). If the water is being chlorinated beyond the breakpoint, then any increase in chlorine dose will result in a corresponding increase in chlorine residual. Use the mg/L to lbs/day equation to determine the expected increase in residual that would result from an increase in the chlorine dose:

$$\text{(mg/}L\text{) (MGD flow) (8.34 lbs/gal)}_{\text{Expected Increase}} = \text{lbs/day Increase in } Cl_2 \text{ Dose}$$

The **actual increase in residual** is simply a comparison of new and old residual data:

$$\text{Actual Increase in Resid., mg/}L = \text{New Residual, mg/}L - \text{Old Residual, mg/}L$$

Example 5: (Chlorine Dose, Demand, Residual)
❑ A chlorinator setting is increased by 4 lbs/day. The chlorine residual before the increased dosage was 0.3 mg/L. After the increased chlorine dose, the chlorine residual was 0.6 mg/L. The average flow rate being chlorinated is 1.45 MGD. Is the water being chlorinated beyond the breakpoint?

First calculate the expected increase in chlorine residual. Remember, this is based directly on lbs/day increase in chlorine dose. Use the mg/L to lbs/day equation:

(mg/L Incr.) (MGD flow) (8.34 lbs/gal) = lbs/day Incr.

(x mg/L) (1.45 MGD) (8.34 lbs/gal) = 4 lbs/day

$$x = \frac{4}{(1.45)\,(8.34)}$$

$$x = \boxed{0.3 \text{ mg/L}}$$

The actual increase in residual was:

0.6 mg/L – 0.3 mg/L = $\boxed{0.3 \text{ mg/L}}$

Because these calculations match, we can conclude that the water is being chlorinated beyond the breakpoint.

Example 6: (Chlorine Dose, Demand, Residual)
❑ A chlorinator setting of 20 lbs chlorine per 24 hours results in a chlorine residual of 0.4 mg/L. The chlorinator setting is increased to 25 lbs per 24 hours. The chlorine residual increased to 0.5 mg/L at this new dosage rate. The average flow being treated is 1.6 MGD. On the basis of this data, is the water being chlorinated past the breakpoint?

First calculate the expected increase in chlorine residual:

(mg/L Incr.) (MGD flow) (8.34 lbs/gal) = lbs/day Incr.

(x mg/L) (1.6 MGD) (8.34 lbs/gal) = 5 lbs/day

$$x = \frac{5}{(1.6)\,(8.34)}$$

$$x = \boxed{0.37 \text{ mg/L}}$$

The actual increase in residual was:

0.5 mg/L – 0.4 mg/L = $\boxed{0.1 \text{ mg/L}}$

The expected residual increase was 0.37 mg/L whereas the actual increase was only 0.1 mg/L. From this analysis it appears that the water is <u>not</u> being chlorinated past the breakpoint.

12.3 DRY HYPOCHLORITE FEED RATE

When chlorinating water with chlorine gas, there is 100% available chlorine. Therefore, if the chlorine demand and residual require 50 lbs/day chlorine, the chlorinator setting would be just that—50 lbs/24 hrs.

Many times, however, a chlorine compound called hypochlorite is used to chlorinate the water. Hypochlorite compounds contain chlorine and are similar to a strong bleach. The most commonly used dry hypochlorite, calcium hypochlorite,* contains about 65-70% available chlorine, depending on the brand. Sodium hypochlorite, or liquid bleach, contains about 12-15% available chlorine in commercial-grade bleaches and 5.25% available chlorine in household bleach.

Because hypochlorites are not 100% pure chlorine, **more lbs/day must be fed into the system to obtain the same amount of chlorine for disinfection.**

The lbs/day hypochlorite required can be calculated in two steps:

1. If the required lbs/day chlorine has not been determined, calculate it:

$$(\text{mg}/L)(\text{MGD})(8.34) = \text{lbs/day}$$
$$\quad\quad \text{Cl}_2 \quad \text{flow} \quad \text{lbs/gal} \quad\quad \text{Cl}_2$$

2. Then calculate the lbs/day hypochlorite needed by dividing the lbs/day chlorine by the percent available chlorine.

$$\frac{\text{Hypochlorite}}{\text{lbs/day}} = \frac{\text{lbs/day Cl}_2}{\dfrac{\text{\% Available Cl}_2}{100}}$$

Example 1: (Dry Hypochlorite Feed Rate)

❑ A chlorine dosage of 105 lbs/day is required to disinfect a flow of 1,640,000 gpd. If the calcium hypochlorite to be used contains 65% available chlorine, how many lbs/day hypochlorite will be required for disinfection?

Since only 65% of the hypochlorite is chlorine, **more than 105 lbs** of hypochlorite will be required:

$$\frac{\text{Hypochlorite,}}{\text{lbs/day}} = \frac{\text{lbs/day Cl}_2}{\dfrac{\text{\% Available Cl}_2}{100}}$$

$$= \frac{105 \text{ lbs/day}}{\dfrac{65}{100}}$$

$$= \frac{105}{0.65}$$

$$= \boxed{\begin{array}{l}161.5 \text{ lbs/day} \\ \text{Hypochlorite}\end{array}}$$

Example 2: (Dry Hypochlorite Feed Rate)

❑ A total chlorine dosage of 35 lbs/day is required to treat a particular water. If the calcium hypochlorite has 65% available chlorine, how many lbs/day of hypochlorite will be required?

Only 65% of the hypochlorite is chlorine. Therefore, <u>more than 35 lbs/day</u> will be required:

$$\frac{\text{Hypochlorite,}}{\text{lbs/day}} = \frac{\text{lbs/day Cl}_2}{\dfrac{\text{\% Available Cl}_2}{100}}$$

$$= \frac{35 \text{ lbs/day Cl}_2}{0.65 \text{ Avail. Cl}_2}$$

$$= \boxed{\begin{array}{l}53.8 \text{ lbs/day} \\ \text{Hypochlorite}\end{array}}$$

---

* The form of calcium hypochlorite typically used for disinfection is called high-test hypochlorite, or HTH.

Example 3: (Dry Hypochlorite Feed Rate)
❑ A water flow of 850,000 gpd requires a chlorine dose of 2.9 mg/L. If calcium hypochlorite (65% available chlorine) is to be used, how many lbs/day of hypochlorite are required?

First, calculate the lbs/day chlorine required:

$$(\text{mg/}L \ Cl_2)(\text{MGD flow})(8.34 \ \text{lbs/gal}) = \text{lbs/day}$$

$$(2.9 \ \text{mg/}L)(0.85 \ \text{MGD})(8.34 \ \text{lbs/gal}) = \boxed{21 \ \text{lbs/day} \atop \text{Chlorine}}$$

Then calculate the lbs/day hypochlorite:

$$\frac{\text{Hypochlorite}}{\text{lbs/day}} = \frac{\text{lbs/day } Cl_2}{\dfrac{\% \text{ Available } Cl_2}{100}}$$

$$= \frac{21 \ \text{lbs/day } Cl_2}{0.65 \ \text{Avail. } Cl_2}$$

$$= \boxed{32 \ \text{lbs/day} \atop \text{Hypochlorite}}$$

In Examples 1 and 2, the required lbs/day chlorine was known, and only the lbs/day hypochlorite had to be determined. In some calculations, however, the lbs/day chlorine will not be known and must first be calculated before the hypochlorite requirement can be determined. For such a problem, the numerator of the hypochlorite equation (as used in Examples 1 and 2) can be expanded to include the mg/L to lbs/day equation:

$$\boxed{\frac{\text{Hypochl.}}{\text{lbs/day}} = \frac{\underset{Cl_2}{(\text{mg/}L)}\underset{\text{flow}}{(\text{MGD})}\underset{\text{lbs/gal}}{(8.34)}}{\dfrac{\% \text{ Available } Cl_2}{100}}}$$

Example 3 illustrates a two-step calculation and Example 4 illustrates the use of the "combined" equation.

Example 4: (Dry Hypochlorite Feed Rate)
❑ A tank contains 575,000 gallons of water. This water is to receive a chlorine dose of 2.2 mg/L. How many pounds of calcium hypochlorite (65% available chlorine) will be required for this disinfection?

Use the "combined" equation to calculate the lbs hypochlorite required:

$$\frac{\text{Hypochlorite,}}{\text{lbs}} = \frac{(\text{mg/}L \ Cl_2)(\text{MG Vol.})(8.34 \ \text{lbs/gal})}{\dfrac{\% \text{ Available } Cl_2}{100}}$$

$$= \frac{(2.2 \ \text{mg/}L)(0.575 \ \text{MG})(8.34 \ \text{lbs/gal})}{\dfrac{65}{100}}$$

$$= \frac{10.6 \ \text{lbs}}{0.65}$$

$$= \boxed{16.3 \ \text{lbs} \atop \text{Hypochlorite}}$$

HYPOCHLORITE DOSING IN A TANK OR PIPELINE

The same basic equation is used when dosing a tank or pipeline. The only difference in the equation is that **instead of MGD flow rate, a MG tank or pipeline volume factor is used:**

$$\boxed{\frac{\text{Hypochl.}}{\text{lbs}} = \frac{\underset{Cl_2}{(\text{mg/}L)}\underset{\text{Vol.}}{(\text{MG})}\underset{\text{lbs/gal}}{(8.34)}}{\dfrac{\% \text{ Available } Cl_2}{100}}}$$

CALCULATING lbs/day or mg/*L* CHLORINE GIVEN HYPOCHLORITE DOSAGE

Occasionally you will know the lbs/day hypochlorite and will want to determine either the lbs/day chlorine or the mg/*L* chlorine dosage. To calculate either of these unknown variables, begin with the hypochlorite equation. This equation enables you to calculate lbs/day chlorine. Then, if calculating mg/*L* chlorine, use the mg/*L* to lbs/day equation as a second step. In effect, this is working the problem "backwards" from the problems shown in Examples 2 and 3. Example 5 illustrates this type of calculation.

Example 5: (Dry Hypochlorite Feed Rate)

❏ A total of 42 lbs of calcium hypochlorite (65% available chlorine) are used in a day. If the flow rate treated is 1,080,000 gpd, what is the chlorine dosage in mg/*L*?

First calculate the lbs/day chlorine dosage:

$$\frac{\text{Hypochlorite}}{\text{lbs/day}} = \frac{\text{lbs/day Cl}_2}{\dfrac{\% \text{ Available Cl}_2}{100}}$$

$$\frac{42 \text{ lbs/day}}{\text{Hypochlorite}} = \frac{x \text{ lbs/day Cl}_2}{0.65}$$

$$(0.65)(42) = x$$

$$\boxed{\begin{array}{c}27.3 \text{ lbs/day} \\ \text{Chlorine}\end{array}} = x$$

Then calculate mg/*L* Cl$_2$, using the mg/*L* to lbs/day equation and filling in the known information:*

$$(x \text{ mg/}L \text{ Cl}_2)(1.08 \text{ MGD})(8.34 \text{ lbs/gal}) = 27.3 \text{ lbs/day Cl}_2$$

$$x = \frac{27.3 \text{ lbs/day}}{(1.08 \text{ MGD})(8.34 \text{ lbs/gal})}$$

$$x = \boxed{\begin{array}{c}3.0 \text{ mg/}L \\ \text{Chlorine}\end{array}}$$

Example 6: (Dry Hypochlorite Feed Rate)

❏ A total of 56 lbs of calcium hypochlorite (65% available chlorine) are used in a day. If the flow rate treated is 2.34 MGD, what is the chlorine dosage in mg/*L*?

First calculate the lbs/day chlorine dosage:

$$\frac{56 \text{ lbs/day}}{\text{Hypochlorite}} = \frac{x \text{ lbs/day Cl}_2}{0.65}$$

$$\boxed{\begin{array}{c}36.4 \text{ lbs/day} \\ \text{Chlorine}\end{array}} = x$$

Then calculate mg/*L* Cl$_2$ using the mg/*L* to lbs/day equation:

$$(x \text{ mg/}L \text{ Cl}_2)(2.34 \text{ MGD})(8.34 \text{ lbs/gal}) = 36.4 \text{ lbs/day}$$

$$x = \boxed{\begin{array}{c}1.9 \text{ mg/}L \\ \text{Chlorine}\end{array}}$$

---

* Refer to Chapter 2 in *Basic Math Concepts* for a review of solving for the unknown value.

Example 7: (Dry Hypochlorite Feed Rate)
❑ A total of 49 lbs of calcium hypochlorite (65% available chlorine) are used in a 24-hour period. If the flow rate treated is 1,960,000 gpd, what is the chlorine dosage in mg/L?

First calculate the lbs/day chlorine dosage:

$$\frac{\text{Hypochlorite}}{\text{lbs/day}} = \frac{\text{lbs/day Cl}_2}{\frac{\text{\% Available Cl}_2}{100}}$$

$$\frac{49 \text{ lbs/day}}{\text{Hypochlorite}} = \frac{x \text{ lbs/day Cl}_2}{0.65}$$

$$(49)\,(0.65) = x$$

$$\boxed{\frac{31.9 \text{ lbs/day}}{\text{Chlorine}}} = x$$

Then calculate mg/L Cl$_2$, using the mg/L to lbs/day equation and filling in the known information:

$$(x \text{ mg/}L \text{ Cl}_2)\,(1.96 \text{ MGD})\,(8.34 \text{ lbs/gal}) = 31.9 \text{ lbs/day Cl}_2$$

$$x = \frac{31.9 \text{ lbs/day}}{(1.96)\,(8.34)}$$

$$x = \boxed{2.0 \text{ mg/}L \text{ Cl}_2}$$

Example 8: (Dry Hypochlorite Feed Rate)
❑ A flow of 2,770,000 gpd is disinfected with calcium hypochlorite (65% available chlorine). If 52 lbs of hypochlorite are used in a 24-hour period, what is the mg/L chlorine dosage?

The lbs/day chlorine dosage is:

$$\frac{52 \text{ lbs/day}}{\text{Hypochlorite}} = \frac{x \text{ lbs/day Cl}_2}{0.65}$$

$$\boxed{\frac{33.8 \text{ lbs/day}}{\text{Chlorine}}} = x$$

The mg/L Cl$_2$ can now be calculated:

$$(x \text{ mg/}L \text{ Cl}_2)\,(2.77 \text{ MGD})\,(8.34 \text{ lbs/gal}) = 33.8 \text{ lbs/day}$$

$$x = \boxed{\frac{1.5 \text{ mg/}L}{\text{Chlorine}}}$$

12.4 HYPOCHLORITE SOLUTION FEED RATE

In Examples 1-4 of Section 12.3, the dry hypochlorite required (lbs/day) was calculated. The same basic equation may be used when calculating gallons per day (gpd) liquid hypochlorite. Only one step must be added to the calculation—**lbs/day hypochlorite required must be converted to gpd hypochlorite required.** The conversion factor of 8.34 lbs/gal is used in this conversion:*

$$\frac{\text{Hypochlorite, lbs/day}}{8.34 \text{ lbs/gal}} = \text{Hypochlorite, gpd}$$

Examples 1-4 illustrate this calculation.

Example 1: (Hypochlorite Solution Feed Rate)
❏ A total of 53 lbs/day sodium hypochlorite are required for disinfection of a 1.4-MGD flow. How many gallons per day hypochlorite is this?

Since lbs/day hypochlorite has already been calculated, simply convert lbs/day to gpd hypochlorite required:

$$\frac{\text{Hypochlorite, lbs/day}}{8.34 \text{ lbs/gal}} = \text{Hypochlorite, gpd}$$

$$\frac{53 \text{ lbs/day}}{8.34 \text{ lbs/gal}} = \boxed{\begin{array}{l}6.4 \text{ gpd} \\ \text{Hypochlorite}\end{array}}$$

Example 2: (Hypochlorite Solution Feed Rate)
❏ A chlorine dose of 1.3 mg/L is required for adequate disinfection. If a total flow of 350,000 gpd will be treated, how many gpd of sodium hypochlorite will be required? The sodium hypochlorite contains 14% available chlorine.

First calculate the lbs/day chlorine required:

$$(\text{mg/}L \text{ Cl}_2)(\text{MGD flow})(8.34 \text{ lbs/gal}) = \text{lbs/day Cl}_2$$

$$(1.3 \text{ mg/}L)(0.35 \text{ MGD})(8.34 \text{ lbs/gal}) = \begin{array}{l}3.8 \text{ lbs/day} \\ \text{Chlorine}\end{array}$$

Next, calculate the lbs/day sodium hypochlorite required:

$$\begin{array}{l}\text{Hypochlorite} \\ \text{lbs/day}\end{array} = \frac{\text{lbs/day Cl}_2}{\dfrac{\% \text{ Available Cl}_2}{100}}$$

$$= \frac{3.8 \text{ lbs/day Cl}_2}{0.14}$$

$$= 27 \text{ lbs/day Hypochlorite}$$

Then, calculate the gpd sodium hypochlorite required:

$$\frac{27 \text{ lbs/day}}{8.34 \text{ lbs/gal}} = \boxed{\begin{array}{l}3.2 \text{ gpd} \\ \text{Sodium Hypochlorite}\end{array}}$$

* For a review of flow conversions, refer to Chapter 8 in *Basic Math Concepts*.

Example 3: (Hypochlorite Solution Feed Rate)
❑ A hypochlorinator is used to disinfect the water pumped from a well. The hypochlorite solution contains 2% available chlorine. A chlorine dose of 1.2 mg/L is required for adequate disinfection throughout the system. If the flow being treated is 0.4 MGD, how many gpd of the hypochlorite solution will be required?

First calculate the lbs/day chlorine required:

$$(1.2 \text{ mg}/L) \ (0.4 \text{ MGD}) \ (8.34 \text{ lbs/gal}) \ = \ 4.0 \text{ lbs/day Cl}_2$$

Next, calculate the lbs/day hypochlorite solution required:

$$\frac{\text{Hypochlorite}}{\text{lbs/day}} \ = \ \frac{4.0 \text{ lbs/day Cl}_2}{0.02}$$

$$= \ 200 \text{ lbs/day Hypochlorite}$$

Then, calculate the gpd hypochlorite solution required:

$$\frac{200 \text{ lbs/day}}{8.34 \text{ lbs/gal}} \ = \ \boxed{24 \text{ gpd} \\ \text{Hypochlorite}}$$

The hypochlorite solution is generally diluted from a 12-15% available chlorine solution, as delivered to the plant, to a solution containing about 2% available chlorine. The calculation of gpd hypochlorite solution required is the same—only a different percent is used. Examples 7 and 8 illustrate this calculation.

Example 4: (Hypochlorite Solution Feed Rate)
❑ A hypochlorite solution with a 3% available chlorine is used to disinfect a water. A chlorine dose of 1.8 mg/L is desired to maintain an adequate chlorine residual. If the flow being treated is 310 gpm, what hypochlorite solution flow (in gpd) will be required?

To determine lbs/day chlorine, the flow rate must be expressed as MGD: [(310 gpm) (1440 min/day)] ÷ 1,000,000 = 0.45 MGD.

$$(1.8 \text{ mg}/L) \ (0.45 \text{ MGD}) \ (8.34 \text{ lbs/gal}) \ = \ 6.8 \text{ lbs/day}$$

Calculate the lbs/day hypochlorite solution required:

$$\frac{\text{Hypochlorite}}{\text{lbs/day}} \ = \ \frac{6.8 \text{ lbs/day Cl}_2}{0.03}$$

$$= \ 227 \text{ lbs/day Hypochlorite}$$

Then, calculate the gpd hypochlorite solution required:

$$\frac{227 \text{ lbs/day}}{8.34 \text{ lbs/gal}} \ = \ \boxed{27 \text{ gpd} \\ \text{Hypochlorite}}$$

A SHORTCUT METHOD FOR SOLUTION FEED CALCULATIONS

Examples 1-4 illustrate one method of calculating the required hypochlorite solution feed rate (gpd). Although this general method is effective when calculating **dry hypochlorite** feed rate, it is perhaps not as good an approach to calculating **hypochlorite solution feed rate**.

An alternative method of calculating hypochlorite solution feed rate is shown to the right. The idea behind this approach is that **the actual chemical dose in a water is a direct result of the chemical dose provided by the solution feeder**. Thus, in the simplified equation, the actual (or desired) dose is always set equal to the dose from the solution feeder. In the expanded equation, note that the lbs/day dose on each side of the equation is simply replaced by an equivalent statement, since (mg/*L*)(MGD flow)(8.34 lbs/gal) = lbs/day. Remember that the actual or desired dosage data goes on the left side of the equation and the solution feed data goes on the right side.

Examples 5-7 illustrate this alternative method of calculating solution feed rate. To compare approaches, Examples 5 and 6 use the **same problems** given in Examples 3 and 4, respectively.

SET THE ACTUAL (OR DESIRED) DOSAGE EQUAL TO THE SOLUTION FEEDER DOSAGE

Simplified Equation:

$$\begin{array}{cc} \text{Actual Dose*} & = & \text{Solution Feeder Dose} \\ \text{lbs/day} & & \text{lbs/day} \end{array}$$

Expanded Equation:

$$\begin{array}{ccccccc} (\text{mg}/L) & (\text{MGD Flow}) & (8.34) & = & (\text{mg}/L) & (\text{MGD}) & (8.34) \\ \text{Dose} & \text{Treated} & \text{lbs/gal} & & \text{Sol'n} & \text{Sol'n} & \text{lbs/gal} \end{array}$$

*Many times solution concentration will be given in terms of percent. Convert percent to mg/L concentration. Remember, 1% = 10,000 mg/L***

*Solution flow may be given as gpm or gpd. If so, convert to MGD flow.***

Example 5: (Hypochlorite Solution Feed Rate)

❑ A hypochlorinator is used to disinfect the water pumped from a well. The hypochlorite solution contains 2% available chlorine. A chlorine dose of 1.2 mg/*L* is required for adequate disinfection throughout the system. If the flow being treated is 0.4 MGD, how many gpd of the hypochlorite solution will be required?
(This problem is the same as Example 3.)

Write the equation and fill in the given data: (Note: a 2% solution concentration = 20,000 mg/*L*)

$$\begin{array}{ccccccc} (\text{mg}/L) & (\text{MGD Flow}) & (8.34) & = & (\text{mg}/L) & (\text{MGD}) & (8.34) \\ \text{Dose} & \text{Treated} & \text{lbs/gal} & & \text{Sol'n} & \text{Sol'n} & \text{lbs/gal} \end{array}$$

$$(1.2 \text{ mg}/L)(0.4 \text{ MGD})(8.34) = (20,000 \text{ mg}/L)(x \text{ MGD})(8.34)$$
$$\text{lbs/gal} \qquad\qquad\qquad\qquad \text{lbs/gal}$$

$$\frac{(1.2)(0.4)(\cancel{8.34})}{(20,000)(\cancel{8.34})} = x$$

$$0.000024 \text{ MGD} = x$$

Then convert MGD flow to gpd flow:**

$$\boxed{24 \text{ gpd}} = x$$

---

* Desired dose can also be put in this position. Data pertaining to solution flow is placed on the right side of the equation.
** For a review of mg/*L* and percent conversions or flow conversions, refer to Chapter 8 in *Basic Math Concepts*.

Example 6: (Hypochlorite Solution Feed Rate)
❏ A hypochlorite solution with a 3% available chlorine is used to disinfect a water. A chlorine dose of 1.8 mg/L is desired to maintain an adequate chlorine residual. If the flow being treated is 310 gpm, what hypochlorite solution flow (in gpd) will be required? (This problem is the same as Example 4.)

Write the equation and fill in the given data: (Note: a 3% solution concentration = 30,000 mg/L and a flow of 310 gpm = 0.45 MGD)

$$\underset{\text{Dose}}{(\text{mg}/L)} \ \underset{\text{Treated}}{(\text{MGD Flow})} \ \underset{\text{lbs/gal}}{(8.34)} = \underset{\text{Sol'n}}{(\text{mg}/L)} \ \underset{\text{Sol'n}}{(\text{MGD})} \ \underset{\text{lbs/gal}}{(8.34)}$$

$$(1.8 \text{ mg}/L)(0.45 \text{ MGD})\underset{\text{lbs/gal}}{(8.34)} = (30,000 \text{ mg}/L)(x \text{ MGD})\underset{\text{lbs/gal}}{(8.34)}$$

$$\frac{(1.8)(0.45)\cancel{(8.34)}}{(30,000)\cancel{(8.34)}} = x$$

$$0.000027 \text{ MGD} = x$$

$$\boxed{27 \text{ gpd}} = x$$

Example 7: (Hypochlorite Solution Feed Rate)
❏ Water from a well is disinfected by a hypochlorinator. The water meter indicated that a total of 2,260,000 gallons of water were pumped during a 7-day period. The 2% hypochlorite solution used to treat the well water is pumped from a 3-ft diameter storage tank. During the same 7-day period, the level in the tank dropped 2.5 ft. What is the chlorine dosage in the treated well water in mg/L?

The water flow rate and solution flow rate must first be expressed as gpd (then MGD) for use in the equation.

Well Water Flow Rate:
$$\frac{2,260,000 \text{ gal}}{7 \text{ days}} = 322,857 \text{ gpd}$$

Solution Feeder Flow Rate:*
$$\frac{(0.785)(3 \text{ ft})(3 \text{ ft})(2.5 \text{ ft})(7.48 \text{ gal/cu ft})}{7 \text{ days}} = 19 \text{ gpd}$$

Now use the "shortcut" equation to determine Cl_2 dose:

$$(x \text{ mg}/L)\underset{\text{MGD}}{(0.323)}\underset{\text{lbs/gal}}{(8.34)} = (20,000 \text{ mg}/L)\underset{\text{MGD}}{(0.000019)}\underset{\text{lbs/gal}}{(8.34)}$$

$$x = \frac{(20,000)(0.000019)\cancel{(8.34)}}{(0.323)\cancel{(8.34)}}$$

$$x = \boxed{1.2 \text{ mg}/L}$$

CALCULATING OTHER UNKNOWN VALUES

In Examples 5 and 6, the shortcut equation has been used to calculate gpd solution feed rate. However, the same equation can be used to calculate any of the four variables: mg/L actual dose, MGD flow treated, mg/L solution concentration or MGD solution flow. Example 7 illustrates one such calculation.

In Example 7, note that the flow rate treated is expressed in gallons per week. This flow rate must first be expressed as gpd then MGD for use in the equation. The solution flow rate is also expressed as gallons per week (using tank volume data) and must be converted to gpd and then MGD before it can be used in the equation.

* For a review of volume calculations, refer to Chapter 11 in *Basic Math Concepts*.

12.5 PERCENT STRENGTH OF SOLUTIONS

PERCENT STRENGTH USING DRY HYPOCHLORITE

The strength of a chlorine solution is a measure of the amount of chlorine (solute) dissolved in the solution. Since percent is calculated as "part over whole,"

$$\% = \frac{Part}{Whole} \times 100$$

percent chlorine strength is calculated as **part chlorine**, in lbs, divided by the **whole solution**, in lbs:

$$\frac{\% \ Cl_2}{Strength} = \frac{Chlorine, \ lbs}{Solution, \ lbs} \times 100$$

Since lbs chlorine is equal to (lbs hypochlorite) (% Avail. Cl_2/100), the numerator of the equation can be replaced. And the denominator of the equation (lbs solution) includes both water (lbs) and chlorine or chlorine compound (lbs). Thus the expanded equation may be written as:

$$\frac{\% \ Cl_2}{Strength} = \frac{\dfrac{(Hypochl,)(\% \ Avail.)}{lbs \quad Cl_2}{100}}{\dfrac{Water + Chl \ Cmpd}{lbs \quad lbs}} \times 100$$

Example 1 illustrates this expanded equation.

As indicated in this last equation, **the hypochlorite added must be expressed in pounds**. If the hypochlorite weight is expressed in ounces (as in Example 1) or grams (as in Example 2), it must first be converted to pounds (to correspond with the other units in the problem) before percent chlorine strength is calculated.

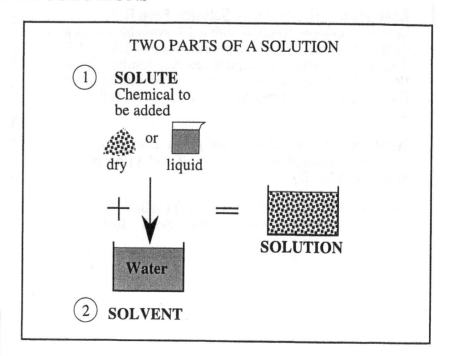

TWO PARTS OF A SOLUTION

① **SOLUTE**
Chemical to be added

dry   or   liquid

+   =   **SOLUTION**

Water

② **SOLVENT**

Example 1: (Percent Strength)
❑ If a total of 64 ounces of calcium hypochlorite (65% available Cl_2) are added to 10 gallons of water, what is the percent chlorine strength (by weight) of the solution?

Before calculating percent chlorine strength, the ounces hypochlorite must be converted to lbs hypochlorite:*

$$\frac{64 \ ounces}{16 \ ounces/pound} = 4 \ lbs \ chemical$$

Now calculate percent chlorine strength:

$$\% \ Cl_2 = \frac{\dfrac{(Hypochl., \ lbs)(\% \ Avail. \ Cl_2)}{100}}{Water, \ lbs + Hypochl., \ lbs} \times 100$$

$$= \frac{(4 \ lbs) (0.65)}{(10 \ gal) (8.34 \ lbs/gal) + 4 \ lbs} \times 100$$

$$= \frac{2.6 \ lbs}{83.4 \ lbs + 4 \ lbs} \times 100$$

$$= \frac{(2.6) (100)}{87.4}$$

$$= \boxed{3.0\% \ Cl_2 \ \ Strength}$$

* To review ounces to pounds conversions refer to Chapter 8 in *Basic Math Concepts*.

Example 2: (Percent Strength)
❏ If 150 grams of calcium hypochlorite (65% available chlorine) are dissolved in 5 gallons of water, what percent chlorine strength is the solution? (1 gram = 0.0022 lbs)

First, convert grams hypochlorite to pounds. Since 1 gram equals 0.0022 lbs, 150 grams is 150 times 0.0022 lbs:

$$(150 \text{ grams}) (0.0022 \text{ lbs/gram}) = 0.33 \text{ lbs}$$
Hypochlorite   Hypochlorite

Now calculate percent strength of the solution:

$$\% \text{ Cl}_2 \text{ Strength} = \frac{\dfrac{(\text{Hypochl., lbs})(\% \text{ Avail. Cl}_2)}{100}}{\text{Water, lbs} + \text{Hypochl., lbs}} \times 100$$

$$= \frac{\dfrac{(0.33 \text{ lbs}) (0.65)}{(5 \text{ gal}) (8.34 \text{ lbs/gal}) + 0.33 \text{ lbs}}}{} \times 100$$

$$= \frac{(0.2) (100)}{42.03}$$

$$= \boxed{0.5\% \text{ Cl}_2 \text{ Strength}}$$

WHEN GRAMS CHEMICAL ARE USED

The chemical (solute) to be used in making a solution may be measured in grams rather than pounds or ounces. When this is the case, convert grams to pounds before calculating percent strength. The following relationship is used for the conversion:

$$\boxed{1 \text{ gram} = 0.0022 \text{ lbs}}$$

Example 3: (Percent Strength)
❏ How many pounds of dry hypochlorite (65% available chlorine) must be added to 25 gallons of water to make a 1% chlorine solution:

First, write the equation as usual and fill in the known information. Then solve for the unknown value.

$$\% \text{ Cl}_2 \text{ Strength} = \frac{\dfrac{(\text{Hypochl., lbs})(\% \text{ Avail. Cl}_2)}{100}}{\text{Water, lbs} + \text{Hypochl., lbs}} \times 100$$

$$1 = \frac{(x \text{ lbs}) (0.65)}{(25 \text{ gal}) (8.34 \text{ lbs/gal}) + x \text{ lbs}} \times 100$$

$$1 = \frac{(x) (0.65) (100)}{208.5 + x}$$

$$208.5 + x = 65 x$$

$$208.5 = 64 x$$

$$\boxed{\begin{array}{c} 3.3 \text{ lbs} \\ \text{Hypochlorite} \end{array}} = x$$

SOLVING FOR OTHER UNKNOWN VARIABLES

In the percent chlorine strength equation there are four variables:

• % Strength

• lbs Hypochlorite

• % Available Cl$_2$

• lbs Water

In Examples 1 and 2, the unknown value was percent chlorine strength. However, the same equation can be used to determine any one of the other variables. Example 3 illustrates this type of calculation.

Note that gallons water can also be the unknown variable in percent strength calculations. In the denominator of the equation, use (X gal) (8.34 lbs/gal) **in place of** "lbs water".

PERCENT STRENGTH USING LIQUID HYPOCHLORITE

When using a liquid chemical to make up a solution, such as liquid polymer, a different calculation is required for percent strength.

The liquid is shipped from the supplier at a certain percent strength—perhaps 10% or 12%. This solution is then added to water to obtain a desired solution of lower percent chlorine—such as 1-3%.

In these percent strength problems, the lbs of chlorine in the liquid hypochlorite (from supplier) are set equal to the lbs of chlorine in the hypochlorite solution. The expanded equations are simply variations of this basic principle. Examples 4-6 illustrate the calculation of percent strength when using liquid hypochlorite.

THE KEY TO THESE CALCULATIONS—POUNDS CHEMICAL REMAINS CONSTANT

LIQUID HYPOCHLORITE (12-15% Chlorine)

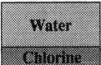

HYPOCHLORITE SOLUTION (1-3% Chlorine)

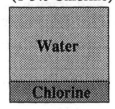

Simplified Equation:

$$\begin{array}{c}\text{lbs Chlorine} \\ \text{in Liquid} \\ \text{Hypochlorite}\end{array} = \begin{array}{c}\text{lbs Chlorine} \\ \text{in Hypochlorite} \\ \text{Solution}\end{array}$$

Expanded Equation:

$$\begin{array}{c}\text{(Liquid)} \\ \text{Hypo.,} \\ \text{lbs}\end{array} \frac{\text{(\% Strength)}}{\text{of Hypo.}}{100} = \begin{array}{c}\text{(Hypochlorite)} \\ \text{Solution,} \\ \text{lbs}\end{array} \frac{\text{(\% Strength)}}{\text{of Hypo.}}{100}$$

Or

$$\begin{array}{c}\text{(Liq. Hypo.)} \\ \text{gal}\end{array} \begin{array}{c}\text{(8.34)} \\ \text{lbs/gal}\end{array} \frac{\text{(\% Strength)}}{\text{of Hypo.}}{100} = \begin{array}{c}\text{(Hypo. Sol'n)} \\ \text{gal}\end{array} \begin{array}{c}\text{(8.34)} \\ \text{lbs/gal}\end{array} \frac{\text{(\% Strength)}}{\text{of Hypo.}}{100}$$

Example 4: (Percent Strength)

❏ A 12% liquid hypochlorite is to be used in making up a hypochlorite solution. If 3.3 gallons of liquid hypochlorite are mixed with water to produce 20 gals of hypochlorite solution, what is the percent strength of the solution?

$$\begin{array}{c}\text{(Liq. Hypo.)} \\ \text{gal}\end{array} \begin{array}{c}\text{(8.34)} \\ \text{lbs/gal}\end{array} \frac{\text{(\% Strength)}}{\text{of Hypo.}}{100} = \begin{array}{c}\text{(Hypo. Sol'n)} \\ \text{gal}\end{array} \begin{array}{c}\text{(8.34)} \\ \text{lbs/gal}\end{array} \frac{\text{(\% Strength)}}{\text{of Hypo.}}{100}$$

$$\text{(3.3 gal)} \begin{array}{c}\text{(8.34)} \\ \text{lbs/gal}\end{array} \frac{(12)}{100} = \text{(20 gal)} \begin{array}{c}\text{(8.34)} \\ \text{lbs/gal}\end{array} \frac{(x)}{100}$$

$$\frac{\cancel{(100)}\,(3.3)\,\cancel{(8.34)}\,(12)}{(20)\,\cancel{(8.34)}\,\cancel{(100)}} = x$$

$$\boxed{2.0\%} = x$$

Example 5: (Percent Strength)
❏ If 10 gallons of 11% liquid hypochlorite are mixed with water to produce 75 gallons of a hypochlorite solution, what is the percent strength of the solution? (Assume the density of both solutions is 8.34 lbs/gal.)

Use the expanded form of the equation, filling in known information:

$$\text{(Liq. Hypo.) gal} \times (8.34)\text{ lbs/gal} \times \frac{\text{(\% Strength) of Hypo.}}{100} = \text{(Hypo. Sol'n) gal} \times (8.34)\text{ lbs/gal} \times \frac{\text{(\% Strength) of Hypo.}}{100}$$

$$(10\text{ gal})(8.34)\text{ lbs/gal}\frac{(11)}{100} = (75\text{ gal})(8.34)\text{ lbs/gal}\frac{(x)}{100}$$

$$\frac{(100)(10)(8.34)(11)}{(75)(8.34)(100)} = x$$

$$\boxed{1.5\%} = x$$

Example 6: (Percent Strength)
❏ If 5 gallons of a 12% liquid hypochlorite are added to a 55-gallon drum, how much water should be added to the drum to produce a 2% hypochlorite solution?

First, use the equation to determine the **total gallons** of hypochlorite solution required:

$$\text{(Liq. Hypo.) gal} \times (8.34)\text{ lbs/gal} \times \frac{\text{(\% Strength) of Hypo.}}{100} = \text{(Hypo. Sol'n) gal} \times (8.34)\text{ lbs/gal} \times \frac{\text{(\% Strength) of Hypo.}}{100}$$

$$(5\text{ gal})(8.34)\text{ lbs/gal}(0.12) = (x\text{ gal})(8.34)\text{ lbs/gal}(0.02)$$

$$\frac{(5)(8.34)(0.12)}{(8.34)(0.02)} = x$$

$$30\text{ gal} = x$$

The hypochlorite solution is a total of 30 gallons. Since 5 gallons have been added to the drum already, only **25 gallons of water** must be added:

$$30\text{ gal solution} - 5\text{ gal added already} = \boxed{25\text{ gal water required}}$$

CALCULATING GALLONS OF WATER REQUIRED

In Example 5, the percent strength of the hypochlorite solution is the unknown value. In fact, <u>any</u> of the four variables can be the unknown value:

- Liquid hypochlorite, gal
- % strength of liquid hypochlorite
- Hypochlorite solution, gal
- % strength of hypochlorite solution.

In addition, there is another calculation that can be made using this equation: **gallons additional water required.** When making this calculation, you will first calculate the total gallons of hypochlorite solution required. Then the gallons additional water required can be calculated as:

$$\begin{array}{ccc}\text{Total gal} & \text{Gallons} & \text{Gallons} \\ \text{Solution} - \text{Liq. Hypo.} = \text{Additional} \\ & \text{Already} & \text{Water} \\ & \text{Added} & \text{Req'd}\end{array}$$

Example 6 illustrates a calculation of this type.

* Refer to Chapter 7, Section 7.1, "Density and Specific Gravity".

12.6 MIXING SOLUTIONS OF DIFFERENT STRENGTHS

There are two types of solution mixture calculations. In one type of calculation, two solutions of different strengths are mixed with no particular target solution strength. The calculation involves determining the percent strength of the solution mixture.

The second type of solution mixture calculation includes a desired or target strength. This type of problem is described later in this section.

WHEN DIFFERENT PERCENT STRENGTH SOLUTIONS ARE MIXED

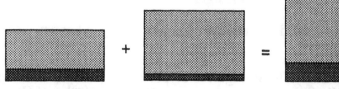

10% Strength Solution $+$ 1% Strength Solution $=$ Solution Mixture (% Strength somewhere between 10% and 1% depending on the quantity contributed by each.)

Simplified Equation:

$$\% \; Cl_2 \text{ Strength of Mixture} = \frac{\text{Chlorine in Mixture, lbs}}{\text{Solution Mixture, lbs}} \times 100$$

Expanded Equations:

$$\% \; Cl_2 \text{ Strength of Mixture} = \frac{\dfrac{\text{lbs } Cl_2 \text{ from Solution 1}} {} + \dfrac{\text{lbs } Cl_2 \text{ from Solution 2}}{}}{\text{lbs Solution 1 + lbs Solution 2}} \times 100$$

$$\% \; Cl_2 \text{ Strength of Mixture} = \frac{\text{(Sol'n 1)} \text{lbs} \dfrac{(\% \text{ Avail. } Cl_2 \text{ of Sol'n 1})}{100} + \text{(Sol'n 2)} \text{lbs} \dfrac{(\% \text{ Avail. } Cl_2 \text{ of Sol'n 2})}{100}}{\text{lbs Solution 1 + lbs Solution 2}} \times 100$$

Example 1: (Solution Mixtures)

❑ If 50 lbs of a hypochlorite solution (10% available chlorine) are mixed with 200 lbs of another hypochlorite solution (1% available chlorine), what is the percent chlorine of the solution mixture?

$$\% \; Cl_2 \text{ Strength of Mixture} = \frac{\text{(Sol'n 1)} \text{lbs} \dfrac{(\% \text{ Avail. } Cl_2 \text{ of Sol'n 1})}{100} + \text{(Sol'n 2)} \text{lbs} \dfrac{(\% \text{ Avail. } Cl_2 \text{ of Sol'n 2})}{100}}{\text{lbs Solution 1 + lbs Solution 2}} \times 100$$

$$= \frac{(50 \text{ lbs})(0.1) + (200 \text{ lbs})(0.01)}{50 \text{ lbs} + 200 \text{ lbs}} \times 100$$

$$= \frac{5 \text{ lbs} + 2 \text{ lbs}}{250 \text{ lbs}} \times 100$$

$$= \boxed{2.8\%}$$

Example 2: (Solution Mixtures)
❏ If 5 gallons of an 8% hypochlorite solution are mixed with 40 gallons of a 0.5% hypochlorite solution, what is the percent strength of the solution mixture? (Assuming the density of both solutions is 8.34 lbs/gal.)

$$\text{% Cl}_2 \text{ Strength of Mixture} = \frac{(\text{Sol'n 1}) \text{ lbs} \dfrac{(\text{% Avail.}) \text{ Chlorine}}{100} + (\text{Sol'n 2}) \text{ lbs} \dfrac{(\text{% Avail.}) \text{ Chlorine}}{100}}{\text{lbs Solution 1} + \text{lbs Solution 2}} \times 100$$

$$= \frac{(5 \text{ gal}) (8.34 \text{ lbs/gal}) (0.08) + (40 \text{ gal}) (8.34 \text{ lbs/gal}) (0.005)}{(5 \text{ gal}) (8.34 \text{ lbs/gal}) + (40 \text{ gal}) (8.34 \text{ lbs/gal})} \times 100$$

$$= \frac{3.3 \text{ lbs Cl}_2 + 1.7 \text{ lbs Cl}_2}{41.7 \text{ lbs Sol'n 1} + 333.6 \text{ lbs Sol'n 2}} \times 100$$

$$= \frac{5 \text{ lbs Chlorine}}{375.3 \text{ lbs Solution}} \times 100$$

$$= \boxed{1.3\% \text{ Strength}}$$

Example 3: (Solution Mixtures)
❏ If 15 gallons of a 10% hypochlorite solution are added to 50 gallons of 2% hypochlorite solution, what is the percent strength of the solution mixture?

$$\text{% Cl}_2 \text{ Strength of Mixture} = \frac{(\text{Sol'n 1}) \text{ gal} \dfrac{(\text{% Avail.}) \text{ Chlorine}}{100} + (\text{Sol'n 2}) \text{ gal} \dfrac{(\text{% Avail.}) \text{ Chlorine}}{100}}{\text{Solution 1, gal} + \text{Solution 2, gal}} \times 100$$

$$= \frac{(15 \text{ gal}) (0.1) + (50 \text{ gal}) (0.02)}{15 \text{ gal} + 50 \text{ gal}} \times 100$$

$$= \frac{1.5 \text{ gal Cl}_2 + 1 \text{ gal Cl}_2}{65 \text{ gal}} \times 100$$

$$= \frac{2.5 \text{ gal Chlorine}}{65 \text{ gal}} \times 100$$

$$= \boxed{3.8\% \text{ Strength}}$$

USE DIFFERENT DENSITY FACTORS WHEN APPROPRIATE

Percent strength is generally expressed in terms of **pounds chemical per pounds solution** in case the solutions have different densities.* Examples 1 and 2 illustrate calculations where the solutions are expressed in pounds.

It is important to know what density factor should be used to convert from gallons to pounds. If the solution has a density the same as water, 8.34 lbs/gal is used. If, however, the solution has a higher density, such as some polymer solutions, then the higher density factor should be used.

When density factors of all solutions in a problem are the same, the equation is sometimes given in terms of gallons rather than lbs as shown in Example 3.

---

* Refer to Chapter 7, Section 7.1, "Density and Specific Gravity."

SOLUTION MIXTURES TARGET PERCENT STRENGTH

In the previous section we examined the first type of solution mixture calculation—a calculation where there is no target percent strength. In this type calculation, two solutions are mixed and the percent strength of the mixture is determined.

In the second type of solution mixture calculation, **two different percent strength solutions are mixed in order to obtain a desired quantity of solution and a target percent strength**. These problems may be solved using the same equation shown in Examples 1-3. An illustration of this approach is given in Example 4.

Another and perhaps preferred approach in solving these problems is by using the dilution rectangle. Although the first use of the dilution rectangle can be confusing, the effort to master its use is rewarded— solution mixture problems are quickly calculated. Example 5 uses the dilution rectangle to solve the problem stated in Example 4. Compare the two methods of calculating this type of mixture problem.

Example 4: (Solution Mixtures)
❏ What weights of a 1% hypochlorite solution and a 12% hypochlorite solution must be mixed to make 850 lbs of a 2% solution?

Use the same equation as shown for Examples 1-3 and fill in given information.* (Note that the lbs of Solution 1 is unknown, x. If lbs of Solution 1 is x, then the lbs of Solution 2 must be the balance of the 850 lbs, or $850 - x$.)

$$\text{% Strength of Mixture} = \frac{(\text{Sol'n 1})\,\text{lbs}\,\dfrac{(\text{% Strength of Sol'n 1})}{100} + (\text{Sol'n 2})\,\text{lbs}\,\dfrac{(\text{% Strength of Sol'n 2})}{100}}{\text{lbs Solution 1 + lbs Solution 2}} \times 100$$

$$2 = \frac{(x\text{ lbs})(0.01) + (850 - x\text{ lbs})(0.12)}{850\text{ lbs}} \times 100$$

$$\frac{(2)}{100}(850) = 0.01x + 102 - 0.12x$$

$$17 = -0.11x + 102$$

$$0.11x = 85$$

$$x = \boxed{773\text{ lbs of 1% Solution}}$$

Then $850 - 773 = \boxed{77\text{ lbs of 12% Solution}}$

THE DILUTION RECTANGLE

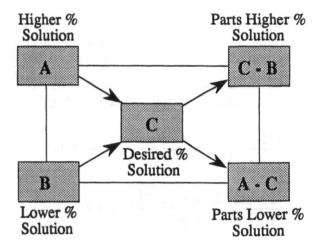

Higher % Solution — A
Parts Higher % Solution — C - B
C — Desired % Solution
Lower % Solution — B
Parts Lower % Solution — A - C

Steps in Using the Dilution Rectangle:

1. Place the % Strength numbers in positions A, B, and C.

2. Calculate parts higher % solution and parts lower % solution, subtracting as indicated.

3. Multiply fractional parts of each solution by the total lbs of solution desired.

---

* Refer to Chapter 2 in *Basic Math Concepts* for a review of solving for the unknown value.

Example 5: (Solution Mixtures)
❏ What weights of a 1% solution and a 12% solution must be mixed to make 850 lbs of a 2% solution?

Use the Dilution Rectangle to solve this problem. First determine the parts required of each solution:

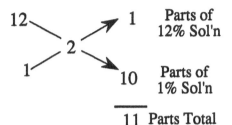

Thus, 1 part of the total 11 parts (1/11) come from the 12% hypochlorite solution, and the other 10 parts (10/11) come from the 1% solution. Now calculate the lbs of 1% and 12% solution, using these fractions:

$$\text{Amt. of}\atop\text{12\% Soln:} \quad \frac{1}{11} \ (850 \text{ lbs}) = \boxed{77 \text{ lbs}}$$

$$\text{Amt. of}\atop\text{1\% Soln:} \quad \frac{10}{11} \ (850 \text{ lbs}) = \boxed{773 \text{ lbs}}$$

Example 6: (Solution Mixtures)
❏ How many lbs of a 10% solution and water should be mixed together to form 425 lbs of a 1% solution?

First calculate the parts of each solution required:

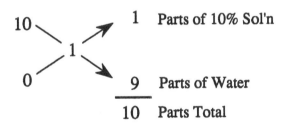

Then calculate the actual lbs of each solution:

$$\text{Amt. of}\atop\text{10\% Sol'n:} \quad \frac{(1)}{10} \ (425 \text{ lbs}) = \boxed{\text{42.5 lbs}\atop\text{10\% Sol'n}}$$

$$\text{Amt. of}\atop\text{Water:} \quad \frac{(9)}{10} \ (425 \text{ lbs}) = \boxed{\text{382.5 lbs}\atop\text{Water}}$$

MIXING A SOLUTION AND WATER

In solution mixing Examples 1-5, two solutions of different strengths are blended. The solution mixing equation and dilution rectangle can also be used when only one solution and water are blended. (Water is considered a 0% strength solution.) Example 6 illustrates such a calculation using the dilution rectangle.

12.7 CHEMICAL USE CALCULATIONS

The lbs/day or gpd chemical use should be recorded each day. From this data, you can calculate the average daily use of the chemical or solution. You can also forecast expected chemical use, compare it with chemical in inventory, and determine when additional chemical supplies will be required.

AVERAGE CHEMICAL USE

First determine the average chemical use:

$$\frac{\text{Average Use}}{\text{lbs/day}} = \frac{\text{Total Chem. Used, lbs}}{\text{Number of Days}}$$

Or

$$\frac{\text{Average Use}}{\text{gpd}} = \frac{\text{Total Chem. Used, gal}}{\text{Number of Days}}$$

Then calculate day's supply in inventory:*

$$\frac{\text{Days' Supply}}{\text{in Inventory}} = \frac{\text{Total Chem. in Inventory, lbs}}{\text{Average Use, lbs/day}}$$

Or

$$\frac{\text{Days' Supply}}{\text{in Inventory}} = \frac{\text{Total Chem. in Inventory, gal}}{\text{Average Use, gpd}}$$

Example 1: (Chemical Use)
❑ The lbs calcium hypochlorite used for each day during a week is given below. Based on this data, what was the average lbs/day hypochlorite chemical use during the week?

Monday—51 lbs/day   Friday—58 lbs/day
Tuesday—57 lbs/day   Saturday—52 lbs/day
Wednesday—53 lbs/day   Sunday—49 lbs/day
Thursday—47 lbs/day

$$\frac{\text{Average Use}}{\text{lbs/day}} = \frac{\text{Total Chem. Used, lbs}}{\text{Number of Days}}$$

$$= \frac{367 \text{ lbs}}{7 \text{ days}}$$

$$= \boxed{\begin{array}{c}52.4 \text{ lbs/day}\\ \text{Aver. Use}\end{array}}$$

---

* Note how similar these equations are to detention time equations. Refer to Chapter 5 for a review of detention time calculations.

Example 2: (Chemical Use)
❑ The average calcium hypochlorite use at a plant is 45 lbs/day. If the chemical inventory in stock is 1200 lbs, how many days' supply is this?

$$\text{Days' Supply in Inventory} = \frac{\text{Total Chem. in Inventory, lbs}}{\text{Average Use, lbs/day}}$$

$$= \frac{1200 \text{ lbs in Inventory}}{45 \text{ lbs/day Aver. Use}}$$

$$= \boxed{\begin{array}{l}26.7 \text{ days'} \\ \text{Supply in Inventory}\end{array}}$$

Example 3: (Chemical Use)
❑ The average gallons of sodium hypochlorite solution used each day at a treatment plant is 50 gpd. A chemical feed tank has a diameter of 3 ft and contains solution to a depth of 4.3 ft. How many days' supply is represented by the solution in the tank?

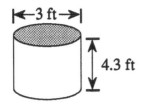

$$\text{Days' Supply in Tank} = \frac{\text{Total Solution in Tank, gal}}{\text{Average Use, gpd}}$$

$$x \text{ days} = \frac{(0.785)\,(3 \text{ ft})\,(3 \text{ ft})\,(4.3 \text{ ft})\,(7.48 \text{ gal/cu ft})}{50 \text{ gpd}}$$

$$= \boxed{\begin{array}{l}4.5 \text{ days'} \\ \text{Supply in Tank}\end{array}}$$

Additional types of chemical use calculations include total lbs chemical use and chlorine cylinders required, as shown in the equations to the right and illustrated in Examples 4-6.

OTHER CHEMICAL USE CALCULATIONS

Total Chemical Use, lbs:

$$\text{Total Chemical Used, lbs} = \frac{(\text{Chem. Use,}) (\text{Days Use})}{\text{lbs/day}}$$

Or

$$\text{Total Chemical Used, lbs} = \frac{(\text{Chem. Use,}) (\text{Hrs Use})}{\text{lbs/hr}}$$

Chlorine Cylinders Required:

$$\text{Required Cylinders/Week} = \frac{\text{Chem. Use, lbs/week}}{\text{lbs Chem./Cylinder}}$$

Or

$$\text{Required Cylinders/Month} = \frac{\text{Chem. Use, lbs/month}}{\text{lbs Chem./Cylinder}}$$

Example 4: (Chemical Use)

❏ An average of 5.8 lbs of chlorine are used each day at a plant. How many pounds of chlorine would be used in one week if the hour meter on the pump indicates 115 hrs operation?

$$\text{Total Chem. Used, lbs} = \frac{(\text{Chem. Use,}) (\text{Hrs Used})}{\text{lbs/hr}}$$

$$= \frac{(5.8 \text{ lbs/day}) (115 \text{ hrs operation})}{24 \text{ hrs/day}}$$

$$= \boxed{27.8 \text{ lbs Used}}$$

Example 5: (Chemical Use)
❑ A chlorine cylinder has 82 lbs chlorine at the beginning of a week. The chlorinator setting is 7.5 lbs per 24 hours. If the pump hour meter indicates the pump has operated a total of 110 hours during the week, how many lbs chlorine should be in the cylinder at the end of the week?

First calculate the lbs/hr chemical use:

$$\frac{7.5 \text{ lbs}}{24 \text{ hrs}} = 0.31 \text{ lbs/hr}$$

The chemical use each hour is 0.31 lbs/hr. For a total of 110 hrs, the chemical use during the week is:

$$\left(0.31 \frac{\text{lbs}}{\text{hr}}\right) (110 \text{ hrs}) = 34.1 \text{ lbs}$$

The lbs chlorine remaining at the end of the week is:

$$82 \text{ lbs} - 34.1 \text{ lbs} = \boxed{\begin{array}{c} 47.9 \text{ lbs chlorine} \\ \text{remaining} \end{array}}$$

Example 6: (Chemical Use)
❑ An average of 35 lbs of chlorine are used each day at a plant. How many 150-lb chlorine cylinders will be required each month? (Assume a 30-day month.)

The lbs chlorine used during the month is:

$$\left(35 \frac{\text{lbs}}{\text{day}}\right) (30 \text{ days}) = 1050 \text{ lbs chlorine}$$

The number of 150-lb containers is:

$$\frac{1050 \text{ lbs Cl}_2}{150 \text{ lbs/cylinder}} = \boxed{\begin{array}{c} 7 \text{ cylinders} \\ \text{required} \end{array}}$$

NOTES:

13 *Fluoridation*

SUMMARY

1. **Methods of Expressing Concentration**

    - mg/*L* chemical
    - Percent chemical
    - lbs chemical/MG treated

2. **Percent Fluoride Ion in a Compound**

$$\text{\% Fluoride Ion in a Compound} = \frac{\text{Molecular Wt. of Fluoride}}{\text{Molecular Wt. of Compound}} \times 100$$

3. **Calculating Dry Feed Rate of a Fluoride Compound, lbs/day**

$$\frac{(\text{mg}/L \text{ F}) \, (\text{MGD flow}) \, (8.34 \text{ lbs/gal})}{\dfrac{(\text{\% Comm. Purity})}{100} \dfrac{(\text{\% F in Comp.})}{100}} = \begin{array}{l}\text{lbs/day}\\ \text{Fluoride}\\ \text{Compound}\end{array}$$

4. **Percent Strength of a Solution**

    <u>Simplified Equation:</u>

$$\text{\% Strength of a Solution} = \frac{\text{lbs Chemical}}{\text{lbs Solution}} \times 100$$

    <u>Expanded Equations:</u>

    When the commercial chemical is 100% pure:

$$\text{\% Strength} = \frac{\text{lbs Chemical}}{\text{lbs Water} + \text{lbs Chemical}} \times 100$$

    When the commercial chemical is less than 100% pure:

$$\text{\% Strength} = \frac{\dfrac{(\text{lbs Chem.}) \, (\text{\% Comm. Purity})}{100}}{\text{lbs Water} + \dfrac{(\text{lbs Chem.}) \, (\text{\% Comm. Pur.})}{100}} \times 100$$

SUMMARY—Cont'd

5. Fluoride Solution Feed Rate Calculations

There are two related methods of calculating feed rates:

• The mg/L to lbs/day method:

Simplified Equation:

$$\begin{array}{ccc} \text{Actual Feed,} & = & \text{Solution Feed,} \\ \text{lbs F/day} & & \text{lbs F/day} \end{array}$$

Expanded Equation:

$$\frac{(\text{mg/}L\ \text{F})\ (\text{MGD})\ (8.34)}{\text{flow}\quad\text{lbs/gal}} = \frac{(\text{mg/L})\ (\text{MGD})\ (8.34)}{\text{Comp. flow}\quad\text{lbs/gal}}\ (\text{sp. gr.})\ \frac{(\%\ \text{F})}{\text{in Comp.}}$$

$$\frac{}{100}$$

When applicable

• The rate/concentration method:

$$R_1 \times C_1 = R_2 \times C_2$$

where: R_1 = feed rate of first solution
C_1 = concentration of first solution
R_2 = feed rate of second solution
C_2 = concentration of second solution

6. Feed Rates Using Charts and Nomographs

Using the Given Chart

• Enter the chart at the indicated flow rate along the left scale.

• Draw a horizontal line to the right until it intersects with the given solution strength line.

• Then move vertically down to the flow rate scale and read the value indicated.

Use a proportion to determine the flow rate corresponding to the desired mg/L dosage:

$$\frac{\text{Desired Dose, mg/}L}{\text{Chart Dose, 1 mg/}L} = \frac{\text{Desired Feed Rate, gpd}}{\text{Chart Feed Rate, gpd}}$$

6. Feed Rates Using Charts and Nomographs—Cont'd

Using the Given Nomograph

- Draw a line from the flow rate scale through the fluoride dosage (mg/L) scale until the line intersects with the lbs/day fluoride scale.

- From the point of intersection, draw a horizontal line to the right to obtain the approximate weight of dry chemical or volume of liquid chemical required.

7. Calculating mg/L Fluoride Dosage

Given lbs/day Fluoride:

$$(mg/L \text{ F}) (MGD \text{ Flow}) (8.34 \text{ lbs/gal}) = \text{lbs/day F}$$

For Dry Feed Rate Calculations:

$$\frac{(mg/L)(MGD)(8.34 \text{ lbs/gal})}{\frac{(\% \text{ Comm. Pur.})}{100} \frac{(\% \text{ F in Comp.})}{100}} = \frac{\text{lbs/day}}{\text{Compound}}$$

For Solution Feed Rate Calculations:

$$(mg/L \text{ F})(MGD)(8.34) = (mg/L)(MGD)(8.34)(\text{sp. gr.})\frac{(\% \text{ F})}{100}$$
flow lbs/gal   Comp. flow lbs/gal   in Comp.

When applicable

SUMMARY—Cont'd

8. Solution Mixtures

<u>Simplified Equation:</u>

$$\text{\% Strength of Mixture} = \frac{\text{lbs Chemical}}{\text{lbs Solution}} \times 100$$

<u>Expanded Equation:</u>

$$\text{\% Strength of Mixture} = \frac{\text{lbs Chem. in Sol'n 1} + \text{lbs Chem. in Sol'n 2}}{\text{lbs Sol'n 1} + \text{lbs Sol'n 2}} \times 100$$

Or

$$\text{\% Strength of Mixture} = \frac{\dfrac{(\text{Sol'n 1, gal})(8.34 \text{ lbs/gal})(\text{\% Strength of Sol'n 1})}{100} + \dfrac{(\text{Sol'n 2, gal})(8.34 \text{ lbs/gal})(\text{\% Strength of Sol'n 2})}{100}}{(\text{Sol'n 1 gal})(8.34 \text{ lbs/gal}) + (\text{Sol'n 2, gal})(8.34 \text{ lbs/gal})} \times 100$$

When mixing solutions with the same density, the lbs/gal factor can be dropped from the equation:

$$\text{\% Strength of Mixture} = \frac{\dfrac{(\text{Sol'n 1, gal})(\text{\% Strength of Sol'n 1})}{100} + \dfrac{(\text{Sol'n 2, gal})(\text{\% Strength of Sol'n 2})}{100}}{\text{Sol'n 1, gal} + \text{Sol'n 2, gal}} \times 100$$

By moving the denominator to the left side of the equation, the equations above can be stated generally as:

$$R_3 C_3 = R_1 C_1 + R_2 C_2$$

where: R_3 = lbs (or gal) of Sol'n Mixture
C_3 = % Strength of Mixture
R_1 = lbs (or gal) of Sol'n 1
C_1 = % Strength of Sol'n 1
R_2 = lbs (or gal) of Sol'n 2
C_2 = % Strength of Sol'n 2

NOTES:

13.1 METHODS OF EXPRESSING CONCENTRATION

In making chemical dosage calculations, it is important to be able to convert back and forth between various expressions of concentration. In this way, should you require a different expression of concentration for a given calculation, this will not present a problem. Three common expressions of concentration are:

- mg/L,
- percent, and
- lbs/MG

PERCENT AND mg/L CONVERSIONS

To convert between mg/L and percent concentration, use the following equation:*

$$\boxed{1\% \;=\; 10{,}000 \text{ mg/}L}$$

The conversion between mg/L and percent therefore involves multiplying or dividing by 10,000, depending on the direction of the conversion. To multiply by 10,000, simply move the decimal point **four places to the right**. To divide by 10,000, move the decimal point **four places to the left**.

Remember that **percent is always the smaller number.** Knowing this will help you decide whether to move the decimal point to the right or to the left for any given conversion. Examples 1-3 illustrate these calculations.

Example 1: (Expressing Concentration)
❑ Express 2.8% concentration in terms of mg/L concentration.

When converting between percent and mg/L, there is always a decimal point move of 4 places (as a result of multiplying or dividing by 10,000). Since percent is always the smaller number in these conversions, the decimal point must be moved **4 places to the right**, thus making 2.8 a larger number:

$$2.8\% \;=\; 2.8000. \;=\; \boxed{28{,}000 \text{ mg/}L}$$

Example 2: (Expressing Concentration)
❑ Convert 5800 mg/L to percent.

From mg/L to percent, the number must get <u>smaller</u>. Therefore, a decimal point move four places <u>to the left</u> is required:

$$5{,}800 \text{ mg/}L \;=\; 0.5800. \;=\; \boxed{0.5\%}$$

This answer makes sense. Since 10,000 mg/L = 1%, about 5000 mg/L is about a half a percent (0.5%).

Example 3: (Expressing Concentration)
❑ Express 24% concentration in terms of mg/L.

From percent to mg/L the number must get larger. Therefore the decimal point must be moved four places to the right:

$$24\% \;=\; 24.0000. \;=\; \boxed{240{,}000 \text{ mg/}L}$$

---

* For a review of mg/L and percent conversions, refer to Chapter 8 in *Basic Math Concepts*.

Example 4: (Expressing Concentration)
❑ Express 11 lbs/MG concentration as mg/*L*.

First convert lbs/MG concentration as lbs/mil. lbs. (The MG is converted to mil. lbs.)

$$\frac{11\ lbs}{(1\ MG)\ (8.34\ lbs/gal)} = \frac{11\ lbs}{8.34\ mil.\ lbs} = \frac{1.3\ lbs}{1\ mil.\ lbs}$$

The expression of lbs/mil. lbs is the same as ppm and mg/*L*:

$$\frac{1.3\ lbs}{1\ mil.\ lbs} = 1.3\ ppm = \boxed{1.3\ mg/L}$$

Example 5: (Expressing Concentration)
❑ Convert 6 mg/*L* to lbs/MG.

First convert mg/*L* to lbs/mil. lbs:

$$6\ mg/L = 6\ ppm = \frac{6\ lbs}{1\ mil.\ lbs}$$

Then convert lbs/mil. lbs to lbs/MG. (The denominator must be converted from mil. lbs to MG.)

$$\frac{6\ lbs}{\dfrac{1\ mil.\ lbs}{8.34\ lbs/gal}} = \frac{6\ lbs}{0.12\ MG} = \boxed{\frac{50\ lbs}{1\ MG}}$$

Example 6: (Expressing Concentration)
❑ Convert 25 lbs/MG concentration to mg/*L* concentration.

First express lbs/MG concentration as lbs/mil. lbs. To do this, convert the denominator from gallons to lbs:

$$\frac{25\ lbs}{(1\ MG)\ (8.34\ lbs/gal)} = \frac{25\ lbs}{8.34\ mil.\ lbs} = \frac{3\ lbs}{1\ mil.\ lbs}$$

Once the concentration has been expressed as lbs/mil. lbs, it has already been converted to mg/*L*:

$$\frac{3\ lbs}{1\ mil.\ lbs} = 3\ ppm = \boxed{3\ mg/L}$$

* The units used in expressing ppm are generally units of weight, so that there is not a problem resulting from different densities. (Refer to Chapter 7 for a discussion of density).

lbs/MG AND mg/*L* CONVERSIONS

The key to these conversions is that both mg/*L* and lbs/MG can be expressed in common terms, lbs/mil. lbs:

$$\boxed{mg/L \leftrightarrow \frac{lbs}{mil.\ lbs} \leftrightarrow \frac{lbs}{MG}}$$

To express mg/*L* concentration as lbs/million lbs requires no concentration conversion at all since mg/*L* is equivalent to parts per million parts (ppm) concentration. Parts per million can be expressed using any units—milligrams, grams, pounds, etc:*

$$mg/L = 6\ ppm = \frac{mg}{1,000,000\ mg}$$

$$or = \frac{grams}{1,000,000\ grams}$$

$$or = \frac{lbs}{1,000,000\ lbs}$$

So, a concentration of 2 mg/*L* may be stated as:

$$2\ mg/L = 2\ ppm = \frac{2\ lbs}{1,000,000\ lbs}$$

And a concentration of 43 lbs/ 1,000,000 lbs may be stated as:

$$\frac{43\ lbs}{1,000,000\ lbs} = 43\ ppm = 43\ mg/L$$

To express lbs/MG concentration as lbs/million lbs, simply requires converting million gallons (MG) in the denominator to million lbs, using the conversion factor 8.34 lbs/gal. In this calculation you multiply only the denominator of the fraction since you are not changing (or increasing) the amount of water represented. You are merely changing how you are reporting that amount of water (lbs instead of gallons).

Examples 4-6 illustrate these calculations.

13.2 PERCENT FLUORIDE ION IN A COMPOUND

Fluoride dosage calculations are based on the amount of fluoride ion added to the water. When fluoride compounds are used, **only a percentage of the compound is fluoride ion**. This percent of fluoride ion should be included in all dosage calculations involving fluoride compounds.

When calculating the percent fluoride ion present in a compound, you will need to know:

- The **chemical formula** for the compound (such as NaF, H_2SiF_6, or Na_2SiF_6), and

- The **atomic weight** of each element in the compound.

The chemical formula of a compound tells you <u>which elements</u> comprise the compound, as well as the <u>number of atoms of each element</u> required in making up a molecule of that compound. For example, the chemical formula for water is H_2O. This means that two atoms of hydrogen combine with one atom of oxygen to form a molecule of water. By using the atomic weights of hydrogen and oxygen, the percent of hydrogen or oxygen (by weight) in a molecule of water can be calculated.

Calculate the percent hydrogen in H_2O: (Use the Periodic Table of Elements shown above to determine the atomic weight of hydrogen and oxygen.)

Element	Number of Atoms	Atomic Wt.	Molec. Wt.
H	2	1.008	2.016
O	1	16.000	16.000

Total Molec. Wt. = 18.016
of H_2O

PARTIAL LISTING OF THE
PERIODIC TABLE OF ELEMENTS*

1 H 1.008					
3 Li 6.940	4 Be 9.02				
11 Na 22.997	12 Mg 24.32				

5 B 10.82	6 C 12.010	7 N 14.008	8 O 16.000	9 F 19.00	10 Ne 20.183
13 Al 26.97	14 Si 28.06	15 P 30.98	16 S 32.06	17 Cl 35.457	18 A 39.944

To calculate the percent fluoride ion in a compound:

1. Calculate the molecular weight of each element in the compound:

Element	No. of Atoms	Atomic Wt.	Molec. Wt.
:	$\Box$ x	$\Box$ =	$\Box$

2. Calculate the percent fluoride in the compound:

$$\text{\% F in Compound} = \frac{\text{Molecular Wt. of Fluoride}}{\text{Molecular Wt. of Compound}} \times 100$$

Example 1: (% F in a Compound)
❑ Calculate the percent fluoride in sodium fluoride (NaF)

First, calculate the molecular weight of each element in the compound: (Note: the atomic weights of both sodium and fluoride are given in the partial Periodic Table above.)

Element	No. of Atoms	Atomic Wt.	Molec. Wt.
Na:	1	22.997	22.997
F:	1	19.00	19.00

Molec. Wt. of NaF: 41.997

Then calculate the percent fluoride in NaF:

$$\text{\% F in NaF} = \frac{\text{Molec. Wt. of F}}{\text{Molec. Wt. of NaF}} \times 100$$

$$= \frac{19.00}{41.997} \times 100$$

$$= \boxed{45.2\%}$$

* For a complete listing of the Periodic Table of Elements, consult a general chemistry text.

Example 2: (% F in a Compound)
❑ Calculate the percent fluoride ion in hydrofluosilicic acid (H_2SiF_6).

First, calculate the molecular weight of each element in the compound:

Element	No. of Atoms		Atomic Wt.		Molec. Wt.
H:	2	x	1.008	=	2.016
Si:	1	x	28.06	=	28.06
F:	6	x	19.00	=	114.00
					144.076

The percent fluoride in the compound can now be calculated:

$$\% \text{ F in } H_2SiF_6 = \frac{\text{Molec. Wt. of F}}{\text{Molec. Wt. of } H_2SiF_6} \times 100$$

$$= \frac{114.00}{144.076} \times 100$$

$$= \boxed{79.1\%}$$

As with all percent calculations,* the numerator of the equation is the part you are interested in (hydrogen in this case), and the denominator is the total amount (total H_2O). To calculate the percent hydrogen in each molecule of H_2O, you would therefore use the following equation:

$$\% \text{ H} = \frac{\text{Molec. Wt. of H}}{\text{Molec. Wt. of } H_2O} \times 100$$

$$= \frac{2.016}{18.016} \times 100$$

$$= 11.2\%$$

The same basic approach is used in calculating the percent fluoride in any given fluoride compound. Examples 1-3 illustrate this calculation.

Example 3: (% F in a Compound)
❑ Calculate the percent fluoride ion present in sodium silicofluoride (Na_2SiF_6).

Calculate the molecular weight of each element in the compound:

Element	No. of Atoms		Atomic Wt.		Molec. Wt.
Na:	2	x	22.997	=	45.994
Si:	1	x	28.06	=	28.06
F:	6	x	19.00	=	114.00
					188.054

Then calculate the percent fluoride in NaF:

$$\% \text{ F in } Na_2SiF_6 = \frac{\text{Molec. Wt. of F}}{\text{Molec. Wt. of } Na_2SiF_6} \times 100$$

$$= \frac{114.00}{188.054} \times 100$$

$$= \boxed{60.6\%}$$

* For a review of percent problems, refer to Chapter 5 in *Basic Math Concepts*.

13.3 CALCULATING DRY FEED RATE, lbs/day

In order to determine the amount of fluoride compound required to accomplish a given fluoride dosage you must consider two factors:

• The percent commercial purity of the compound, and

• The percent fluoride ion in the compound.

When a fluoride compound is manufactured, it contains various impurities. The amount of these impurities depends both on the compound being manufactured and the process used in manufacturing that compound. The commercial purity of three common fluoride compounds is commonly as follows:

Sodium Fluoride, NaF = 90-98%

Sodium Silicofluoride, Na_2SiF_6 = 98-99%

Hydrofluosilicic Acid, H_2SiF_6 = 22-30%

Even if the compound is 100% pure (i.e. pure NaF, pure Na_2SiF_6 or pure H_2SiF_6), only a portion of that chemical is fluoride ion. The rest is sodium, hydrogen, or silica.**

Therefore, when you are calculating fluoride compound feed rate (lbs/day), you must incorporate the two factors of percent compound purity and percent fluoride ion present in the compound. These calculations are similar to hypochlorite calculations described in Chapter 3.

Examples 1-3 illustrate this calculation.

FLUORIDE COMPOUND DOSAGE CALCULATIONS MUST INCORPORATE TWO CONSIDERATIONS

1. Percent commercial purity of the compound, and

2. Percent fluoride ion in the compound.

First calculate the lbs/day fluoride required:*

$$\text{(mg/}L\text{ F) (MGD flow) (8.34 lbs/gal)} = \text{lbs/day Fluoride}$$

Then calculate the lbs/day fluoride compound required:

$$\frac{\text{Fluoride, lbs/day}}{\dfrac{\text{(\% Comm. Purity)}}{100}\dfrac{\text{(\% F in Comp.)}}{100}} = \begin{array}{l}\text{lbs/day} \\ \text{Fluoride} \\ \text{Compound}\end{array}$$

These two equations can be combined into one equation:

$$\frac{\text{(mg/}L\text{ F) (MGD flow) (8.34 lbs/gal)}}{\dfrac{\text{(\% Comm. Purity)}}{100}\dfrac{\text{(\% F in Comp.)}}{100}} = \begin{array}{l}\text{lbs/day} \\ \text{Fluoride} \\ \text{Compound}\end{array}$$

Example 1: (Dry Feed Rate, lbs/day)

❏ A fluoride dosage of 1.1 mg/L is desired. How many lbs/day dry sodium silicofluoride (Na_2SiF_6) will be required if the flow to be treated is 1.9 MGD? The commercial purity of the sodium silicofluoride is 98% and the percent of fluoride ion in the compound is 60.6%. (Assume the water to be treated contains no fluoride.)

$$\frac{\text{(mg/}L\text{ F) (MGD flow) (8.34 lbs/gal)}}{\dfrac{\text{(\% Comm. Purity)}}{100}\dfrac{\text{(\% F in Comp.)}}{100}} = \begin{array}{l}\text{lbs/day} \\ \text{Fluoride} \\ \text{Compound}\end{array}$$

$$\frac{\text{(1.1 mg/}L\text{) (1.9 MG) (8.34 lbs/gal)}}{\dfrac{(98)}{100}\dfrac{(60.6)}{100}} = \boxed{\begin{array}{l}\text{29.4 lbs/day} \\ Na_2SiF_6\end{array}}$$

---

* For a review of mg/L to lbs/day calculations, refer to Chapter 3.
** Section 13.2 describes how to calculate the percent fluoride in various fluoride compounds.

Example 2: (Dry Feed Rate, lbs/day)
❑ A fluoride dosage of 1.2 mg/L is desired. The flow to be treated is 2,760,000 gpd. How many lbs/day dry sodium silicofluoride (Na_2SiF_6) will be required if the commercial purity of the sodium silicofluoride is 98% and the percent of fluoride ion in the compound is 60.6%? (Assume the water to be treated contains no fluoride.)

$$\frac{(mg/L\ F)\ (MGD\ flow)\ (8.34\ lbs/gal)}{\dfrac{(\%\ Comm.\ Purity)}{100}\ \dfrac{(\%\ F\ in\ Comp.)}{100}} = \begin{array}{l} lbs/day \\ Fluoride \\ Compound \end{array}$$

$$\frac{(1.2\ mg/L)\ (2.76\ MG)\ (8.34\ lbs/gal)}{\dfrac{(98)}{100}\ \dfrac{(60.6)}{100}} = \boxed{\begin{array}{l} 46.5\ lbs/day \\ Na_2SiF_6 \end{array}}$$

Example 3: (Dry Feed Rate, lbs/day)
❑ The fluoride level desired in a water is 1.3 mg/L. The flow to be treated is 1,925,000 gpd. If sodium silicofluoride (Na_2SiF_6) will be used to provide the fluoride with 98% commercial purity and 60.6% fluoride ion, how many lbs/day sodium silicofluoride will be required? The natural level of fluoride in the water is 0.2 mg/L.

The desired fluoride level in the water is 1.3 mg/L. Since the level of fluoride in the water before treatment is 0.2 mg/L, the desired fluoride dosage is 1.3 mg/L – 0.2 mg/L = 1.1 mg/L. The sodium silicofluoride required to add 1.1 mg/L fluoride to the water is:

$$\frac{(mg/L\ F)\ (MGD\ flow)\ (8.34\ lbs/gal)}{\dfrac{(\%\ Comm.\ Purity)}{100}\ \dfrac{(\%\ F\ in\ Comp.)}{100}} = \begin{array}{l} lbs/day \\ Fluoride \\ Compound \end{array}$$

$$\frac{(1.1\ mg/L)\ (1.925\ MG)\ (8.34\ lbs/gal)}{\dfrac{(98)}{100}\ \dfrac{(60.6)}{100}} = \boxed{\begin{array}{l} 29.7\ lbs/day \\ Na_2SiF_6 \end{array}}$$

WHEN THERE IS A NATURAL FLUORIDE LEVEL IN THE WATER

Occasionally there will be a natural level of fluoride present in the water to be treated. In this case, you will need to reduce the desired fluoride dosage by the amount of fluoride already present in the water. For example, if the desired fluoride dosage for a water is 1.5 mg/L, and the natural fluoride level of the water is 0.4 mg/L, the adjusted fluoride dosage should be:

$$\boxed{\begin{array}{l} Desired \\ F, mg/L \end{array} - \begin{array}{c} Natural \\ Level\ of\ F, \\ mg/L \end{array} = \begin{array}{c} Adjusted \\ F\ Dosage, \\ mg/L \end{array}}$$

1.5 mg/L – 0.4 mg/L = 1.1 mg/L F

Example 3 illustrates this calculation.

COMBINING PERCENT COMMERCIAL PURITY AND PERCENT FLUORIDE DATA

In some cases the percent commercial purity and percent fluoride ion data have been combined.* In other words the fluoride content of the chemical has already been calculated. This combined fluoride content may be expressed as:

- Percent fluoride ion in each pound of compound, or

- Pounds of fluoride ion in each pound of compound.

These are related terms. For example, if each pound of compound contains 0.6 lbs fluoride ion, the percent fluoride ion in each lb is:

$$\frac{0.6 \text{ lbs F}}{1 \text{ lb Compound}} = \begin{array}{l} 60\% \text{ F in the} \\ \text{Compound,} \\ \text{as delivered} \end{array}$$

Using this example, regardless of whether the fluoride content of the compound is expressed as 0.6 lbs F/lb compound or % F in the compound, you would divide the lbs/day fluoride by 0.6. The general structure of the dry feed rate calculation remains unchanged.

Examples 4-7 illustrate the calculation of feed rate when fluoride ion content data has already been combined.

Example 4: (Dry Feed Rate, lbs/day)

❏ A flow of 1.8 MGD is to be treated with sodium silicofluoride, Na_2SiF_6. The raw water contains no fluoride. If the desired fluoride concentration in the water is 1.4 mg/L, what should be the chemical feed rate (in lbs/day)? (The manufacturer's data indicates that each pound of Na_2SiF_6 contains 0.6 lbs of fluoride ion.)

The compound contains 0.6 lbs fluoride ion for each pound of compound. This may be expressed as a percent fluoride ion content of:

$$\frac{0.6 \text{ lbs F}}{1 \text{ lb Compound}} \times 100 = 60\%$$

Continue the problem as in Examples 1-3:

$$\frac{(1.4 \text{ mg/}L)(1.8 \text{ MGD})(8.34 \text{ lbs/gal})}{\frac{60}{100}} = \boxed{\begin{array}{l} 35.0 \text{ lbs/day} \\ Na_2SiF_6 \end{array}}$$

Example 5: (Dry Feed Rate, lbs/day)

❏ A flow of 2.96 MGD is to be treated with sodium silicofluoride, Na_2SiF_6. The raw water contains no fluoride. If the desired fluoride concentration in the water is 1.1 mg/L, what should be the chemical feed rate (in lbs/day)? (The manufacturer's data indicates that each pound of Na_2SiF_6 contains 0.6 lbs of fluoride ion.)

0.6 lbs of F/lb of compound may be converted to a percent, if desired: (Whether converted to a percent or not, you will be dividing lbs/day fluoride by 0.6.)

$$\frac{0.6 \text{ lbs F}}{1 \text{ lb Compound}} \times 100 = 60\%$$

Continue with the dry feed rate calculation as usual:

$$\frac{(1.1 \text{ mg/}L)(2.96 \text{ MGD})(8.34 \text{ lbs/gal})}{\frac{60}{100}} = \boxed{\begin{array}{l} 45.3 \text{ lbs/day} \\ Na_2SiF_6 \end{array}}$$

---

* The commercial purity of sodium silicofluoride, Na_2SiF_6, generally ranges between 98-99%. When this is combined with the fact that the chemical Na_2SiF_6 has only 60.6% fluoride ion, this means that only about 60% (98/100 x 60.6/100 = 0.59 or 59%) of the compound is fluoride ion.

Example 6: (Dry Feed Rate, lbs/day)
❏ A flow of 3.7 MGD is to be treated with sodium fluoride, NaF. The raw water contains no fluoride and the desired fluoride concentration in the finished water is 1.2 mg/L. What should be the chemical feed rate in lbs/day? (The manufacturer's data indicates that each pound of NaF contains 0.44 lbs of fluoride ion.)

The fluoride ion content can be expressed as a percent:

$$\frac{0.44 \text{ lbs F}}{1 \text{ lb Compound}} \times 100 = 44\% \text{ F}$$

The feed rate can now be calculated:

$$\frac{(1.2 \text{ mg/L})(3.7 \text{ MGD})(8.34 \text{ lbs/gal})}{\dfrac{44}{100}} = \boxed{\begin{array}{c} 84.2 \text{ lbs/day} \\ \text{NaF} \end{array}}$$

Example 7: (Dry Feed Rate, lbs/day)
❏ A flow of 1,020,000 gpd is to be treated with sodium fluoride, NaF. The raw water contains 0.08 mg/L fluoride and the desired fluoride level in the finished water is 1.5 mg/L. What should be the chemical feed rate in lbs/day? (The manufacturer's data indicates that each pound of NaF contains 0.45 lbs of fluoride ion.)

It is not necessary to convert the fluoride ion content to a percent before using it in the denominator of the feed rate calculation, since the two terms are equivalent (0.45/1 = 45/100).

The fluoride ion to be added to the water is
1.5 mg/L – 0.08 mg/L = 1.42 mg/L

$$\frac{(1.42 \text{ mg/L})(1.02 \text{ MGD})(8.34 \text{ lbs/gal})}{0.45} = \boxed{\begin{array}{c} 26.8 \text{ lbs/day} \\ \text{NaF} \end{array}}$$

13.4 PERCENT STRENGTH OF A SOLUTION

Many times fluoride solutions are used in dosing a water with fluoride. The appropriate chemical feed rate of such a solution depends on the concentration or percent strength of that solution.

As illustrated in the diagram to the right, the percent strength of a solution depends on how much chemical is dissolved in the solution. The more chemical dissolved, the higher the percent strength or concentration.

The percent strength calculation is simply a percent calculation:*

$$\% = \frac{\text{Part}}{\text{Whole}} \times 100$$

In this case, the "part" of interest is the pounds chemical in the solution, and the "whole" is the total pounds solution, as shown in the equation to the right.

Examples 1-3 illustrate the calculation of percent strength.

PERCENT STRENGTH OF A SOLUTION DEPENDS ON HOW MUCH CHEMICAL IS DISSOLVED IN THE SOLUTION

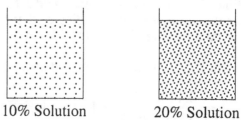

10% Solution   20% Solution

Simplified Equation:

$$\% \text{ Strength} = \frac{\text{lbs Chemical}}{\text{lbs Solution}} \times 100$$

Expanded Equation:

When the commercial chemical is 100% pure:

$$\% \text{ Strength} = \frac{\text{lbs Chemical}}{\text{lbs Water ** + lbs Chem. Cmpd.}} \times 100$$

When the commercial chemical is less than 100% pure:

$$\% \text{ Strength} = \frac{\dfrac{(\text{lbs Chem.}) (\% \text{ Comm. Purity})}{100}}{\text{lbs Water + lbs Chem. Cmpd.}} \times 100$$

Example 1: (% Strength of a Solution)
❑ If 5 lbs of sodium fluoride, NaF, are dissolved in 70 gallons of water, what is the percent strength of the solution? (Assume the chemical used is 100% pure NaF.)

$$\% \text{ Strength} = \frac{\text{lbs Chemical}}{\text{lbs Water + lbs Chem. Cmpd.}} \times 100$$

$$= \frac{5 \text{ lbs}}{(70 \text{ gal}) (8.34 \text{ lbs/gal}) + 5 \text{ lbs}} \times 100$$

$$= \frac{5 \text{ lbs}}{588.8 \text{ lbs}} \times 100$$

$$= \boxed{0.85\% \text{ Strength NaF}}$$

---

 * For a review of percent calculations, refer to Chapter 5 in *Basic Math Concepts.*
** To review the conversion of gallons to pounds, refer to Chapter 8 in *Basic Math Concepts.*

Example 2: (% Strength of a Solution)
❑ If 7 lbs of sodium fluoride are mixed with 50 gal of water, what is the percent strength of the solution? (The commercially available sodium fluoride is 98% pure.)

The expanded equation including percent commercial purity must be used in this calculation:

$$\% \text{ Strength} = \frac{(\text{lbs Chem.}) \dfrac{(\% \text{ Comm. Purity})}{100}}{\text{lbs Water} + \text{lbs Chem. Cmpd.}} \times 100$$

$$= \frac{(7 \text{ lbs}) \dfrac{(98)}{100}}{(50 \text{ gal}) (8.34 \text{ lbs/gal}) + 7 \text{ lbs}} \times 100$$

$$= \frac{6.86}{417 + 7} \times 100$$

$$= \boxed{1.6\% \text{ Strength NaF}}$$

WHEN A CHEMICAL IS NOT 100% PURE

In Example 1, it was assumed that the sodium fluoride (NaF) compound used was 100% pure. However, as discussed in Section 13.3, the chemical compounds are not 100% pure, due to the source of the chemical, the manufacturing processes, etc. Therefore, when calculating the percent strength of a solution, the commercial purity must be included in the calculation to reflect the actual amount of compound present in the solution. The expanded equation including commercial purity is shown on the opposite page. Examples 2 and 3 illustrate this calculation.

Example 3: (% Strength of a Solution)
❑ How many pounds of sodium fluoride (98% pure) must be added to 100 gallons of water to make a 1.1% solution of sodium fluoride?

Use the same equation, filling in given data:

$$\% \text{ Strength} = \frac{(\text{lbs Chem.}) \dfrac{(\% \text{ Comm. Purity})}{100}}{\text{lbs Water} + \text{lbs Chem. Cmpd.}} \times 100$$

$$1.1 = \frac{(x \text{ lbs}) \dfrac{(98)}{100}}{(100 \text{ gal}) (8.34 \text{ lbs/gal}) + x \text{ lbs}} \times 100$$

$$1.1 = \frac{0.98x}{834 + x} \times 100$$

$$1.1 = \frac{98x}{834 + x}$$

$$1.1 (834 + x) = 98x$$

$$917.4 + 1.1x = 98x$$

$$917.4 = 96.9x$$

$$\boxed{9.5 \text{ lbs NaF}} = x$$

SOLVING FOR A DIFFERENT UNKNOWN VARIABLE

There are four variables in the expanded percent strength equation:

- lbs water

- lbs chemical

- % commercial purity of the chemical

- percent strength

In Examples 1 and 2, percent strength was the unknown variable. However, the same equation may be used to calculate <u>any</u> of the variables, provided data is known for the other three variables. Simply fill in the given data in the equation and solve for the unknown variable.* Example 3 illustrates this calculation.

* For a review of percent problems, refer to Chapter 5 in *Basic Math Concepts.*

EXPRESSING PERCENT STRENGTH AS mg/L FLUORIDE

The top number in the table to the right shows the approximate percent strength of the sodium fluoride solution that would result from adding the indicated pounds of NaF (left column) and gallons of water (top row). For example, if 7 lbs of NaF (98% pure) (left column) is added to 50 gallons of water (top row), the resulting solution would have a **1.62% NaF strength** (read the number where 7 lbs and 50 gal intersect on the table).

Compare this reading from the table (1.62% strength) with the value calculated in Example 2 (1.6% strength). The upper values shown in each box of the table were generated from calculations such as that in Example 2.**

What about the lower numbers in each box? These numbers represent the mg/L fluoride present in the solution.

In several different calculations we have converted from percent to mg/L.* **The conversion involves a decimal point move four places to the right** (multiplication by 10,000). For example,

$$1\% = 1.0000. \text{ mg/}L$$

PREPARATION OF SOLUTIONS OF SODIUM FLUORIDE*

lbs NaF (98% pure)	Gallons of Water									
	10	20	30	40	50	60	70	80	90	100
1	1.16 5250	0.58 2650	0.39 1750	0.29 1350						
2	2.30 10400	1.16 5250	0.78 3500	0.58 2650	0.47 2100	0.39 1750	0.33 1500			
3		1.73 7850	1.16 5250	0.87 3950	0.70 3150	0.58 2650	0.50 2250	0.44 2000	0.39 1750	0.35 1600
4		2.30 10400	1.54 7000	1.16 5250	0.93 4200	0.78 3500	0.67 3000	0.58 2650	0.52 2350	0.47 2100
5			1.92 8700	1.45 6550	1.16 5250	0.97 4400	0.83 3750	0.73 3300	0.65 2950	0.58 2650
6			1.73 7850	1.39 6300	1.16 5250	1.00 4500	0.87 3950	0.78 3500	0.70 3150
7				2.01 9100	1.62 7300	1.35 6100	1.16 5250	1.02 4600	0.91 4100	0.82 3700
8				1.85 8350	1.54 7000	1.33 6000	1.16 5250	1.03 4700	0.93 4200
9				2.07 9400	1.73 7850	1.49 6750	1.30 5900	1.16 5250	1.05 4750
10					1.92 8700	1.65 7450	1.45 6550	1.29 5850	1.16 5250

The table gives strengths of solutions prepared by dissolving the given weights of sodium fluoride in various volumes of water. The upper figures represent approximate solution strength in percent sodium fluoride (100% pure NaF), while the lower figures represent approximate solution concentration in mg/L fluoride.

Example 4: (% Strength of a Solution)

❏ A 1.92% sodium fluoride solution has been prepared by dissolving 5 lbs of NaF (98% pure) in 30 gallons of water. What is the mg/L fluoride concentration of this solution? (Compare this calculated value with the value shown on the table above.)

The 1.92% NaF concentration can be expressed as mg/L NaF concentration as follows:

$$1.92\% \text{ NaF} = 1.9200. \text{ mg/}L \text{ NaF}$$

Then, using the percent fluoride ion, the mg/L fluoride concentration can be calculated:

$$\frac{(19,200 \text{ mg/}L)}{\text{NaF}} \frac{(45.25)}{100} = \boxed{8688 \text{ mg/}L \text{ F}}$$

The table indicates a fluoride concentration of 8700 mg/L.

---

* For a review of percent and mg/L conversions, refer to Chapter 8 in *Basic Math Concepts*.

** Example 1 assumes the sodium fluoride is 100% pure, so it is not precisely like the calculations using 98% pure NaF, as shown in Example 2.

Example 5: (% Strength of a Solution)
❑ A 1.03% sodium fluoride solution results from dissolving 8 lbs of NaF (98% pure) in 90 gallons of water. What is the mg/L fluoride concentration of this solution? (Compare this calculated value with the value shown in the table.)

A 1.03% NaF can be expressed in terms of mg/L as follows:

$$1.03\% = 1.0300. \text{ mg/}L \text{ NaF}$$

Then, using percent fluoride ion present, the mg/L F can be calculated:

$$\frac{(10,300 \text{ mg/}L)}{\text{NaF}} \frac{(45.25)}{100} = \boxed{4661 \text{ mg/}L \text{ F}}$$

Let's examine the two values shown on the table for a solution prepared using 7 lbs NaF dissolved in 50 gallons water:

$\boxed{1.62}$ ← % Na F strength (100% pure NaF)
$\boxed{7300}$ ← mg/L F in Solution

The upper number indicates a sodium fluoride solution strength of 1.62%. Remember that the percent commercial purity (98%) of the sodium fluoride has already been considered in this calculation (similar to the calculation shown in Example 2). The <u>percent strength</u> of sodium fluoride (100% pure NaF) can now be expressed as <u>mg/L NaF</u> concentration:

$$1.62\% = 1.6200. \text{ mg/}L$$
$$= 16,200 \text{ mg/}L \text{ NaF}$$

Now that the mg/L NaF concentration has been calculated, **the mg/L fluoride concentration** in the solution can be calculated. Simply multiply the mg/L NaF by the percent fluoride ion content:

$$\frac{\begin{array}{c}(\text{mg/}L \text{ NaF}) \\ \text{at 100\% pure}\end{array} \frac{\begin{array}{c}(\% \text{ F}) \\ \text{in Comp}\end{array}}{100} = \begin{array}{c}\text{mg/}L \\ \text{F}\end{array}}$$

The percent fluoride in NaF is always 45.25% (refer to Section 13.2). Therefore, a concentration of 16,200 mg/L NaF is equal to a fluoride ion concentration of:

Example 6: (% Strength of a Solution)
❑ A total of 6 lbs of sodium fluoride (98% pure) is dissolved in 50 gallons of water. (a) What is the percent sodium fluoride strength of the solution, and (b) what is the mg/L fluoride concentration of the solution? Compare these values to those given in the table.

(a) First calculate the percent strength of the solution:

$$\begin{array}{c}\% \text{ Strength} \\ \text{NaF}\end{array} = \frac{(6 \text{ lbs}) \dfrac{(98)}{100}}{(50 \text{ gal}) (8.34 \text{ lbs/gal}) + 6 \text{ lbs}} \times 100$$

$$= \frac{5.88 \text{ lbs}}{417 \text{ lbs} + 6 \text{ lbs}} \times 100$$

$$= \boxed{1.39\% \text{ NaF}}$$

(b) To calculate mg/L fluoride concentration, the mg/L sodium fluoride must first be determined:

$$1.39\% = 1.3900. \text{ mg/}L$$
$$= 13,900 \text{ mg/}L \text{ NaF}$$

Now the mg/L F can be calculated:

$$\frac{(13,900 \text{ mg/}L)}{\text{NaF}} \frac{(45.25)}{100} = \boxed{6300 \text{ mg/}L \text{ F}}$$

The table indicates 1.39% NaF solution and 6300 mg/L F.

$$\frac{(16,200)}{\begin{array}{c}\text{mg/}L \\ \text{NaF}\end{array}} \frac{(45.25)}{100} = \boxed{\begin{array}{c}7331 \text{ mg/}L \\ \text{Fluoride}\end{array}}$$

$$\text{or rounded to} = \boxed{\begin{array}{c}7350 \text{ mg/}L \\ \text{Fluoride}\end{array}}$$

The value shown in the table (7300) is slightly different due to difference in rounding.

CONSTRUCTING A DOSAGE TABLE

Tables such as that shown on the previous page are simply a compilation of many separate calculations. Examples 7 and 8 illustrate the calculations required to construct the sodium fluoride solution preparation table.

Example 7: (% Strength of a Solution)
❑ If 7 lbs of sodium fluoride (98% pure) are dissolved in 70 gallons of water, (a) What is the percent strength of the NaF solution? (b) What is the mg/L fluoride ion concentration of this solution? (Compare these calculated values with the values shown on the table given on the previous page.)

(a) Calculate the percent strength of the sodium fluoride solution:

$$\text{\% Strength NaF} = \frac{(7 \text{ lbs}) \frac{(98)}{100}}{(70 \text{ gal}) (8.34 \text{ lbs/gal}) + 7 \text{ lbs}} \times 100$$

$$= \frac{6.86 \text{ lbs}}{583.8 \text{ lbs} + 7 \text{ lbs}} \times 100$$

$$= \frac{6.86 \text{ lbs}}{590.8 \text{ lbs}} \times 100$$

$$= \boxed{1.16\% \text{ NaF}}$$

(b) To determine the mg/L fluoride concentration, first determine the mg/L NaF concentration of the solution. Converting percent NaF to mg/L NaF:

$$1.16\% = 1.1600. \text{ mg/}L$$

$$= 11,600 \text{ mg/}L \text{ NaF}$$

Now the mg/L F can be calculated:

$$\frac{(11,600 \text{ mg/}L) (45.25)}{\text{NaF} \quad 100} = \boxed{5249 \text{ mg/}L \text{ F}}$$

Example 8: **(% Strength of a Solution)**
❑ A sodium fluoride solution is prepared using 10 lbs of NaF (98% pure) in 80 gallons water. (a) What is the percent strength of the NaF solution? (b) What is the mg/*L* fluoride ion concentration of this solution? (Compare these calculated values with those shown on the table given on the previous page.)

(a) First, calculate the percent NaF strength of the solution:

$$\text{\% Strength NaF} = \frac{(10 \text{ lbs}) \dfrac{(98)}{100}}{(80 \text{ gal}) (8.34 \text{ lbs/gal}) + 10 \text{ lbs}} \times 100$$

$$= \frac{9.8 \text{ lbs}}{667.2 \text{ lbs} + 10 \text{ lbs}} \times 100$$

$$= \frac{9.8 \text{ lbs}}{677.2 \text{ lbs}} \times 100$$

$$= \boxed{1.45\% \text{ NaF}}$$

(b) In order to calculate mg/*L* fluoride concentration, the mg/*L* sodium fluoride must be determined:

$$1.45\% = 1.4500. \text{ mg/}L$$

$$= 14,500 \text{ mg/}L \text{ NaF}$$

Then the mg/*L* fluoride ion can be determined:

$$\text{(14,500 mg/}L\text{)} \atop \text{NaF} \quad \frac{(45.25)}{100} = \boxed{6561 \text{ mg/}L \text{ F}}$$

13.5 CALCULATING SOLUTION FEED RATES

There are many approaches to calculating solution feed rates. Two related methods are shown in the box to the right:

* the mg/L to lbs/day method, and
* the rate/concentration method.

The mg/L to lbs/day* method is based on the concept that the dosage rate of the solution feeder directly affects the actual dosage accomplished. The structure of these problems, therefore, is to put the actual (or desired) dosage information on the left side of the equation and the solution feeder data on the right side of the equation as shown in the box to the right.

There are two important considerations in these calculations:

* Solution concentration is generally expressed as a percent and must be converted to mg/L concentration before continuing with the calculation.

* Commercial purity factor is not used in these calculations. This is because percent concentration of the solution is essentially the percent commercial purity (the actual amount of chemical present).

Examples 1-3 illustrate calculations of solution of feed rate for hydrofluosilicic acid. Examples 4-7 are calculations of solution feed rate for sodium fluoride. Examples 8-11 illustrate how to calculate solution feed rate expressed in mL/min. And Examples 12-15 illustrate the rate/concentration method of calculating solution feed problems.

TWO METHODS OF CALCULATING SOLUTION FEED RATES

1. The mg/L to lbs/day method:

Simplified Equation:

$$\begin{array}{ll} \text{Actual Feed,} & = \text{Solution Feed,} \\ \text{lbs/day} & \text{mg/L} \end{array}$$

Expanded Equation:

$$\underset{\text{F}}{(mg/L)} \underset{\text{flow}}{(MGD)} \underset{\text{lbs/gal}}{(8.34)} = \underset{\text{Comp.}}{(mg/L)} \underset{\text{flow}}{(MGD)} \underset{\text{lbs/gal}}{(8.34)} \underset{}{(sp.\ gr.)} \underset{\frac{\text{in}}{\text{Comp.}}}{(\%\ F)}$$
$$\frac{}{100}$$

2. The rate/concentration method: *When applicable*

$$R_1 \times C_1 = R_2 \times C_2$$

where: R_1 = feed rate of first solution
C_1 = concentration of first solution
R_2 = feed rate of second solution
C_2 = concentration of second solution

Example 1: (Solution Feed Rates)

❑ A flow of 3.1 MGD is to be treated with a 20% solution of hydrofluosilicic acid (H_2SiF_6). The desired fluoride concentration is 1.3 mg/L. What should be the solution feed rate (in gpd)? The hydrofluosilicic acid weighs 9.8 lbs/gal. (Assume the water to be treated contains no natural fluoride level.) The percent fluoride ion content of H_2SiF_6 is 79.2%.

Desired Feed, lbs F/day = Solution Feed, lbs F/day

$$\underset{\text{F}}{(mg/L)} \underset{\text{flow}}{(MGD)} \underset{\text{lbs/gal}}{(8.34)} = \underset{\text{Comp.}}{(mg/L)} \underset{\text{flow}}{(MGD)} \underset{\text{lbs/gal}}{(Sol'n)} \underset{\frac{\text{Comm.}}{100}}{(\%\ F\ in)}$$

$$\underset{mg/L}{(1.3)} \underset{\text{flow}}{(3.1)} \underset{\text{lbs/gal}}{(8.34)} = \underset{mg/L}{(200,000)} \underset{}{(x\ MGD)} \underset{\text{lbs/gal}}{(9.8)} \underset{100}{\frac{(79.2)}{}}$$

$$\frac{(1.3)\ (3.1)\ (8.34)}{(200,000)\ (9.8)\ (0.79)} = x\ MGD$$

$$0.0000217\ MGD = x$$

$$\boxed{21.7\ gpd} = \text{or}$$

---

* For a review of mg/L to lbs/day calculations, refer to Chapter 3.

Example 2: (Solution Feed Rates)

❑ A flow of 6 MGD is to be treated with a 20% solution of hydrofluosilicic acid (H_2SiF_6). The acid has a specific gravity of 1.18. If the desired fluoride level in the water is 1.5 mg/*L*, what should be the solution feed rate (in gpd)? The fluoride level in the water to be treated is 0.1 mg/*L*. (The percent fluoride ion content of H_2SiF_6 is 79.2%.)

Since there is a fluoride concentration of 0.1 mg/*L* in the raw water, only 1.4 mg/*L* fluoride must be added:

$$\begin{array}{l} \underset{\text{F}}{(mg/L)} \underset{\text{flow}}{(MGD)} \underset{\text{lbs/gal}}{(8.34)} = \underset{\text{Comp.}}{(mg/L)} \underset{\text{flow}}{(MGD)} \underset{\text{lbs/gal}}{(8.34)} (sp.\ gr.) \underset{\substack{\text{in Comp.}}}{\dfrac{(\%\ F)}{100}} \end{array}$$

$$\underset{mg/L}{(1.4)} \ \underset{MGD}{(6)} \ \underset{lbs/gal}{(8.34)} = \underset{mg/L}{(200{,}000)} \ (x\ MGD) \ \underset{lbs/gal}{(8.34)} \ (1.18) \ \dfrac{(79.2)}{100}$$

$$\frac{(1.4)\,(6)\,\cancel{(8.34)}}{(200{,}000)\cancel{(8.34)}(1.18)(0.792)} = x\ MGD$$

$$0.0000448\ MGD = x$$

$$\boxed{44.8\ \text{gpd}} = \text{ or}$$

Example 3: (Solution Feed Rates)

❑ A water flow rate of 1100 gpm is to be treated with a 25% solution of hydrofluosilicic acid (H_2SiF_6). If the desired concentration of fluoride in the water is 1 mg/*L*, what should be the solution feed rate (in gpd)? The specific gravity of the acid is 1.22. Assume there is no fluoride in the water to be treated. (The fluoride ion content of H_2SiF_6 is 79.2%.)

Before the solution feed rate can be calculated, the flow rate to be treated must be expressed as MGD:**

$$(1100\ gpm)\,(1440\ min/day) = 1{,}584{,}000\ gpd$$

$$\text{or} = 1.584\ MGD$$

Then calculate the solution feed rate required:

$$\underset{mg/L}{(1.0)} \ \underset{MGD}{(1.584)} \ \underset{lbs/gal}{(8.34)} = \underset{mg/L}{(250{,}000)} \ (x\ MGD) \ \underset{lbs/gal}{(8.34)} \ (1.22) \ \dfrac{(79.2)}{100}$$

$$\frac{(1.0)\,(1.584)\,\cancel{(8.34)}}{(250{,}000)\cancel{(8.34)}(1.22)(0.792)} = x\ MGD$$

$$0.00000656\ MGD = x$$

$$\boxed{6.6\ \text{gpd}} = \text{ or}$$

SOLUTION FEED RATE —HYDROFLUOSILICIC ACID (H_2SiF_6)

In calculations of solution feed rate, the solution feed rate data is placed on the right side of the equation and actual (or desired) feed rate on the left side of the equation. Many solutions are fed at a relatively low concentration (4% or less) and therefore the density* of the solution is considered to be the same as that of water (8.34 lbs/gal). However, when feeding hydrofluosilicic acid solutions, the density is often greater than 8.34 lbs/gal and must be included in the calculation.

To account for a solution with a different density, you must either include the higher density factor (such as 9.17 lbs/gal) or use 8.34 lbs/gal and a specific gravity factor. (Such as 8.34 lbs/gal x 1.1 sp. gr.).

---

* For a discussion of density and specific gravity, refer to Chapter 7.

** For a review of flow conversions, refer to Chapter 8 in *Basic Math Concepts*.

SOLUTION FEED RATE— SODIUM FLUORIDE (NaF)

Sodium fluoride is available in a powder or crystalline form. It can be added to water using a dry chemical feeder, as done for sodium silicofluoride (described in Section 13.3), or it may be dissolved in a saturator and added as a solution.

When calculating the dry chemical feed rate of sodium fluoride, a calculation similar to that described in Section 13.3 must be used.

When calculating the feed rate of a sodium fluoride solution feeder, the solution feed rate equation must be used (such as shown in Examples 1-3). Examples 4-7 illustrate this calculation.

Example 4: (Solution Feed Rates)

❏ A flow of 432,000 gpd is to be treated with a 2.5% solution of sodium fluoride, NaF. If the desired fluoride ion concentration is 1.4 mg/L, what should be the sodium fluoride solution feed rate (in gpd)? Sodium fluoride has a fluoride ion content of 45.25%. The water to be treated contains 0.2 mg/L of fluoride ion. (Assume the solution density is 8.34 lbs/gal).

The desired fluoride ion concentration is 1.4 mg/L. Since the natural fluoride level is 0.2 mg/L, only 1.2 mg/L fluoride ion must be added to the water.

$$\underset{\substack{F}}{(mg/L)} \underset{\substack{flow}}{(MGD)} \underset{\substack{lbs/gal}}{(8.34)} = \underset{\substack{NaF}}{(mg/L)} \underset{\substack{flow}}{(MGD)} \underset{\substack{lbs/gal}}{(8.34)} \underset{\substack{Comp.}}{(\% \ F \ in)} \frac{}{100}$$

$$\underset{\substack{mg/L}}{(1.2)} \underset{\substack{MGD}}{(0.432)} \underset{\substack{lbs/gal}}{(8.34)} = \underset{\substack{mg/L}}{(25,000)} (x \ MGD) \underset{\substack{lbs/gal}}{(8.34)} \frac{(45.25)}{100}$$

$$\frac{(1.2) \ (0.432) \ \cancel{(8.34)}}{(25,000) \ \cancel{(8.34)} \ (0.4525)} = x \ MGD$$

$$0.0000457 \ MGD = x$$

$$\boxed{45.7 \ gpd} = x$$

Example 5: (Solution Feed Rates)

❏ A fluoride feed solution containing 1.6% fluoride ion is fed to a flow of 460,000 gpd. If the desired fluoride level in the water is 1.3 mg/L and the fluoride content of the raw water is 0.07 mg/L, what should be the solution feed rate, in gpd? (Assume the solution density is 8.34 lbs/gal.)

The fluoride concentration to be added to the water is 1.3 mg/L − 0.07 mg/L = 1.23 mg/L.

In this problem the fluoride ion content of the solution is given. Therefore, the percent fluoride ion present in the compound has already been accounted for and is not required on the right side of the equation:

$$\underset{\substack{mg/L}}{(1.23)} \underset{\substack{MGD}}{(0.46)} \underset{\substack{lbs/gal}}{(8.34)} = \underset{\substack{mg/L}}{(16,000)} (x \ MGD) \underset{\substack{lbs/gal}}{(8.34)}$$

$$\frac{(1.23) \ (0.46) \ \cancel{(8.34)}}{(16,000) \ \cancel{(8.34)}} = x \ MGD$$

$$0.0000354 \ MGD = x$$

$$\boxed{35.4 \ gpd} = x$$

Example 6: (Solution Feed Rates)
❏ A flow of 1,940,000 gpd is to be treated with a 2.6% solution of sodium fluoride. The desired fluoride level in the water is 1.1 mg/*L*. What should be the sodium fluoride solution feed rate (in gpd)? Sodium fluoride has a fluoride ion content of 45.25%. The water to be treated contains no fluoride. (Assume the solution density is 8.34 lbs/gal.)

$$
\underset{\substack{\text{F}}}{(\text{mg}/L)} \underset{\substack{\text{flow}}}{(\text{MGD})} \underset{\substack{\text{lbs/gal}}}{(8.34)} = \underset{\substack{\text{NaF}}}{(\text{mg}/L)} \underset{\substack{\text{flow}}}{(\text{MGD})} \underset{\substack{\text{lbs/gal}}}{(8.34)} \underset{\substack{\text{Comp.}\\100}}{(\%\ \text{F in})}
$$

$$
\underset{\substack{\text{mg}/L}}{(1.1)} \underset{\substack{\text{MGD}}}{(1.94)} \underset{\substack{\text{lbs/gal}}}{(8.34)} = \underset{\substack{\text{mg}/L}}{(26{,}000)} (x\ \text{MGD}) \underset{\substack{\text{lbs/gal}}}{(8.34)} \underset{100}{(45.25)}
$$

$$
\frac{(1.1)\,(1.94)\,\cancel{(8.34)}}{(26{,}000)\,\cancel{(8.34)}\,(0.4525)} = x\,\text{MGD}
$$

$$
0.000181\ \text{MGD} = x
$$

$$
\boxed{181\ \text{gpd}} = x
$$

Example 7: (Solution Feed Rates)
❏ The feed from a saturator contains 4% sodium fluoride, NaF. If the flow to be treated is 2.5 MGD, what should be the solution feed (in gpd) when the desired fluoride ion content of the finished water is 1.5 mg/*L*? The fluoride level of the raw water is 0.2 mg/*L*. The fluoride ion content of NaF is 45.25%. (Assume the density of the solution is 8.34 lbs/gal.)

The fluoride ion to be added to the water is 1.5 mg/*L* – 0.2 mg/*L* = 1.3 mg/*L*:

$$
\underset{\substack{\text{F}}}{(\text{mg}/L)} \underset{\substack{\text{flow}}}{(\text{MGD})} \underset{\substack{\text{lbs/gal}}}{(8.34)} = \underset{\substack{\text{NaF}}}{(\text{mg}/L)} \underset{\substack{\text{flow}}}{(\text{MGD})} \underset{\substack{\text{lbs/gal}}}{(8.34)} \underset{\substack{\text{Comp.}\\100}}{(\%\ \text{F in})}
$$

$$
\underset{\substack{\text{mg}/L}}{(1.3)} \underset{\substack{\text{MGD}}}{(2.5)} \underset{\substack{\text{lbs/gal}}}{(8.34)} = \underset{\substack{\text{mg}/L}}{(40{,}000)} (x\ \text{MGD}) \underset{\substack{\text{lbs/gal}}}{(8.34)} \underset{100}{(45.25)}
$$

$$
\frac{(1.3)\,(2.5)\,\cancel{(8.34)}}{(40{,}000)\,\cancel{(8.34)}\,(0.4525)} = x\,\text{MGD}
$$

$$
0.000179\ \text{MGD} = x
$$

$$
\boxed{179\ \text{gpd}} = x
$$

EXPRESSING SOLUTION FEED RATE AS mL/min

The solution feed rate is
sometimes expressed as mL/min.
To calculate the mL/min feed rate,
first calculate the gpd feed rate as
illustrated in Examples 1-7. Then
convert gpd feed rate to mL/min
feed rate as follows:

$$\frac{(gpd)(3785 \text{ m}L/gal)}{1440 \text{ min/day}} = \text{m}L/\text{min}$$

To verify the mathematical setup
of this conversion, use
dimensional analysis:*

$$\frac{\cancel{gal}}{\cancel{day}} \cdot \frac{3785 \text{ m}L}{1 \cancel{gal}} \cdot \frac{1 \cancel{day}}{1440 \text{ min}} = \frac{\text{m}L}{\text{min}}$$

Example 8: (Solution Feed Rates)
❑ The desired solution feed rate has been determined to be
98 gpd. What is this feed rate expressed as mL/min?

$$\frac{(gpd) (3785 \text{ m}L/gal)}{1440 \text{ min/day}} = \text{m}L/\text{min}$$

$$\frac{(98 \text{ gpd}) (3785 \text{ m}L/gal)}{1440 \text{ min/day}} = \boxed{258 \text{ m}L/\text{min}}$$

Example 9: (Solution Feed Rates)
❑ The desired feed rate for a solution feeder has been
calculated to be 65 gpd. What is this feed rate expressed as
mL/min?

$$\frac{(gpd) (3785 \text{ m}L/gal)}{1440 \text{ min/day}} = \text{m}L/\text{min}$$

$$\frac{(65 \text{ gpd}) (3785 \text{ m}L/gal)}{1440 \text{ min/day}} = \boxed{171 \text{ m}L/\text{min}}$$

---

* Dimensional analysis is described in Chapter 15 of *Basic Math Concepts.*

Example 10: (Solution Feed Rates)
❏ A flow of 2.12 MGD is to be treated with a 25% solution of hydrofluosilicic acid, H_2SiF_6. The raw water contains no fluoride and the desired fluoride concentration is 1.4 mg/L. The hydrofluosilicic acid weighs 9.8 lbs/gal. What should be the mL/min solution feed rate? (The percent fluoride content of H_2SiF_6 is 79.2%.)

First calculate the gpd solution feed rate:

$$\underset{\substack{\text{mg/}L}}{(1.4)}\ \underset{\substack{\text{MGD}}}{(2.12)}\ \underset{\substack{\text{lbs/gal}}}{(8.34)} = \underset{\substack{\text{mg/}L}}{(250,000)}\ (x\,\text{MGD})\ \underset{\substack{\text{lbs/gal}}}{(9.8)}\ \frac{(79.2)}{100}$$

$$\frac{(1.4)\,(2.12)\,(8.34)}{(250,000)\,(9.8)\,(0.792)} = x\,\text{MGD}$$

$$0.0000128\,\text{MGD} = x$$

$$12.8\,\text{gpd} = x$$

Then convert gpd feed rate to mL/min feed rate:

$$\frac{(12.8\,\text{gpd})\,(3785\,\text{m}L/\text{gal})}{1440\,\text{min/day}} = \boxed{33.6\,\text{m}L/\text{min}}$$

Example 11: (Solution Feed Rates)
❏ A flow of 1.06 MGD is to be treated with a 2.8% solution of sodium fluoride. The raw water contains no fluoride and the desired fluoride level in the finished water is 1.1 mg/L. What should be the sodium fluoride feed rate in mL/min? Sodium fluoride has a fluoride ion content of 45.25%. (Assume the solution density is 8.34 lbs/gal.)

First calculate the gpd solution feed rate:

$$\underset{\substack{\text{mg/}L}}{(1.1)}\ \underset{\substack{\text{MGD}}}{(1.06)}\ \underset{\substack{\text{lbs/gal}}}{(8.34)} = \underset{\substack{\text{mg/}L}}{(28,000)}\ (x\,\text{MGD})\ \underset{\substack{\text{lbs/gal}}}{(8.34)}\ \frac{(45.25)}{100}$$

$$\frac{(1.1)\,(1.06)\,\cancel{(8.34)}}{(28,000)\,\cancel{(8.34)}\,(0.4525)} = x\,\text{MGD}$$

$$0.0000919\,\text{MGD} = x$$

$$91.9\,\text{gpd} = x$$

Then convert gpd feed rate to mL/min feed rate:

$$\frac{(91.9\,\text{gpd})\,(3785\,\text{m}L/\text{gal})}{1440\,\text{min/day}} = \boxed{242\,\text{m}L/\text{min}}$$

In Examples 1-11, the first method of calculating solution feed rates was used—**the mg/L to lbs/day method.**

A second method may be used to calculate solution feed rates—**the rate/concentration method.**

As with the mg/L to lbs/day method of calculating solution feed rate, the actual (or desired) feed data is placed on the left side of the equation, and the solution feed data is placed on the right side of the equation:*

Actual or Desired Feed Rate, as F	=	Solution Feed Rate, as F

$$(R_1)(C_1) = (R_2)(C_2)(\text{sp. gr.})\frac{(\% \text{ F})}{\text{in Comp.}}{100}$$

where: R_1 = The feed rate of the water to treated, MGD

C_1 = The desired fluoride ion concentration, in mg/L

R_2 = The feed rate of the fluoride solution, MGD

C_2 = The concentration of second solution in percent or mg/L

The rate R_1 can be expressed in other terms, such as gpd or lbs/day. However, whichever terms R_1 is expressed in, R_2 must be expressed in the same terms. Likewise, C_1 may be expressed as a percent as well as mg/L; however, C_2 must be expressed in the same terms.

Example 12: (Solution Feed Rates)
❑ A flow of 3.1 MGD is to be treated with a 20% solution of hydrofluosilicic acid (H_2SiF_6). The desired fluoride concentration is 1.3 mg/L. What should be the solution feed rate (in gpd)? The hydrofluosilicic acid weighs 9.8 lbs/gal. (Assume the water to be treated contains no natural fluoride level.) The percent fluoride ion content of H_2SiF_6 is 79.2%. (This is the same problem given in Example 1.)[6]

$$(R_1)(C_1) = (R_2)(C_2)(\text{sp. gr.})\frac{(\% \text{ F in Comp.})}{100}$$

Before filling in data in the equation, the density given for the acid solution must be expressed as specific gravity. Remember, specific gravity is a comparison (or ratio) of the density of a substance to that of water:**

$$\frac{9.8 \text{ lbs/gal}}{8.34 \text{ lbs/gal}} = 1.18$$

Now continue with the solution feed calculation:

$$(3.1 \text{ MGD})(1.3 \text{ mg/L}) = (x \text{ MGD})(200,000 \text{ mg/L})(1.18)\frac{(79.2)}{100}$$

$$\frac{(3.1)(1.3)}{(200,000)(1.18)\frac{(79.2)}{100}} = x \text{ MGD}$$

$$0.0000214 \text{ MGD} = x$$

$$\boxed{21.4 \text{ gpd}} = x$$

(The answer to Example 1 was 21.7 gpd. The difference in answers is due to rounding from density 9.8 lbs/gal to specific gravity, 1.18.)

---

* This equation is sometimes written as $Q_1 C_1 = Q_2 C_2$.
** For a review of density and specific gravity, refer to Chapter 7.

Example 13: (Solution Feed Rates)
❏ A flow of 1,940,000 gpd is to be treated with a 2.6% solution of sodium fluoride. The desired fluoride level in the water is 1.1 mg/*L*. What should be the sodium fluoride solution feed rate (in gpd)? Sodium fluoride has a fluoride ion content of 45.25%. The water to be treated contains no fluoride. (Assume the solution density is 8.34 lbs/gal.) (This is the same problem given in Example 6.)

$$(R_1)(C_1) = (R_2)(C_2)(\text{sp. gr.})\frac{(\% \text{ F in Comp.})}{100}$$

$$(1.94 \text{ MGD})(1.1 \text{ mg/}L) = (x \text{ MGD})(26,000 \text{ mg/}L)(1.0)\frac{(45.25)}{100}$$

$$\frac{(1.94)\,(1.1)}{(26,000)\,(1.0)\,(0.4525)} = x \text{ MGD}$$

$$0.000181 \text{ MGD} = x$$

$$\boxed{181 \text{ gpd}} = x$$

(The answer to Example 6 was 181 gpd.)

Examples 12-14 illustrate this method of calculation solution feed rate. These examples are repeat problems from Examples 1, 6, and 7, respectively, so that you can compare the two methods of calculation solution feed rates.

Example 14: (Solution Feed Rates)
❏ The feed from a saturator contains 4% sodium fluoride, NaF. If the flow to be treated is 2.5 MGD, what should be the solution feed (in gpd) when the desired fluoride ion content of the finished water is 1.5 mg/*L*? The fluoride level of the raw water is 0.2 mg/*L*. The fluoride ion content of NaF is 45.25%. (Assume the density of the solution is 8.34 lbs/gal.) (This is the same problem given in Example 7.)

The fluoride ion to be added to the water is 1.5 mg/*L* – 0.2 mg/*L* = 1.3 mg/*L*:

$$(R_1)(C_1) = (R_2)(C_2)(\text{sp. gr.})\frac{(\% \text{ F in Comp.})}{100}$$

$$(2.5 \text{ MGD})(1.3 \text{ mg/}L) = (x \text{ MGD})(40,000 \text{ mg/}L)(1.0)\frac{(45.25)}{100}$$

$$\frac{(2.5)\,(1.3)}{(40,000)\,(1.0)\,(0.4525)} = x \text{ MGD}$$

$$0.000179 \text{ MGD} = x$$

$$\boxed{179 \text{ gpd}} = x$$

(The answer to Example 7 was 179 gpd.)

13.6 FEED RATES USING CHARTS AND NOMOGRAPHS

FEED RATES USING CHARTS

Feed rate charts are a shortcut method of determining chemical feed rates. Four such charts are given in this section.

Feed rate charts are constructed on the basis of a series of solution feed calculations (such as described on Section 13.5) which are then consolidated on one chart.

The steps for determining solution feed rate using the charts are described to the right. Examples 1-7 illustrate this process.

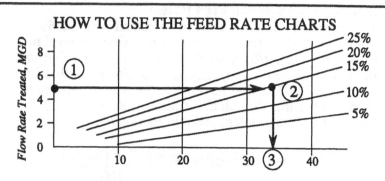

HOW TO USE THE FEED RATE CHARTS

Step 1: Enter the chart at the indicated flow rate along the left scale (Point 1).

Step 2: Draw a horizontal line to the right until it intersects with the given solution strength line (Point 2).

Step 3: Move vertically down to the flow rate scale and read the value indicated (Point 3).

Step 4: Use a proportion to determine the flow rate corresponding to the desired mg/L dosage:

$$\frac{\text{Desired Dose, mg/L}}{\text{Chart Dose, 1 mg/L}} = \frac{\text{Desired Feed Rate, gpd}}{\text{Chart Feed Rate, gpd}}$$

Example 1: (Charts and Nomographs)

❑ A flow of 6 MGD is to be treated with a 20% solution of hydrofluosilicic acid (H_2SiF_6). The fluoride ion concentration to be added is 1.4 mg/L. The raw water contains no fluoride. Use the treatment chart given on page 348 to determine the appropriate solution feed rate. (This problem is the same as Example 2, where the problem was calculated longhand.)

Use Treatment Chart I since Chart I is for lower flows. Enter the chart at the 6 MGD mark along the left scale. Follow the 6 MGD horizontal line to the point at which it intersects with the 20% line. From that point, move directly down to the flow rate scale and read the value: <u>30 gpd</u>. The charts are based on 1 mg/L fluoride. Therefore a proportion must be used to determine the flow rate corresponding to 1.4 mg/L:

$$\frac{\text{Desired Dose, mg/L}}{\text{Chart Dose, 1 mg/L}} = \frac{\text{Desired Feed Rate, gpd}}{\text{Chart Feed Rate, gpd}}$$

$$\frac{1.4\ \text{mg/L}}{1.0\ \text{mg/L}} = \frac{x\ \text{gpd}}{30\ \text{gpd}}$$

$$\frac{(1.4)(30)}{1} = x$$

$$\boxed{42\ \text{gpd}} = x$$

(The answer in Example 2 was 44.8 gpd. The difference in answers is due to the fact that answers read from graphs are only approximations.)

Example 2: (Charts and Nomographs)
❑ A flow of 5 MGD is to be treated with a 15% solution of hydrofluosilicic acid. The desired fluoride concentration is 1.5 mg/*L*. The raw water contains no fluoride. Use the treatment chart on page 348 to determine the appropriate solution feed rate.

Enter Treatment Chart I at 5 MGD along the left scale. Follow a horizontal line to the right until it intersects with the 15% line. From the point of intersection, move directly down to the flow rate scale and read the value: approximately 32.5 gpd

The chart is based on 1 mg/*L* fluoride. Therefore, a proportion must be used to determine the flow rate corresponding to 1.5 mg/*L* fluoride concentration:

$$\frac{\text{Desired Dose, mg/}L}{\text{Chart Dose, 1 mg/}L} = \frac{\text{Desired Feed Rate, gpd}}{\text{Chart Feed Rate, gpd}}$$

$$\frac{1.5 \text{ mg/}L}{1.0 \text{ mg/}L} = \frac{x \text{ gpd}}{32.5 \text{ gpd}}$$

$$\frac{(1.5)(32.5)}{1} = x$$

$$\boxed{48.8 \text{ gpd}} = x$$

Example 3: (Charts and Nomographs)
❑ A flow of 30 MGD is to be treated with a 25% hydrofluosilicic acid. The desired fluoride concentration in the finished water is 1.2 mg/L and the raw water contains no fluoride. Use the treatment chart on page 348 to determine the appropriate gpd solution feed rate.

Use Treatment Chart II. From the 30 MGD along the left scale, move to the right until it intersects with the 25% line. Since the intersection falls directly on a gph dashed line, use the gph flow rate and convert to gpd. This will be more accurate than trying to estimate a reading from the gpd scale.

$$(5 \text{ gph})(24 \text{ hrs/day}) = 120 \text{ gpd}$$

Use a proportion to determine the flow rate corresponding to 1.2 mg/*L* fluoride concentration:

$$\frac{\text{Desired Dose, mg/}L}{\text{Chart Dose, 1 mg/}L} = \frac{\text{Desired Feed Rate, gpd}}{\text{Chart Feed Rate, gpd}}$$

$$\frac{1.2 \text{ mg/}L}{1 \text{ mg/}L} = \frac{x \text{ gpd}}{120 \text{ gpd}}$$

$$(1.2)(120) = x$$

$$\boxed{144 \text{ gpd}} = x$$

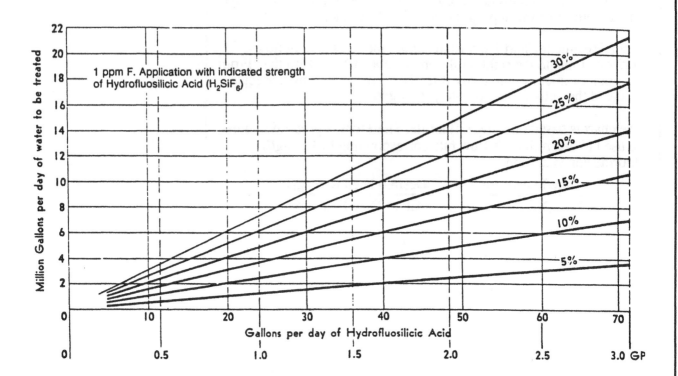

TREATMENT CHART I
Hydrofluosilicic Acid

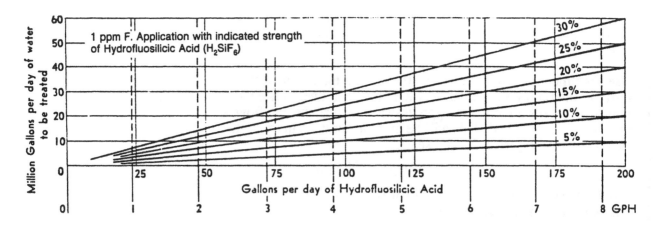

TREATMENT CHART II
Hydrofluosilicic Acid

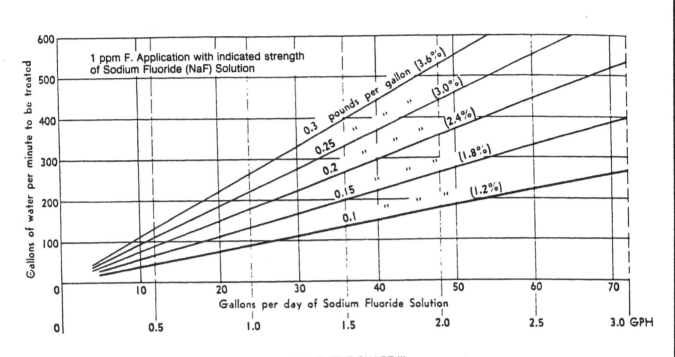

TREATMENT CHART III
Sodium Fluoride

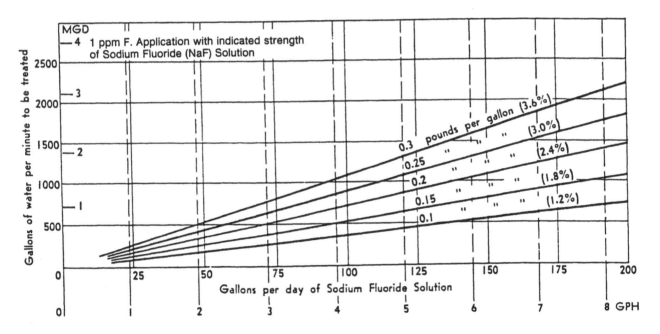

TREATMENT CHART IV

Treatment Charts Courtesy of Wallace & Tiernan

Example 4: (Charts and Nomographs)
❑ A flow of 300 gpm is to be treated with a 2.4% solution of sodium fluoride, NaF. The water to be treated contains no fluoride and the desired fluoride level in the finished water is 1.4 mg/L. Use the treatment charts on page 349 to determine the sodium fluoride solution feed rate, in gpd.

Use Treatment Chart III. From 300 gpm on the left scale, move horizontally to the right until it intersects with the 2.4% line. From the point of intersection, move directly down to the flow rate scale and read the value: 41 gpd.

Use a proportion to determine the flow rate corresponding to a fluoride dose of 1.4 mg/L:

$$\frac{\text{Desired Dose, mg/L}}{\text{Chart Dose, 1 mg/L}} = \frac{\text{Desired Feed Rate, gpd}}{\text{Chart Feed Rate, gpd}}$$

$$\frac{1.4 \text{ mg/L}}{1.0 \text{ mg/L}} = \frac{x \text{ gpd}}{41 \text{ gpd}}$$

$$(1.4)(41) = x$$

$$\boxed{57.4 \text{ gpd}} = x$$

Example 5: (Charts and Nomographs)
❑ A flow of 200 gpm is to be treated with a 3% solution of sodium fluoride. The water to be treated contains no fluoride and the desired fluoride level in the finished water is 1.6 mg/L. Use the treatment charts to determine the sodium fluoride solution feed rate, in gpd.

Use Treatment Chart III. From 200 gpm on the left scale, make a line horizontally to the right until it intersects with the 3% line. From the point of intersection, move directly down to the flow rate scale and read the value: 21.5 gpd.

Use a proportion to determine the flow rate that corresponds to 1.6 mg/L fluoride:

$$\frac{\text{Desired Dose, mg/L}}{\text{Chart Dose, 1 mg/L}} = \frac{\text{Desired Feed Rate, gpd}}{\text{Chart Feed Rate, gpd}}$$

$$\frac{1.6 \text{ mg/L}}{1.0 \text{ mg/L}} = \frac{x \text{ gpd}}{21.5 \text{ gpd}}$$

$$\frac{(1.6)(21.5)}{1} = x$$

$$\boxed{34.4 \text{ gpd}} = x$$

Example 6: (Charts and Nomographs)
❑ A flow of 1 MGD is to be treated with a 3% solution of sodium fluoride. The water to be treated contains a natural fluoride level of 0.2 mg/*L* and the desired fluoride level in the finished water is 1.1 mg/*L*. Use the treatment charts on page 349 to determine the sodium fluoride solution feed rate, in gpd.

Use Treatment Chart IV. From 1 MGD on the left scale, draw a horizontal line until it intersects with the 3% line. Then, from the point of intersection, move directly down to the flow rate scale and read the value: approximately 83 gpd.

Then use a proportion to determine the flow rate that corresponds to a fluoride level to be added of 1.1 mg/*L* – 0.2 mg/*L* = 0.9 mg/*L*:

$$\frac{\text{Desired Dose, mg/}L}{\text{Chart Dose, 1 mg/}L} = \frac{\text{Desired Feed Rate, gpd}}{\text{Chart Feed Rate, gpd}}$$

$$\frac{0.9 \text{ mg/}L}{1.0 \text{ mg/}L} = \frac{x \text{ gpd}}{83 \text{ gpd}}$$

$$(0.9)(83) = x$$

$$\boxed{74.7 \text{ gpd}} = x$$

Example 7: (Charts and Nomographs)
❑ A flow of 1000 gpm is to be treated with a 2.4% solution of sodium fluoride. The water to be treated contains no fluoride and the desired fluoride level in the finished water is 1.5 mg/*L*. Use the treatment charts on page 349 to determine the sodium fluoride solution, in gpd.

Use Treatment Chart IV. Follow the horizontal line from 1000 gpm until it intersects with the 2.4% line. From the point of intersection, move directly down to the flow rate scale and read the value: about 137 gpd.

Then use a proportion to determine the flow rate that corresponds to a fluoride level of 1.5 mg/*L*:

$$\frac{\text{Desired Dose, mg/}L}{\text{Chart Dose, 1 mg/}L} = \frac{\text{Desired Feed Rate, gpd}}{\text{Chart Feed Rate, gpd}}$$

$$\frac{1.5 \text{ mg/}L}{1.0 \text{ mg/}L} = \frac{x \text{ gpd}}{137 \text{ gpd}}$$

$$(1.5)(137) = x$$

$$\boxed{205.5 \text{ gpd}} = x$$

FEED RATES USING NOMOGRAPHS

Approximate feed rates may be determined using a nomograph, such as that shown on the opposite page. Nomographs such as this are not often used in determining fluoride compound feed rates since a more accurate determination of feed rates is required. Nomographs would be used only for a very general estimate of feed rate.

To determine the approximate fluoride compound feed rate:

1. If there is any natural level of fluoride in the water to be treated, the natural level must be subtracted from the desired fluoride level in the finished water. The result will be the mg/*L* fluoride ion that must be <u>added</u> to the water.

2. Draw a line from the flow rate to be treated (gpm or MGD) on the far left scale through the diagonal mg/*L* fluoride added scale to the first of the grouped scales (lbs/day fluoride ion).

3. This point indicates the desired lbs/day fluoride ion to be added. From this scale, draw a horizontal line intersecting the other four scales. The points of intersection on these scales indicate the amounts (lbs/day or gpd) of the listed chemical required to provide the desired fluoride dose.

Examples 8-11 illustrate the use of the fluoride nomograph.

Example 8: (Charts and Nomographs)

❑ A flow of 0.8 MGD is to be treated with a saturated solution of sodium fluoride, NaF. Use the fluoridation nomograph to determine the gpd sodium fluoride required for a fluoride level of 0.9 mg/*L*. (Assume the raw water contains no fluoride.)

Find 0.8 MGD on the left scale. From that point, draw a line through the diagonal fluoride dose scale (at 0.9 mg/*L* F) to the lbs/day fluoride ion scale. Then draw a horizontal line through all the scales to the right. Read the NaF feed rate:

> 42 gpd NaF

Example 9: (Charts and Nomographs)

❑ A flow of 600 gpm is to be treated with a 30% solution of hydrofluoric acid, H_2SiF_6. If the desired fluoride concentration of the finished water is 1.0 mg/*L*, what is the required acid solution feed rate (in gpd)? Use the nomograph to determine the feed rate. (Assume the raw water contains no fluoride.)

Find 600 gpm on the left scale and draw a line through the diagonal fluoride dose scale (at 1.0 mg/*L*) to the lbs/day fluoride feed rate scale. Then draw a horizontal line through all the scales to the right. Read the H_2SiF_6 feed rate:

> 2.85 lbs/day H_2SiF_6

Example 10: (Charts and Nomographs)

❑ A flow of 550,000 gpd is to be treated with sodium silicofluoride, Na_2SiF_6. The desired fluoride ion concentration in the finished water is 1.2 mg/*L*. The fluoride concentration in the raw water is 0.05 mg/*L*. Use the nomograph to determine the approximate sodium silicofluoride feed rate in lbs/day.

The fluoride to be added to the water is 1.2 mg/*L* − 0.05 mg/*L* = 1.15 mg/*L*. First find the 0.55 MGD mark on the left scale. (It is halfway between the 0.5 and 0.6 marks.) Then draw a line from the flow scale through the 1.15 mark on the mg/*L* scale (halfway between the 1.1 and 1.2 marks) to the lbs/day fluoride scale. And from the lbs/day fluoride ion scale, draw a horizontal line to the right, intersecting all scales. Read the feed rate indicated on the Na_2SiF_6 feed rate scale:

> 8.9 lbs/day Na_2SiF_6

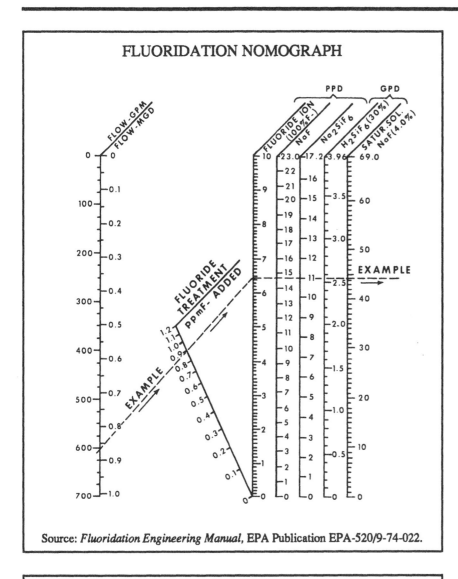

FLUORIDATION NOMOGRAPH

Source: *Fluoridation Engineering Manual*, EPA Publication EPA-520/9-74-022.

Example 11: (Charts and Nomographs)
❑ A flow of 6 MGD is to be treated with sodium fluoride, NaF to obtain a fluoride level of 1 mg/*L* in the finished water. The raw water contains no fluoride. Using the nomograph, determine the approximate lbs/day sodium fluoride feed rate required.

Since 6 MGD is not on the flow scale, divide 6 MGD by 10 (Resulting in 0.6 MGD) and use that value on the flow scale. Then proceed as in Examples 8-10 to determine the lbs/day sodium fluoride required. The answer read from the lbs/day NaF scale is about 11.5 lbs/day NaF. However, this answer must be adjusted since we temporarily changed one of the values. The value was divided by 10. So to reverse the process (and restore it to its proper value), the answer must be multiplied by 10:

$$(11.5 \text{ lbs/day}) (10) = 115 \text{ lbs/day NaF}$$

WHEN THE READING IS OFF THE SCALE

When a desired value is not shown on the flow rate or mg/*L* F scales, or when the alignment of a line drawn between the flow rate and mg/*L* fluoride scales result in a reading off the lbs/day fluoride scale, values can be temporarily adjusted to remedy this situation. Simply multiply or divide one of the values by 10. Then, **when the answer is determined, reverse the process** (i.e., if the value was multiplied by 10, then divide the answer by 10; or if the value was divided by 10, then multiply the answer by 10). Example 11 illustrates this type of problem.

13.7 CALCULATING mg/*L* FLUORIDE DOSAGE

To calculate the mg/*L* fluoride dose that results from adding a particular quantity of chemical, one of the feed rate equations must be used, such as those shown to the right. (These are the equations used in sections 13.3 and 13.5.)

It is suggested that you leave the feed rate equation in the form shown, fill in the given information, then solve for the unknown value*—mg/*L* fluoride. Examples 1-3 illustrate this calculation.

TO CALCULATE FLUORIDE DOSAGE (mg/*L*), USE A FEED RATE EQUATION

Given lbs/day Fluoride:

$$(\text{mg}/L \text{ F}) \ (\text{MGD Flow}) \ (8.34 \text{ lbs/gal}) = \text{lbs/day F}$$

For Dry Feed Rate Calculations:

$$\frac{(\text{mg}/L) \ (\text{MGD}) \ (8.34 \text{ lbs/gal})}{\dfrac{(\% \text{ Comm. Pur.})}{100} \ \dfrac{(\% \text{ F in Comp.})}{100}} = \frac{\text{lbs/day}}{\text{Compound}}$$

For Solution Feed Rate Calculations:**

$$\underset{\text{flow} \quad \text{lbs/gal}}{(\text{mg}/L \text{ F}) \ (\text{MGD}) \ (8.34)} = \underset{\text{Comp. flow} \quad \text{lbs/gal}}{(\text{mg}/L) \ (\text{MGD}) \ (8.34)} \ (\text{sp. gr.}) \ \frac{\underset{\text{in Comp.}}{(\% \text{ F})}}{100}$$

↑ When applicable

Example 1: (mg/*L* Fluoride Dosage)
❑ A flow of 0.1 MGD is treated with 3 lbs/day of sodium fluoride, NaF. The commercial purity of the chemical is 98%. The fluoride ion content of NaF is 45.25%. What was the fluoride dosage in mg/*L*?

First select the appropriate feed rate equation. Since this problem involves dry feed rate, the following equation is used:

$$\frac{(\text{mg}/L \text{ F}) \ (\text{MGD}) \ (8.34 \text{ lbs/gal})}{\dfrac{(\% \text{ Comm. Pur.})}{100} \ \dfrac{(\% \text{ F in Comp.})}{100}} = \frac{\text{lbs/day}}{\text{Compound}}$$

Fill in the given information and solve for the unknown value:

$$\frac{(x \text{ mg}/L) \ (0.1 \text{ MGD}) \ (8.34 \text{ lbs/gal})}{\dfrac{(98)}{100} \ \dfrac{(45.25)}{100}} = \frac{3 \text{ lbs/day}}{\text{NaF}}$$

$$(x) \ (0.1) \ (8.34) = (3) \ (0.98) \ (0.4525)$$

$$x = \frac{(3) \ (0.98) \ (0.4525)}{(0.1) \ (8.34)}$$

$$x = \boxed{1.6 \text{ mg}/L \text{ F}}$$

---

* For a review of solving for the unknown value, refer to Chapter 2 in *Basic Math Concepts*.

** The shortened solution feed equation ($R_1 \times C_1 = R_2 \times C_2$) can also be used to calculate mg/*L* fluoride ion.

Example 2: (mg/L Fluoride Dosage)
❑ A total of 25 lbs of sodium silicofluoride (Na $_2$ SiF $_6$) is added to a flow of 1,750,000 gpd. The commercial purity of the sodium silicofluoride is 98% and the percent fluoride ion content of Na $_2$ SiF $_6$ is 60.7. What is the concentration of fluoride ion in the treated water?

This is a calculation involving dry feed rate. Therefore, the equation to be used is:

$$\frac{(mg/L\ F)\ (MGD)\ (8.34\ lbs/gal)}{\dfrac{(\%\ Comm.\ Pur.)}{100}\ \dfrac{(\%\ F\ in\ Comp.)}{100}} = \frac{lbs/day}{Compound}$$

Fill in the equation with the given data, then solve for the unknown value:

$$\frac{(x\ mg/L)\ (1.75\ MGD)\ (8.34\ lbs/gal)}{\dfrac{(98)}{100}\ \dfrac{(60.7)}{100}} = \frac{25\ lbs/day}{Na_2SiF_6}$$

$$(x)\ (1.75)\ (8.34) = (25)\ (0.98)\ (0.607)$$

$$x = \frac{(25)\ (0.98)\ (0.607)}{(1.75)\ (8.34)}$$

$$x = \boxed{1.02\ mg/L\ F}$$

Example 3: (mg/L Fluoride Dosage)
❑ A flow of 3.07 MGD is treated with a 20% solution of hydrofluosilicic acid, H $_2$ SiF $_6$. If the solution feed rate is 20 gpd, what is the calculated fluoride ion concentration of the finished water? The acid weighs 9.8 lbs/gal and the percent fluoride ion in H $_2$ SiF $_6$ is 79.2%. (Assume the raw water contains no fluoride.)

Use the solution feed equation, fill in given data and solve for the unknown:

$$\underset{flow\quad lbs/gal}{(mg/L\ F)\ (MGD)\ (8.34)} = \underset{Comp.\ flow\quad lbs/gal\quad 100}{(mg/L)\ (MGD)\ (Sol'n)\ \left(\frac{\%\ F\ in\ Comp}{}\right)}$$

$$\underset{MGD\ lbs/gal}{(x\ mg/L)\ (3.07)\ (8.34)} = \underset{mg/L\qquad MGD\ lbs/gal\ 100}{(200,000)\ (0.000020)\ (9.8)\ \frac{(79.2)}{}}$$

$$x = \frac{(200,000)\ (0.000020)\ (9.8)\ (0.792)}{(3.07)\ (8.34)}$$

$$x = \boxed{1.2\ mg/L\ F}$$

13.8 SOLUTION MIXTURES

Reviewing various texts concerning solution mixture calculations, you will encounter several different equations, depending on the equation style preference of the author, how the data is expressed, etc. However, all these equations simplify down to about two basic approaches to solution mixture problems—and even these two equations are actually related:

$$\begin{array}{c} \% \\ \text{Strength} \\ \text{of Mixture} \end{array} = \frac{\text{Amt. of Chem.}}{\text{Amt. of Sol'n}} \times 100$$

and

$$R_3\, C_3 = R_1\, C_1 + R_2\, C_2$$

where: R is the feed rate for solutions 1, 2 or 3, and

C is the concentration of solutions 1, 2 or 3

In the first approach to solution mixtures, the focus is solely on the end product—the mixture. The simplified equation for this method is precisely the same as any other percent strength calculation,* as shown in the simplified equation to the right.

In the **first expanded equation,** the numerator and denominator of the equation have been defined or broken down further. The <u>total lbs of chemical</u> in a solution mixture comes from two sources: the lbs chemical in the first solution and the lbs chemical in the second solution. <u>The total lbs solution</u> also comes from two sources: the lbs of the first solution and the lbs of the second solution used to make the mixture.

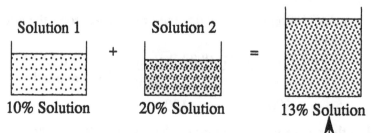

BLENDING TWO SOLUTIONS WILL ALWAYS RESULT IN A SOLUTION WHOSE PERCENT STRENGTH IS SOMEWHERE <u>BETWEEN</u> THE STRENGTHS OF THE ORIGINAL TWO SOLUTIONS

The solution strength of this mixture must be somewhere <u>between</u> 10% and 20% strength, depending on how much solution is contributed by each source.

ONE APPROACH TO SOLUTION MIXTURE CALCULATIONS FOCUSES PRIMARILY ON THE SOLUTION MIXTURE

Solution Mixture

As with all percent strength calculations, the percent strength of the solution mixture depends on two factors:*

- *The total amount of chemical in the solution mixture, and*

- *The total amount of solution.*

Simplified Equation:

$$\begin{array}{c} \% \text{ Strength} \\ \text{of Mixture} \end{array} = \frac{\text{Total lbs Chemical}}{\text{Total lbs Solution}} \times 100$$

Expanded Equation:

①

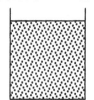

$$\begin{array}{c} \% \text{ Strength} \\ \text{of Mixture} \end{array} = \frac{\begin{array}{c}\text{lbs Chem. in} \\ \text{Sol'n 1}\end{array} + \begin{array}{c}\text{lbs Chem. in} \\ \text{Sol'n 2}\end{array}}{\text{lbs Sol'n 1} + \text{lbs Sol'n 2}} \times 100$$

②   Or

$$\begin{array}{c} \% \\ \text{Strength} \\ \text{of Mixture} \end{array} = \frac{\dfrac{(\text{Sol'n 1,})(8.34)\;(\%)}{100} + \dfrac{(\text{Sol'n 2,})(8.34)\;(\%)}{100}}{(\text{Sol'n 1, gal})(8.34) + (\text{Sol'n 2, gal})(8.34)} \times 100$$

---

* Refer to Section 13.4 for a review of percent strength calculations.

Expanded Equations (Cont'd):

When mixing solutions with the same density, the 8.34 lbs/gal factor can be divided out of all four terms, resulting in this equation:

③

$$\text{% Strength of Mixture} = \frac{\dfrac{(\text{Sol'n 1,})(\text{% Strength})}{\text{gal} \quad \text{of Sol'n 1}} + \dfrac{(\text{Sol'n 2,})(\text{% Strength})}{\text{gal} \quad \text{of Sol'n 2}}}{100 \qquad\qquad 100}{\dfrac{\text{Sol'n 1,}}{\text{gal}} + \dfrac{\text{Sol'n 2,}}{\text{gal}}} \times 100$$

Example 1: (Solution Mixtures)

❏ A tank contains 400 lbs of 20% hydrofluosilicic acid, $H_2 SiF_6$. If 3000 lbs of 25% hydrofluosilicic acid are added to the tank, what is the percent strength of the solution mixture?

$$\text{% Strength of Mixture} = \frac{\text{lbs Chem. in Sol'n 1} + \text{lbs Chem. in Sol'n 2}}{\text{lbs Sol'n 1} + \text{lbs Sol'n 2}} \times 100$$

$$= \frac{(400 \text{ lbs}) \dfrac{(20)}{100} + (3000 \text{ lbs}) \dfrac{(25)}{100}}{400 \text{ lbs} + 3000 \text{ lbs}} \times 100$$

$$= \frac{80 \text{ lbs} + 750 \text{ lbs}}{3400 \text{ lbs}} \times 100$$

$$= \boxed{24.4\%}$$

The **second expanded equation** is to be used when solution data is expressed in <u>gallons</u> rather than pounds. Each term of the previous equation has been replaced with an equivalent expression. Let's examine the first term:

$$\boxed{(\text{lbs Chem. in Sol'n 1})}$$

is replaced by

$$\boxed{(\text{Sol'n 1,}) (8.34) \dfrac{(\text{% Strength})}{100}}{\text{gal} \quad \text{lbs/gal}}$$

These two terms are equivalent terms. If you multiply the gallons of solution by 8.34 lbs/gal*, you will have calculated the lbs of solution. If you then multiply by the percent strength of the solution, you will obtain the <u>lbs chemical</u> in that solution. A similar replacement is made for the second term of the numerator.

Examining the denominator, we find that each term expressed in lbs has been replaced with an equivalent statement in terms of gallons:**

$$\boxed{\text{lbs Sol'n 1}}$$

is replaced by

$$\boxed{(\text{Sol'n 1, gal}) (8.34 \text{ lbs/gal})}$$

When the density is the same for both solutions being mixed, then the 8.34 lbs/gal factor can be divided out of all four terms of the equation, as shown in the **third expanded equation.**

---

* This assumes the density of the solution is 8.34 lbs/gal. If the density is different, use that figure instead of 8.34 lbs/gal.

** For a review of lbs to gallons conversions, refer to Chapter 8 in *Basic Math Concepts*.

Example 2: (Solution Mixtures)
❑ If 1000 lbs of 25% hydrofluosilicic acid were added to a tank containing 200 lbs of 15% hydrofluosilicic acid, what would be the percent strength of the solution mixture?

$$\frac{\% \text{ Strength}}{\text{of Mixture}} = \frac{\frac{\text{lbs Chem. in}}{\text{Sol'n 1}} + \frac{\text{lbs Chem. in}}{\text{Sol'n 2}}}{\text{lbs Sol'n 1 + lbs Sol'n 2}} \times 100$$

$$= \frac{(1000 \text{ lbs}) \frac{(25)}{100} + (200 \text{ lbs}) \frac{(15)}{100}}{1000 \text{ lbs} + 200 \text{ lbs}} \times 100$$

$$= \frac{250 \text{ lbs} + 30 \text{ lbs}}{1200 \text{ lbs}} \times 100$$

$$= \boxed{23.3\%}$$

Example 3: (Solution Mixtures)
❑ A tank of hydrofluosilicic acid ($H_2 SiF_6$) contains 300 gallons with a strength of 18%. If a truck delivers 2000 gallons of 20% hydrofluosilicic acid to be added to the tank, what is the percent strength of the solution mixture? Assume the 20% solution weighs 9.8 lbs/gal and the 18% solution weighs 9.6 lbs/gal.

$$\frac{\% \text{ Str.}}{\text{of Mix.}} = \frac{\frac{(\text{Sol'n 1,}) (\text{Dens.}) (\% \text{ Str.,})}{\text{gal} \quad \text{lbs/gal} \quad \text{Sol'n 1}}}{100} + \frac{(\text{Sol'n 2,}) (\text{Dens.}) (\% \text{ Str.,})}{\text{gal} \quad \text{lbs/gal} \quad \text{Sol'n 2}}{100}}{\frac{(\text{Sol'n 1,}) (\text{Dens.,})}{\text{gal} \quad \text{lbs/gal}} + \frac{(\text{Sol'n 2,}) (\text{Dens.,})}{\text{gal} \quad \text{lbs/gal}}} \times 100$$

$$= \frac{(300 \text{ gal}) (9.6 \text{ lbs/gal}) \frac{(18)}{100} + (2000) (9.8 \text{ lbs/gal}) \frac{(20)}{100}}{(300 \text{ gal}) (9.6 \text{ lbs/gal}) + (2000) (9.8 \text{ lbs/gal})} \times 100$$

$$= \frac{518.4 \text{ lbs} + 3920 \text{ lbs}}{2880 \text{ lbs} + 19,600 \text{ lbs}} \times 100$$

$$= \frac{4438.4 \text{ lbs}}{22,480 \text{ lbs}} \times 100$$

$$= \boxed{19.7\%}$$

THE SECOND APPROACH TO SOLUTION MIXTURE
CALCULATIONS FOCUSES ON THE
ENTIRE MIXTURE PROCESS

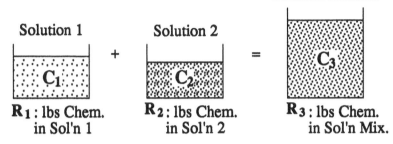

Solution Mixture

Solution 1   +   Solution 2   = C_3

C_1 C_2

R_1: lbs Chem. R_2: lbs Chem. R_3: lbs Chem.
in Sol'n 1   in Sol'n 2   in Sol'n Mix.

$$R_1 C_1 + R_2 C_2 = R_3 C_3$$

Where: R_1 = lbs (or gal) of Sol'n
        C_1 = % Strength of Sol'n 1
        R_2 = lbs (or gal) of Sol'n 2
        C_2 = % Strength of Sol'n 2
        R_3 = lbs (or gal) of Sol'n Mixture
        C_3 = % Strength of Mixture

The second approach to solution mixture calculations is shown to the left.* Rather than focusing on the solution mixture to structure the equation, this method focuses on the entire mixture process.

The equation can be paraphrased as follows:

• The lbs chemical in the first solution ($R_1 C_1$) plus the lbs chemical in the second solution ($R_2 C_2$) equals the lbs chemical in the mixture ($R_3 C_3$).

Examples 4-8 illustrate solution mixture calculations using this equation. The use of boxes to denote lbs chemical and concentration of each component in the calculation, as shown in the examples, can help you organize the data for calculation.

Example 4: (Solution Mixtures)
❑ A tank contains 400 lbs of 20% hydrofluosilicic acid, $H_2 SiF_6$. If 3000 lbs of 25% hydrofluosilicic acid are added to the tank, what is the percent strength of the solution mixture? (This is the same problem as Example 1 so that you can compare the two methods of calculation.)

When using this equation, remember that the rate of the solution mixture (R_3) is a combination of R_1 and R_2:

$$\boxed{\frac{20}{100}} + \boxed{\frac{25}{100}} = \boxed{\frac{x}{100}}$$

400 lbs   3000 lbs   400 lbs + 3000 lbs

$$R_1 C_1 + R_2 C_2 = R_3 C_3$$

$$(400 \text{ lbs})\frac{(20)}{100} + (3000 \text{ lbs})\frac{(25)}{100} = (3400 \text{ lbs})\frac{(x)}{100}$$

$$80 \text{ lbs} + 750 \text{ lbs} = (34)(x)$$

$$\frac{830}{34} = x$$

$$\boxed{24.4\%} = x$$

* This equation is related to the equations described for Examples 1-3. By moving the denominator of either Expanded Equation 1 or Expanded Equation 2 to the left side of the equation, the resulting equation is the same structure as that described here.

Example 5: (Solution Mixtures)
❑ If 1000 lbs of 25% hydrofluosilicic acid were added to a tank containing 200 lbs of 15% hydrofluosilicic acid, what would be the percent strength of the solution mixture? (This is the same problem as Example 2.)

$$\left[\frac{25}{100}\right] + \left[\frac{15}{100}\right] = \left[\frac{x}{100}\right]$$

1000 lbs   200 lbs   1000 lbs + 200 lbs

$$R_1 C_1 + R_2 C_2 = R_3 C_3$$

$$(1000 \text{ lbs})\frac{(25)}{100} + (200 \text{ lbs})\frac{(15)}{100} = (1200 \text{ lbs})\frac{(x)}{100}$$

$$250 \text{ lbs} + 30 \text{ lbs} = 12\,x$$

$$\frac{280}{12} = x$$

$$\boxed{23.3\%} = x$$

Example 6: (Solution Mixtures)
❑ A tank of hydrofluosilicic acid ($H_2 SiF_6$) contains 300 gallons with a strength of 18%. If a truck delivers 2000 gallons of 20% hydrofluosilicic acid to be added to the tank, what is the percent strength of the solution mixture? Assume the 20% solution weighs 9.8 lbs/gal and the 18% solution weighs 9.6 lbs/gal. (This is the same problem as Example 3.)

$$\left[\frac{18}{100}\right] + \left[\frac{20}{100}\right] = \left[\frac{x}{100}\right]$$

(300 gal)(9.6 lbs/gal)   (2000 gal)(9.8 lbs/gal)   2880 lbs + 19,600 lbs

= 2880 lbs   = 19,600 lbs

$$R_1 C_1 + R_2 C_2 = R_3 C_3$$

$$(2880 \text{ lbs})\frac{(18)}{100} + (19{,}600 \text{ lbs})\frac{(20)}{100} = (22{,}480 \text{ lbs})\frac{(x)}{100}$$

$$518.4 \text{ lbs} + 3920 \text{ lbs} = 224.8\,x$$

$$\frac{4438.4}{224.8} = x$$

$$\boxed{19.7\%} = x$$

Example 7: (Solution Mixtures)
❑ How many lbs of water must be mixed with 200 lbs of a 20% solution of hydrofluosilicic acid to make a 15% solution of hydrofluosilicic acid?

$$\boxed{0} + \boxed{\frac{20}{100}} = \boxed{\frac{15}{100}}$$

x lbs   200 lbs $x + 200$ lbs

$$R_1 C_1 + R_2 C_2 = R_3 C_3$$

$$(x \text{ lbs})(0) + (200 \text{ lbs})\frac{(20)}{100} = (x + 200)\frac{(15)}{100}$$

$$0 + 40 = (x + 200)(0.15)$$

$$40 = 0.15x + 30$$

$$30 = 0.15x$$

$$\boxed{66.7 \text{ lbs} \atop \text{Water}} = x$$

Example 8: (Solution Mixtures)
❑ How many gallons of water must be mixed with 100 gallons of 20% hydrofluosilicic acid to result in a 12% solution of hydrofluosilicic acid? The 20% solution of acid weighs 9.8 lbs/gal.

$$\boxed{0} + \boxed{\frac{20}{100}} = \boxed{\frac{12}{100}}$$

$(x$ gal$)(8.34$ lbs/gal$)$ $(100$ gal$)(9.8$ lbs/gal$)$ $8.34x + 980$

$$R_1 C_1 + R_2 C_2 = R_3 C_3$$

$$(8.34x \text{ lbs})(0) + (980 \text{ lbs})\frac{(20)}{100} = (8.34x + 980)\frac{(12)}{100}$$

$$0 + 196 = (8.34x + 980)(0.12)$$

$$196 = 1.0008x + 117.6$$

$$78.4 = 1.0008x$$

$$\boxed{78.3 \text{ lbs} \atop \text{Water}} = x$$

---

* Quoted from *Fluoridation Engineering Manual*, EPA-520/9-74-022.

WHEN WATER IS MIXED WITH A SOLUTION

Is some cases, water is mixed with a solution in order to dilute the concentration of that solution. This type of problem can be calculated like other mixture problems, as illustrated in Examples 7 and 8.

When making these calculations, remember that **the percent strength of water will always be zero.**

When making percent strength or solution mixture calculations, the amount of solution is often expressed in pounds, so that there will not be a miscalculation resulting from solutions having different densities. Gallons can only be used in calculations where the density for all solutions in the calculation is the same.

*Note:** The dilution of hydrofluosilicic acid should be made by carefully pouring the measured volume (or weight) of acid into a measured volume of water. (While adding water to acid would not be particularly hazardous in this case, because of the low strength of commercial acid, it is always good practice to add acid to water rather than the reverse.)

Dilution of hydrofluosilicic acid can result in the formation of a precipitate (silica) when the dilution is in the range of ten to twenty parts of water to one part acid. Such precipitation, which can result in clogged feeders, valves, and orifices, can be avoided by using fortified acid (acid to which a small amount of hydrofluoric acid has been added by the supplier) or by using acid manufactured from hydrofluoric acid rather than from phosphate rock.

NOTES:

14 *Softening*

SUMMARY

1. **Equivalent Weight and Equivalents**

$$\text{Equivalent Weight} = \frac{\text{Atomic Weight}}{\text{Valence}}$$

Or

$$\text{Equivalent Weight} = \frac{\text{Molecular Weight}}{\text{Net Positive Valence}}$$

Or

$$\text{Equivalent Weight} = \frac{\text{Molecular Weight}}{\text{Number of Equivalents}}$$

The number of equivalents or milliequivalents of a chemical may be calculated as:

$$\text{Number of Equivalents} = \frac{\text{Chemical, grams}}{\text{Equivalent Wt.}}$$

And

$$\text{Number of Milliequivalents} = \frac{\text{Chemical, mg}}{\text{Equivalent Wt.}}$$

2. **Hardness, as $CaCO_3$**

    Calcium Hardness, as $CaCO_3$:

$$\frac{\text{Calcium Hardness, mg/}L \text{ as } CaCO_3}{\text{Equivalent Wt. of } CaCO_3} = \frac{\text{Calcium, mg/}L}{\text{Equivalent Wt. of Calcium}}$$

    Magnesium Hardness, as $CaCO_3$:

$$\frac{\text{Magnesium Hardness, mg/}L \text{ as } CaCO_3}{\text{Equivalent Wt. of } CaCO_3} = \frac{\text{Magnesium, mg/}L}{\text{Equivalent Wt. of Magnesium}}$$

* Another method of expressing total hardness is: Total Hardness = Carbonate Hardness + Noncarbonate Hardness. This method is described in Section 14.3.

SUMMARY—Cont'd

2. Hardness, as CaCO₃ —Cont'd

Total Hardness, as CaCO₃ *:

$$\begin{array}{c} \text{Total Hardness,} \\ \text{mg/}L \text{ as CaCO}_3 \end{array} = \begin{array}{c} \text{Calcium Hardness,} \\ \text{mg/}L \text{ as CaCO}_3 \end{array} + \begin{array}{c} \text{Magnesium Hardness,} \\ \text{mg/}L \text{ as CaCO}_3 \end{array}$$

3. Carbonate and Noncarbonate Hardness

The alkalinity of the water dictates the type of hardness present:

If the alkalinity (as CaCO₃) is greater than the total hardness, then all the hardness is carbonate hardness:

$$\begin{array}{c} \text{Total Hardness,} \\ \text{mg/}L \text{ as CaCO}_3 \end{array} = \begin{array}{c} \text{Carbonate Hardness,} \\ \text{mg/}L \text{ as CaCO}_3 \end{array}$$

If the alkalinity (as CaCO₃) is less than the total hardness, the alkalinity is carbonate hardness and noncarbonate hardness is present as well:

$$\begin{array}{c} \text{Total Hardness,} \\ \text{mg/}L \text{ as CaCO}_3 \end{array} = \begin{array}{c} \text{Carbonate Hardness,} \\ \text{as CaCO}_3 \end{array} + \begin{array}{c} \text{Noncarbonate} \\ \text{Hardness,} \\ \text{mg/}L \text{ as CaCO}_3 \end{array}$$

These terms represent the same number

$$\begin{array}{c} \text{Total Hardness,} \\ \text{mg/}L \text{ as CaCO}_3 \end{array} = \begin{array}{c} \text{Alkalinity,} \\ \text{mg/}L \text{ as CaCO}_3 \end{array} + \begin{array}{c} \text{Noncarbonate} \\ \text{Hardness,} \\ \text{mg/}L \text{ as CaCO}_3 \end{array}$$

4. Phenolphthalein and Total Alkalinity

$$\begin{array}{c} \text{Phenolphthalein} \\ \text{Alkalinity} \\ \text{mg/}L \text{ as CaCO}_3 \end{array} = \frac{(A)(N)(50,000)}{\text{m}L \text{ of Sample}}$$

$$\begin{array}{c} \text{Total Alkalinity} \\ \text{mg/}L \text{ as CaCO}_3 \end{array} = \frac{(B)(N)(50,000)}{\text{m}L \text{ of Sample}}$$

Where: A = mL titrant used to pH 8.3
B = Total mL of titrant used to pH 4.5
N = normality of the acid

SUMMARY—Cont'd

5. Bicarbonate, Carbonate and Hydroxide Alkalinity

The phenophthalein and total alkalinity values can be used to determine bicarbonate, carbonate and hydroxide alkalinity. Use the table given below:

Alkalinity, mg/L as CaCO3			
Results of Titration	Bicarbonate Alkalinity	Carbonate Alkalinity	Hydroxide Alkalinity
P = O	T	O	O
P is less than 1/2 T	T – 2P	2P	O
P = 1/2 T	O	2P	O
P is greater then 1/2 T	O	2T – 2P	2P – T
P = T	O	O	T

Where   P = Phenolphthalein alkalinity
        T = Total alkalinity

6. Lime Dosage for Softening

Using Quicklime, CaO:

$$\text{Quicklime (CaO) Feed, mg/L} = \frac{(A + B + C + D)\,1.15}{\dfrac{\%\text{ Purity of Lime}}{100}}$$

Where   A: CO_2 in Source water—
            (mg/L as CO_2)(56/44)

        B: Bicarbonate alkalinity removed in softening
            (mg/L as $CaCO_3$)(56/100)

        C: Hydroxide alkalinity in softener effluent
            (mg/L as $CaCO_3$)(56/100)

        D: Magnesium removed in softening
            (mg/L as Mg^{+2})(56/24.3)

        1.15: Excess lime dosage
               (using 15 percent excess)

SUMMARY—Cont'd

<u>Using hydrated Lime, $Ca(OH)_2$:</u>

To calculate hydrate lime required, use the same equation as given for quicklime except substitute 74 for 56 in A, B, C & D.

7. Soda Ash Dosage for Removal of Noncarbonate Hardness

1. First calculate noncarbonate hardness. (Review Section 14.3)

2. Then calculate soda ash dosage required:

$$\begin{array}{l}\text{Soda Ash } (Na_2CO_3)\\ \text{Feed, mg/L}\end{array} = \begin{array}{l}\text{(Noncarbonate)}\\ \text{Hardness}\\ \text{mg/L as } CaCO_3\end{array} \frac{(106)}{100}$$

8. Carbon Dioxide Required for Recarbonation

1. First calculate mg/L excess lime:

$$\begin{array}{l}\text{Excess Lime,}\\ \text{mg/L}\end{array} = (A + B + C + D)(0.15)$$

Where A, B, C and D are the same factors described in Section 14.6.

2. Then calculate total carbon dioxide dosage (mg/L) required:

$$\begin{array}{l}\text{Total } CO_2\\ \text{Feed, mg/L}\end{array} = \begin{array}{l}\left[Ca(OH)_2\right]\\ \text{Excess,}\\ \text{mg/L}\end{array} \frac{(44)}{74} + \begin{array}{l}(Mg^{+2})\\ \text{Residual}\\ \text{mg/L}\end{array} \frac{(44)}{24.3}$$

9. Chemical Feeder Settings

To calculate lbs/day chemical required:

$$(\text{mg/L Chemical})(\text{MGD Flow})(8.34 \text{ lbs/gal}) = \begin{array}{l}\text{lbs/day}\\ \text{Chemical}\end{array}$$

To express this feed rate as lbs/min:

$$\frac{\text{Chemical, lbs/day}}{1440 \text{ min/day}} = \begin{array}{l}\text{Chemical,}\\ \text{lbs/min}\end{array}$$

SUMMARY—Cont'd

10. Expressions of Hardness – mg/*L* and gpg

$$1 \text{ gpg} = 17.12 \text{ mg/}L$$

When using the "box method" of conversion, the following diagram is used:

11. Ion Exchange Capacity

$$\begin{array}{l}\text{Exchange} \\ \text{Capacity,} \\ \text{grains}\end{array} = \begin{array}{c}\text{(Removal Capacity,)(Media Vol., cu ft)} \\ \text{grains/cu ft}\end{array}$$

12. Water Treatment Capacity
(before Regeneration)

$$\begin{array}{l}\text{Water Treatment} \\ \text{Capacity, gal}\end{array} = \frac{\text{Exchange Capacity, grains}}{\text{Hardness Removed, gpg}}$$

13. Operating Time (Until Regeneration Required)

$$\begin{array}{l}\text{Operating} \\ \text{Time, hrs}\end{array} = \frac{\text{Water Treated, gal}}{\text{Flow Rate, gph}}$$

14. Salt and Brine Required for Regeneration

$$\begin{array}{l}\text{Salt Required,} \\ \text{lbs}\end{array} = \begin{array}{c}\text{(Salt Req'd,) (Hardness Removed,)} \\ \text{lbs/1000 grains \quad kilograins}\end{array}$$

$$\begin{array}{l}\text{Brine Required,} \\ \text{gal}\end{array} = \frac{\text{Salt Required, lbs}}{\text{Brine Solution, lbs Salt/gal Brine}}$$

14.1 EQUIVALENT WEIGHT AND EQUIVALENTS OF ELEMENTS AND COMPOUNDS

When making calculations of the hardness of a water, the concentrations of various constituents of the water are expressed in terms of a common compound—calcium carbonate, $CaCO_3$.

To make these calculations (as described in the next section), it is essential that you are familiar with the concepts of equivalent weight and equivalents.

EQUIVALENT WEIGHT OF AN ELEMENT

In simplified terms, each atom of an element is comprised of protons, neutrons and electrons. The protons (positively charged particles) and neutrons (neutral particles) make up the center or nucleus of the atom. The electrons revolve around the nucleus like planets around the sun. It is the electrons which interact with electrons of other elements to form new substances called compounds. The electrons which interact are called **valence electrons**. The elements listed under Groups I, II and III in the Periodic Table of Elements (listed in any general chemistry text) have 1, 2 and 3 valence electrons, respectively.

By using valence information and atomic weight (also given on the Periodic Table), the equivalent weight for each element can be calculated. The equivalent weight of the element is the atomic weight of that element divided by the valence of that element, as shown in the equation to the right.

Stated another way, it is the portion of the atomic weight of an element associated with <u>each valence electron.</u>

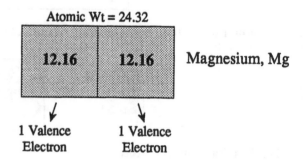

EQUIVALENT WEIGHT IS THE ATOMIC WEIGHT ASSOCIATED WITH EACH VALENCE ELECTRON

Atomic Wt = 24.32

Magnesium, Mg

1 Valence Electron   1 Valence Electron

In this example, the equivalent weight of magnesium would be:

$$\text{Equivalent Weight of Mg} = \frac{24.32}{2}$$

$$= \boxed{12.16}$$

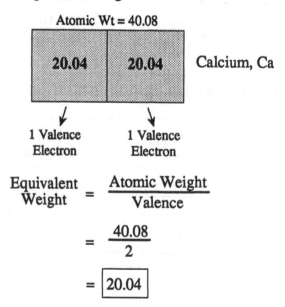

Example 1: (Equivalent Weight and Equivalents)
❑ If calcium has an atomic weight of 40.08 and a valence of 2, what is the equivalent weight of calcium?

Atomic Wt = 40.08

Calcium, Ca

1 Valence Electron   1 Valence Electron

$$\text{Equivalent Weight} = \frac{\text{Atomic Weight}}{\text{Valence}}$$

$$= \frac{40.08}{2}$$

$$= \boxed{20.04}$$

Example 2: (Equivalent Weight and Equivalents)
❑ The atomic weight listed in the Periodic Table for sodium (Na) is 22.997, for oxygen (O) is 16.000 and for hydrogen is 1.008. What is the equivalent weight of NaOH?

First calculate the molecular weight of NaOH:

$$
\begin{array}{l}
\text{Na: } 22.997 \\
\text{O: } 16.000 \\
\text{H: } \underline{1.008} \\
\phantom{\text{H: }} 40.005
\end{array}
$$

The net positive valence of NaOH is 1. Therefore the equivalent weight for NaOH is:

$$
\begin{array}{c}
\text{Equivalent Weight} \\
\text{of NaOH}
\end{array} = \frac{\text{Molecular Weight}}{\text{Net Positive Valence}}
$$

$$
= \frac{40.005}{1}
$$

$$
= \boxed{40.005}
$$

Valences and Atomic Weights of Selected Elements

Element	Valence	Atomic Wt.
Hydrogen, H	+1	1.008
Sodium, Na	+1	22.997
Potassium, K	+1	39.096
Calcium, Ca	+2	40.08
Magnesium, Mg	+2	24.32

Example 1 illustrates the calculation of equivalent weight of elements.

EQUIVALENT WEIGHT OF A COMPOUND

The equivalent weight of a compound is the molecular weight of the compound divided by the net positive valence:

$$
\begin{array}{c}
\text{Equivalent} \\
\text{Weight}
\end{array} = \frac{\text{Molecular Wt.}}{\text{Net Positive Valence}}
$$

To determine the net positive valence of the compound:

1. Write the name of the compound. For Example:

$$ \text{NaCl or CaCO}_3 $$

2. Draw a line between the positive and negative parts of the compound, such as:

$$
\begin{array}{cc}
\text{Na} \mid \text{Cl} & \text{Ca} \mid \text{CO} \\
\text{pos} \mid \text{neg} & \text{pos} \mid \text{neg}
\end{array}
$$

3. Consider the positive (left) side of the formula only. The valence of the left side of the formula equals the **net positive valence**.

	Valence	Subscript		Net Pos. Valence
Na	1	x 1	=	1
Ca	2	x 1	=	2

Examples 2 and 3 illustrate the calculation of equivalent weight of compounds.

Occasionally a different equation must be used for calculating the equivalent weight of a substance.*

Example 3: (Equivalent Weight and Equivalents)
❑ What is the equivalent weight of $CaCO_3$ given the atomic weights listed below?

$$
\begin{array}{llll}
\text{Ca} = & 40.08 & \text{x 1} = & 40.08 \\
\text{C} = & 12.010 & \text{x 1} = & 12.010 \\
\text{O}_3 = & 16.000 & \text{x 3} = & \underline{48.000} \\
& & & 100.090
\end{array}
$$

The net positive valence of $CaCO_3$ is 2. Therefore the equivalent weight of $CaCO_3$ is:

$$
\begin{array}{c}
\text{Equivalent Weight} \\
\text{of CaCO}_3
\end{array} = \frac{\text{Molecular Weight}}{\text{Net Positive Valence}}
$$

$$
= \frac{100.090}{2}
$$

$$
= \boxed{50.045}
$$

* One such case is when determining the equivalent weight of a gas, such as carbon dioxide (CO_2) the following equation is used:

$$
\text{Equivalent Wt} = \frac{\text{Molecular Weight}}{\text{Number of Equivalents}}
$$

Elements and compounds react with one another based on several considerations. One such consideration is that of equivalents.

An **equivalent*** is the amount of an element or compound equal to its equivalent weight. For example, the equivalent weight of calcium is 20.04 (as calculated in Example 1). One equivalent of calcium would therefore be 20.04 grams of calcium. The equivalent weight of calcium carbonate ($CaCO_3$) is 50.045 (as calculated in Example 3). One equivalent of $CaCO_3$ is therefore 50.045 grams of $CaCO_3$.

A **milliequivalent** is similar to an equivalent; however, instead of measuring the chemical in grams, **the chemical is measured in milligrams**.

Examples 4-6 illustrate the calculation of equivalents and milliequivalents. This calculation is important since it is the basis of the calculations presented in Section 14.2—expressing calcium and magnesium in terms of $CaCO_3$

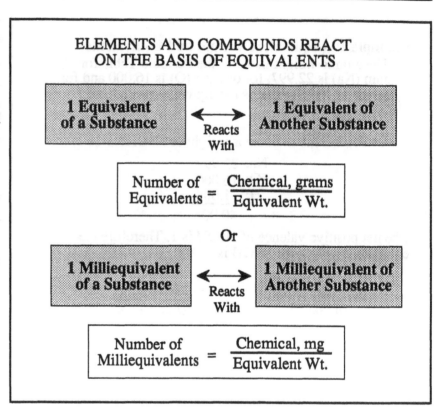

ELEMENTS AND COMPOUNDS REACT
ON THE BASIS OF EQUIVALENTS

1 Equivalent of a Substance ⟷ (Reacts With) 1 Equivalent of Another Substance

$$\text{Number of Equivalents} = \frac{\text{Chemical, grams}}{\text{Equivalent Wt.}}$$

Or

1 Milliequivalent of a Substance ⟷ (Reacts With) 1 Milliequivalent of Another Substance

$$\text{Number of Milliequivalents} = \frac{\text{Chemical, mg}}{\text{Equivalent Wt.}}$$

Example 4: (Equivalent Weight and Equivalents)
❑ 300 milligrams of calcium is how many equivalents of calcium? (The equivalent weight of calcium is 20.04.)

Since equivalents are desired, the mg calcium must first be converted to grams:**

$$300 \text{ mg} = 0.300.$$

$$= 0.3 \text{ grams Ca}$$

Now the number of equivalents can be determined:

$$\text{Number of Equivalents} = \frac{\text{Chemical, grams}}{\text{Equivalent Wt.}}$$

$$= \frac{0.3 \text{ grams}}{20.04}$$

$$= \boxed{0.015 \text{ equivalents}}$$

As indicated by the result of this calculation, the concentration of constituents measured in water is generally too small to be measured as equivalents. These constituents are therefore expressed as milliequivalents, as illustrated in Examples 5 and 6.

---

* Equivalents can be measured in various measures of weight, such as grams or pounds. However, unless stated otherwise, an "equivalent" is understood to mean a "gram-equivalent."
** For a review of metric conversions, refer to Chapter 8 in *Basic Math Concepts*.

Example 5: (Equivalent Weight and Equivalents)
❑ The magnesium content of a water is 25 mg/*L*. How many milliequivalents/liter of magnesium is this? (The equivalent weight of magnesium is 12.15.)

$$\text{Number of Milliequivalents}/L = \frac{\text{Chemical, mg}/L}{\text{Equivalent Wt.}}$$

$$= \frac{25 \text{ mg}/L}{12.15}$$

$$= \boxed{2.06 \text{ milliequivalents}/L}$$

Example 5: (Equivalent Weight and Equivalents)
❑ The calcium content of a water is 37 mg/*L*. How many milliequivalents/liter of calcium is this? (The equivalent weight of calcium is 20.04.)

$$\text{Number of Milliequivalents}/L = \frac{\text{Chemical, mg}/L}{\text{Equivalent Wt.}}$$

$$= \frac{37 \text{ mg}/L}{20.04}$$

$$= \boxed{1.85 \text{ milliequivalents}/L}$$

14.2 HARDNESS, AS CaCO₃

Two primary constituents of water that determine the hardness of water are calcium and magnesium. Once the concentration of these elements in the water is known, the total hardness of the water can be calculated. Before the concentration of these elements can be added, however, they must be expressed in common terms—as calcium carbonate. By expressing both calcium and magnesium as milliequivalents of $CaCO_3$, you have described each hardness constituent with similar reacting characteristics. (Remember, one milliequivalent of one substance will react with one milliequivalent of another substance.)

As shown in the equations given to the right, the milliequivalents of $CaCO_3$ are set equal to the milliequivalents of either calcium or magnesium. The expanded form of each equation is the equation to be used in calculating hardness as $CaCO_3$. First, write the equation and fill in the known values. Then solve for the unknown value (either "Calcium Hardness as $CaCO_3$ " or "Magnesium Hardness as $CaCO_3$ ".

Equivalent weights of calcium, magnesium, and calcium carbonate are given below.

Equivalent Weights	
Calcium, Ca	20.04
Magnesium, Mg	12.15
Calcium Carbonate, CaCO₃	50.045

WHEN EXPRESSING HARDNESS CONSTITUENTS AS $CaCO_3$, YOU ARE ACTUALLY CALCULATING MILLIEQUIVALENTS

Calcium Hardness, as CaCO₃ :

$$\text{Milliequivalents of } CaCO_3 = \text{Milliequivalents of Calcium}$$

In expanded form, the equation given above can be written as:

$$\frac{\text{Calcium Hardness, } mg/L \text{ as } CaCO_3}{\text{Equivalent Wt. of } CaCO_3} = \frac{\text{Calcium, } mg/L}{\text{Equivalent Weight of Calcium}}$$

Magnesium Hardness, as CaCO₃ :

$$\text{Milliequivalents of } CaCO_3 = \text{Milliequivalents of Magnesium}$$

In expanded form, the equation given above can be written as:

$$\frac{\text{Magnesium Hardness, } mg/L \text{ as } CaCO_3}{\text{Equivalent Wt. of } CaCO_3} = \frac{\text{Magnesium, } mg/L}{\text{Equivalent Wt. of Magnesium}}$$

A Commonly Used Form of These Equations*:

$$\text{Ca (or Mg) Hardness, } mg/L \text{ as } CaCO_3 = \frac{(\text{Ca (or Mg), } mg/L)(\text{Equiv. Wt. of } CaCO_3)}{\text{Equiv. Wt. of Ca (or Mg)}}$$

---

* The equation for calculating calcium or magnesium hardness as $CaCO_3$ is generally given in the form indicated. Although this equation isolates the desired term on the left side of the equation, the student often does not recognize how the equation was derived. By keeping the equation shown above, it is often easier for the student to remember the equation since it is based on the calculation of milliequivalents.

Example 1: (Hardness as CaCO3)
❑ A water sample has a calcium content of 48 mg/*L*. What is this calcium hardness expressed as CaCO3 ?

First write the equation and fill in the known information:

$$\frac{\text{Calcium Hardness, mg/}L \text{ as CaCO}_3}{\text{Equivalent Wt. of CaCO}_3} = \frac{\text{Calcium, mg/}L}{\text{Equivalent Wt. of Calcium}}$$

$$\frac{x \text{ mg/}L}{50.045} = \frac{48 \text{ mg/}L}{20.04}$$

Then solve for the unknown value:*

$$x = \frac{(48)(50.045)}{20.04}$$

$$x = \boxed{\begin{array}{c}119.9 \text{ mg/}L \text{ Ca} \\ \text{as CaCO}_3\end{array}}$$

CALCIUM HARDNESS, AS CACO3

To calculate calcium hardness as calcium carbonate (CaCO3), simply set the milliequivalents of CaCO3 equal to the milliequivalents of calcium. Use the expanded equation shown on the facing page. Examples 1 and 2 illustrate this calculation.

Example 2: (Hardness as CaCO3)
❑ The calcium content of a water sample is 28 mg/*L*. What is this calcium hardness expressed as CaCO3 ?

$$\frac{\text{Calcium Hardness, mg/}L \text{ as CaCO}_3}{\text{Equivalent Wt. of CaCO}_3} = \frac{\text{Calcium, mg/}L}{\text{Equivalent Wt. of Calcium}}$$

$$\frac{x \text{ mg/}L}{50.045} = \frac{28 \text{ mg/}L}{20.04}$$

$$x = \frac{(28)(50.045)}{20.04}$$

$$x = \boxed{\begin{array}{c}69.9 \text{ mg/}L \text{ Ca} \\ \text{as CaCO}_3\end{array}}$$

* For a review of solving for the unknown value, refer to Chapter 2 in *Basic Math Concepts.*

MAGNESIUM HARDNESS, AS $CaCO_3$

To calculate magnesium hardness as calcium carbonate ($CaCO_3$), set the milliequivalents of $CaCO_3$ equal to the milliequivalents of magnesium. Use the expanded equation shown on the previous page.

Examples 3 and 4 illustrate this calculation.

Example 3: (Hardness as $CaCO_3$)

❑ A sample of water contains 25 mg/*L* magnesium. Express this magnesium hardness as $CaCO_3$.

First, write the equation and fill in the given information:

$$\frac{\text{Magnesium Hardness, mg/}L \text{ as } CaCO_3}{\text{Equivalent Wt. of } CaCO_3} = \frac{\text{Magnesium, mg/}L}{\text{Equivalent Wt of Magnesium}}$$

$$\frac{x \text{ mg/}L}{50.045} = \frac{25 \text{ mg/}L}{12.15}$$

$$x = \frac{(25)(50.045)}{12.15}$$

$$x = \boxed{103.0 \text{ mg/}L \text{ Mg as } CaCO_3}$$

Example 4: (Hardness as $CaCO_3$)

❑ The magnesium content of a water sample is 15 mg/*L*. Express this magnesium hardness as $CaCO_3$.

$$\frac{\text{Magnesium Hardness, mg/}L \text{ as } CaCO_3}{\text{Equivalent Wt. of } CaCO_3} = \frac{\text{Magnesium, mg/}L}{\text{Equivalent Wt. of Magnesium}}$$

$$\frac{x \text{ mg/}L}{50.045} = \frac{15 \text{ mg/}L}{12.15}$$

$$x = \frac{(15)(50.045)}{12.15}$$

$$x = \boxed{61.8 \text{ mg/}L \text{ Mg as } CaCO_3}$$

Example 5: (Hardness as CaCO₃)
❏ A sample of water has a calcium content of 80 mg/L as $CaCO_3$ and a magnesium content of 92 mg/L as $CaCO_3$. What is the total hardness (as $CaCO_3$) of the sample?

$$\begin{array}{c}\text{Total Hardness,} \\ \text{mg/}L \text{ as } CaCO_3\end{array} = \begin{array}{c}\text{Calcium Hardness,} \\ \text{mg/}L \text{ as } CaCO_3\end{array} + \begin{array}{c}\text{Magnesium Hardness,} \\ \text{mg/}L \text{ as } CaCO_3\end{array}$$

$$= 80 \text{ mg/}L + 92 \text{ mg/}L$$

$$= \boxed{172 \text{ mg/}L \text{ as } CaCO_3}$$

Example 6: (Hardness as CaCO₃)
❏ Determine the total hardness as $CaCO_3$ of a sample of water that has a calcium content of 30 mg/L and a magnesium content of 8 mg/L.

First calcium and magnesium must be expressed in terms of $CaCO_3$:

$$\frac{\begin{array}{c}\text{Calcium Hardness,} \\ \text{mg/}L \text{ as } CaCO_3\end{array}}{\text{Equivalent Wt. of } CaCO_3} = \frac{\text{Calcium, mg/}L}{\text{Equivalent Weight of Calcium}}$$

$$\frac{x \text{ mg/}L}{50.045} = \frac{30 \text{ mg/}L}{20.04}$$

$$x = 74.9 \text{ mg/}L \text{ Ca as } CaCO_3$$

$$\frac{\begin{array}{c}\text{Magnesium Hardness,} \\ \text{mg/}L \text{ as } CaCO_3\end{array}}{\text{Equivalent Wt. of } CaCO_3} = \frac{\text{Magnesium, mg/}L}{\text{Equivalent Wt. of Magnesium}}$$

$$\frac{x \text{ mg/}L}{50.045} = \frac{8 \text{ mg/}L}{12.15}$$

$$x = 33.0 \text{ mg/}L \text{ Mg as } CaCO_3$$

Then total hardness can be calculated:

$$\begin{array}{c}\text{Total Hardness,} \\ \text{mg/}L \text{ as } CaCO_3\end{array} = \begin{array}{c}\text{Calcium Hardness,} \\ \text{mg/}L \text{ as } CaCO_3\end{array} + \begin{array}{c}\text{Magnesium Hardness,} \\ \text{mg/}L \text{ as } CaCO_3\end{array}$$

$$= 74.9 \text{ mg/}L + 33.0 \text{ mg/}L$$

$$= \boxed{107.9 \text{ mg/}L \text{ as } CaCO_3}$$

TOTAL HARDNESS, AS CaCO₃

The two constituents that are the primary cause of hardness of a water are calcium and magnesium ions. And although ions of other metals such as aluminum, iron, manganese, strontium and zinc may contribute to the hardness of a water, they are generally not present to any significant level in natural water.

For this reason, total hardness is generally calculated by adding the concentrations of calcium and magnesium ions, expressed in terms of calcium carbonate, $CaCO_3$:*

$$\begin{array}{c}\text{Total} \\ \text{Hardness,} \\ \text{mg/}L \text{ as} \\ CaCO_3\end{array} = \begin{array}{c}\text{Calcium} \\ \text{Hardness,} \\ \text{mg/}L \text{ as} \\ CaCO_3\end{array} + \begin{array}{c}\text{Magnesium} \\ \text{Hardness,} \\ \text{mg/}L \text{ as} \\ CaCO_3\end{array}$$

Examples 5 and 6 illustrate the calculation of total hardness, as $CaCO_3$.

* According to *Standard Methods* section on "Hardness," if any of the other metallic ions are present in significant amounts, they should also be included in the hardness calculation.

14.3 CARBONATE AND NONCARBONATE HARDNESS

Total hardness is comprised of calcium and magnesium hardness, as described in the previous section. Once total hardness has been calculated, it is sometimes used to determine another expression of hardness—carbonate and noncarbonate hardness.

As shown in the equation to the right, total hardness includes both carbonate hardness (temporary hardness) and noncarbonate hardness (permanent hardness). The amount of hardness contributed by each source depends on the alkalinity of the water.

Examples 1-3 illustrate calculations involving carbonate and noncarbonate hardness.

TOTAL HARDNESS IS COMPRISED OF CARBONATE HARDNESS AND NONCARBONATE HARDNESS

$$\text{Total Hardness,} = \text{Carbonate Hardness,} + \text{Noncarbonate Hardness,}$$

When the alkalinity (as $CaCO_3$) is greater than the total hardness, all the hardness is carbonate hardness:

$$\begin{array}{l}\text{Total Hardness,} \\ \text{mg/}L \text{ as } CaCO_3\end{array} = \begin{array}{l}\text{Carbonate Hardness,} \\ \text{mg/}L \text{ as } CaCO_3\end{array}$$

When the alkalinity (as $CaCO_3$) is less than the total hardness, then the alkalinity represents carbonate hardness and the balance of the hardness is noncarbonate hardness:

$$\begin{array}{l}\text{Total Hardness,} \\ \text{mg/}L \text{ as } CaCO_3\end{array} = \begin{array}{l}\text{Carbonate Hardness,} \\ \text{mg/}L \text{ as } CaCO_3\end{array} + \begin{array}{l}\text{Noncarbonate Hardness,} \\ \text{mg/}L \text{ as } CaCO_3\end{array}$$

Or | *Carbonate hardness is represented by the alkalinity*

$$\begin{array}{l}\text{Total Hardness,} \\ \text{mg/}L \text{ as } CaCO_3\end{array} = \begin{array}{l}\text{Alkalinity,} \\ \text{mg/}L \text{ as } CaCO_3\end{array} + \begin{array}{l}\text{Noncarbonate Hardness,} \\ \text{mg/}L \text{ as } CaCO_3\end{array}$$

Example 1: (Carbonate and Noncarbonate Hardness)
❏ A sample of water contains 115 mg/L alkalinity as $CaCO_3$ and 103 mg/L total hardness as $CaCO_3$. What is the carbonate and noncarbonate hardness of this water?

Since the alkalinity is greater than the total hardness, all the hardness is carbonate hardness:

$$\begin{array}{l}\text{Total Hardness,} \\ \text{mg/}L \text{ as } CaCO_3\end{array} = \begin{array}{l}\text{Carbonate Hardness,} \\ \text{mg/}L \text{ as } CaCO_3\end{array}$$

$$\boxed{\begin{array}{l}103 \text{ mg/}L \\ \text{as } CaCO_3\end{array}} = \begin{array}{l}\text{Carbonate} \\ \text{Hardness}\end{array}$$

There is no noncarbonate hardness in this water.

Example 2: (**Carbonate and Noncarbonate Hardness**)
❏ A water has an alkalinity of 88 mg/L as CaCO₃ and a
total hardness of 110 mg/L. What is the carbonate and
noncarbonate hardness of the water?

Since the alkalinity is less than the total hardness, there will
be both carbonate <u>and</u> noncarbonate hardness present in the
water. The alkalinity content of the water represents the
carbonate hardness of the water.

Using both carbonate and total hardness data, the
noncarbonate hardness can be determined:

$$\begin{array}{l} \text{Total Hardness,} \\ \text{mg/}L\text{ as CaCO}_3 \end{array} = \begin{array}{l} \text{Carbonate Hardness,} \\ \text{mg/}L\text{ as CaCO}_3 \end{array} + \begin{array}{l} \text{Noncarbonate} \\ \text{Hardness,} \\ \text{mg/}L\text{ as CaCO}_3 \end{array}$$

$$110 \text{ mg/}L = 88 \text{ mg/}L + x \text{ mg/}L$$

$$110 \text{ mg/}L - 88 \text{ mg/}L = x \text{ mg/}L$$

$$\boxed{\begin{array}{c} 22 \text{ mg/}L \\ \text{noncarbonate} \\ \text{hardness} \end{array}} = x$$

Example 3: (**Carbonate and Noncarbonate Hardness**)
❏ The alkalinity of a water is 90 mg/L as CaCO₃ . If the total
hardness of the water is 115 mg/L as CaCO₃ , what is the
carbonate and noncarbonate hardness in mg/L as CaCO₃ ?

Alkalinity is less than total hardness. Therefore, both
carbonate and non carbonate hardness will be present in the
water. The alkalinity content represents the carbonate
hardness of the water.

Using both carbonate and total hardness data, the
noncarbonate hardness can be calculated:

$$\begin{array}{l} \text{Total Hardness,} \\ \text{mg/}L\text{ as CaCO}_3 \end{array} = \begin{array}{l} \text{Carbonate Hardness,} \\ \text{mg/}L\text{ as CaCO}_3 \end{array} + \begin{array}{l} \text{Noncarbonate} \\ \text{Hardness,} \\ \text{mg/}L\text{ as CaCO}_3 \end{array}$$

$$115 \text{ mg/}L = 90 \text{ mg/}L + x \text{ mg/}L$$

$$115 \text{ mg/}L - 90 \text{ mg/}L = x \text{ mg/}L$$

$$\boxed{\begin{array}{c} 25 \text{ mg/}L \\ \text{noncarbonate} \\ \text{hardness} \end{array}} = x$$

14.4 PHENOLPHTHALEIN AND TOTAL ALKALINITY

The alkalinity of a water is the capacity of that water to neutralize, or buffer, acids. The higher the alkalinity, the greater the capacity of the water to neutralize acids; and conversely, the lower the alkalinity, the less the neutralizing capacity.

Alkalinity in the water results from the presence of the following radicals:*

- Bicarbonates, HCO_3^{-1},

- Carbonates, CO_3^{-2}, and

- Hydroxides, OH^{-1}

To detect the presence of these different types of alkalinity, the water is tested for phenolphthalein and total alkalinity.

Phenolphthalein alkalinity is the alkalinity that exists above a pH of 8.3. To determine the presence of phenolphthalein alkalinity, therefore, the pH of a 100-mL sample is measured. If the pH is 8.3 or less, there is no phenolphthalein alkalinity present. If, however, the pH is greater than 8.3, the sample is titrated with a 0.02N solution of sulfuric acid (H_2SO_4) until the pH lowers to 8.3. Then the amount of acid required in the titration is recorded.

Total alkalinity is the alkalinity that exists above pH 4.5. It is determined by continuing the titration described above until the pH reaches 4.5. The total amount of acid from the beginning of the titration procedure to the end is then recorded.

ALKALINITY, mg/L $CaCO_3$

$$\text{Phenolphthalein Alkalinity mg/L as } CaCO_3 = \frac{(A)(N)(50,000)}{\text{mL of Sample}}$$

and

$$\text{Total Alkalinity mg/L as } CaCO_3 = \frac{(B)(N)(50,000)}{\text{mL of Sample}}$$

Where: A = mL titrant used to pH 8.3

B = total mL of titrant used to titrate to pH 4.5

N = normality of the acid (0.02 N H_2SO_4 for this alkalinity test)

Example 1: (Alkalinity)

❑ A 100-mL water sample is tested for phenolphthalein alkalinity. If 1.1 mL titrant is used to pH 8.3 and the sulfuric acid solution has a normality of 0.02N, what is the phenolphthalein alkalinity of the water? (mg/L as $CaCO_3$)

First write the equation for phenolphthalein alkalinity, then fill in the known data:

$$\text{Phenolphthalein Alkalinity mg/L as } CaCO_3 = \frac{(A)(N)(50,000)}{\text{mL of Sample}}$$

$$= \frac{(1.1 \text{ mL}) (0.02N) (50,000)}{100 \text{ mL}}$$

$$= \boxed{\begin{array}{l}11 \text{ mg/L as } CaCO_3 \\ \text{Phenolphthalein Alkalinity}\end{array}}$$

---

* A radical is a cluster of elements held together by covalent bonds which behave as if they were a single element. That is, they combine and dissociate with other elements as if they were a single element.

Example 2: (Alkalinity)
❏ A 100-m*L* sample of water is tested for phenolphthalein and total alkalinity. A total of 0.6 m*L* titrant is used to pH 8.3 and a total of 7.5 m*L* titrant is used to titrate to pH 4.5. The normality of the acid used for titrating is 0.02 *N*. What is the phenolphthalein and total alkalinity of the sample? (mg/*L* as $CaCO_3$)

$$\text{Phenolphthalein Alkalinity mg/}L \text{ as } CaCO_3 = \frac{(0.6 \text{ m}L) (0.02N) (50{,}000)}{100 \text{ m}L}$$

$$= \boxed{6 \text{ mg/}L}$$

$$\text{Total Alkalinity mg/}L \text{ as } CaCO_3 = \frac{(7.5 \text{ m}L) (0.02N) (50{,}000)}{100 \text{ m}L}$$

$$= \boxed{75 \text{ mg/}L}$$

The titration results are then used to calculate phenolphthalein and total alkalinity, using the equations shown on the facing page. Letters have been used in the equation so that it could be written briefly. The 50,000 factor is a combined term derived from 50 times 1000, as shown to the right.

The basic equation from which the two equations have been simplified is as follows:

$$\frac{\text{m}L \text{ Titrant}(0.02 \text{ Equiv.})(50 \text{ g})^*(1000 \text{ mg})}{\text{M}L \text{ Sample} \quad 1 \text{ liter} \quad 1 \text{ Equiv} \quad 1 \text{ g}}$$

$$= \text{Alkalinity mg/}L \text{ as } CaCO_3$$

When the acid normality is 0.02*N* (as called for in the *Standard Methods* procedure) and the sample size is 100 m*L*, note that the calculated answer is exactly <u>10 times the m*L* titrant used</u>. For instance, in Example 1, 1.1 m*L* titrant is used and the final answer is 10 times that amount: 11 m*L*. Multiplying by ten is the same as <u>moving the decimal point one place to the right</u>.

Example 3: (Alkalinity)
❏ A 100-m*L* sample of water is tested for alkalinity. The normality of the sulfuric acid used for titrating is 0.02*N*. If 0 m*L* titrant is used to pH 8.3, and 7.8 m*L* titrant is used to pH 4.5, what is the phenolphthalein and total alkalinity of the sample?

$$\text{Phenolphthalein Alkalinity mg/}L \text{ as } CaCO_3 = \frac{(0 \text{ m}L) (0.02N) (50{,}000)}{100 \text{ m}L}$$

$$= \boxed{0 \text{ mg/}L}$$

$$\text{Total Alkalinity mg/}L \text{ as } CaCO_3 = \frac{(7.8 \text{ m}L) (0.02N) (50{,}000)}{100 \text{ m}L}$$

$$= \boxed{78 \text{ mg/}L}$$

---

* 1 equivalent of CaCO3 is equal to 50 grams/liter. Note also that the normality of the acid is expressed as equivalents per liter.

14.5 BICARBONATE, CARBONATE AND HYDROXIDE ALKALINITY

Bicarbonate, carbonate and hydroxide alkalinity can be calculated using the results of the phenolphthalein and total alkalinity tests.* To calculate bicarbonate, carbonate and hydroxide alkalinity, use the values indicated in the table to the right. Examples 1-3 illustrate this type of calculation.

USE PHENOLPHTHALEIN AND TOTAL ALKALINITY TO CALCULATE BICARBONATE, CARBONATE, AND HYDROXIDE ALKALINITY

Alkalinity, mg/L as $CaCO_3$			
Results of Titration	Bicarbonate Alkalinity	Carbonate Alkalinity	Hydroxide Alkalinity
P = O	T	O	O
P is less than 1/2 T	T – 2P	2P	O
P = 1/2 T	O	2P	O
P is greater then 1/2 T	O	2T – 2P	2P – T
P = T	O	O	T

Where P = Phenolphthalein alkalinity
T = Total alkalinity

Example 1: (Bicarbonate, Carbonate and Hydroxide Alk.)
❑ A water sample is tested for phenolphthalein and total alkalinity. If the phenolphthalein alkalinity is 11 mg/L as $CaCO_3$ and the total alkalinity is 54 mg/L as $CaCO_3$, what is the bicarbonate, carbonate, and hydroxide alkalinity of the water?

Based on the titration test results, P alkalinity (11 mg/L) is less than half of the T alkalinity (54 mg/L ÷ 2 = 25 mg/L). Therefore, each type of alkalinity is calculated as follows:

$$\text{Bicarbonate Alkalinity} = T - 2P$$

$$= 54 \text{ mg/L} - 2 (11 \text{ mg/L})$$

$$= 54 \text{ mg/L} - 22 \text{ mg/L}$$

$$= \boxed{32 \text{ mg/L as } CaCO_3}$$

$$\text{Carbonate Alkalinity} = 2P$$

$$= 2 (11 \text{ mg/L})$$

$$= \boxed{22 \text{ mg/L as } CaCO_3}$$

$$\text{Hydroxide Alkalinity} = \boxed{0 \text{ mg/L as } CaCO_3}$$

---

* Refer to Section 14.4 for a review of phenolphthalein and total alkalinity tests.

Example 2: (Bicarbonate, Carbonate and Hydroxide Alk.)
❑ A water sample is found to have a phenolphthalein alkalinity of 0 mg/L and a total alkalinity of 76 mg/L. What is the bicarbonate, carbonate and hydroxide alkalinity of the water?

First compare P and T alkalinity values to determine which row to use in the table. Since P = O, use the first row of the table:

Bicarbonate Alkalinity = T

$$= \boxed{76 \text{ mg/}L \text{ as CaCO}_3}$$

Carbonate Alkalinity = $\boxed{0 \text{ mg/}L \text{ as CaCO}_3}$

Hydroxide Alkalinity = $\boxed{0 \text{ mg/}L \text{ as CaCO}_3}$

Example 3: (Bicarbonate, Carbonate and Hydroxide Alk.)
❑ Alkalinity titrations on a water sample resulted as follows: sample—100 mL; 1.5 mL titrant used to pH 8.3; 2.6 mL total titrant used to pH 4.5. Acid normality was 0.02N H$_2$SO$_4$. What is the phenolphthalein, total bicarbonate, carbonate and hydroxide alkalinity?

$$\text{Phenolphthalein Alkalinity} \atop \text{mg/}L \text{ as CaCO}_3 = \frac{(1.5 \text{ m}L)(0.02N)(50{,}000)}{100 \text{ m}L}$$

$$= \boxed{15 \text{ mg/}L \text{ as CaCO}_3}$$

$$\text{Total Alkalinity} \atop \text{mg/}L \text{ as CaCO}_3 = \frac{(2.6 \text{ m}L)(0.02N)(50{,}000)}{100 \text{ m}L}$$

$$= \boxed{26 \text{ mg/}L \text{ as CaCO}_3}$$

Now use the table to calculate the other alkalinity constituents: (P is greater than 1/2 T)

Bicarbonate Alkalinity = $\boxed{0 \text{ mg/}L \text{ as CaCO}_3}$

Carbonate Alkalinity = 2T – 2P

$$= 2(26 \text{ mg/}L) - 2(15 \text{ mg/}L)$$

$$= \boxed{22 \text{ mg/}L \text{ as CaCO}_3}$$

Hydroxide Alkalinity = 2P – T

$$= 2(15 \text{ mg/}L) - 26 \text{ mg/}L$$

$$= \boxed{4 \text{ mg/}L \text{ as CaCO}_3}$$

14.6 LIME DOSAGE FOR REMOVAL OF CARBONATE HARDNESS

Lime-soda ash softening is used to reduce the hardness of a water. Sections 14.6-14.9 describe various calculations which pertain to the lime-soda ash softening process.

When the hardness in the water is a result of calcium and magnesium bicarbonates (carbonate hardness), the addition of lime only is generally sufficient to soften the water to an acceptable level.

The molecular weights of various chemicals and compounds used in lime-soda ash softening calculations are as follows:

Quicklime, CaO_2	= 56
Hydrated Lime, $Ca(OH)_2$	= 74
Magnesium, Mg^{+2}	= 24.3
Carbon Dioxide, CO_2	= 44
Magnesium Hydroxide, $Mg(OH)_2$	= 58.3
Soda Ash, Na_2CO_3	= 106
Alkalinity, as $CaCO_3$	= 100
Hardness, as $CaCO_3$	= 100

TO ESTIMATE QUICKLIME OR HYDRATED LIME DOSAGE, mg/L

For Quicklime Dosage, mg/L:

$$\text{Quicklime (CaO)} \atop \text{Feed, mg/L} = \frac{(A + B + C + D)\,1.15}{\dfrac{\%\text{ Purity of Lime}}{100}}$$

Where   A: CO_2 in Source water— (mg/L as CO_2)(56/44)

B: Bicarbonate alkalinity removed in softening (mg/L as $CaCO_3$)(56/100)

C: Hydroxide alkalinity in softener effluent (mg/L as $CaCO_3$)(56/100)

D: Magnesium removed in softening (mg/L as Mg^{+2})(56/24.3)

1.15: Excess lime dosage (using 15 percent excess)

For Hydrated Lime Dosage:

To calculate hydrate lime required, use the same equation as given for quicklime except substitute 74 for 56 in A, B, C & D.

Example 1: (Lime Dosage)

❑ A water sample has a carbon dioxide content of 5 mg/L as CO_2, total alkalinity of 135 mg/L as $CaCO_3$, and magnesium content of 28 mg/L as MG^{+2}. Approximately how much quicklime (CaO) (90% purity) will be required for softening? (Assume 15% excess lime.)

First calculate the A – D factors:

$A = (CO_2, \text{mg/L})(56/44)$

   $= (5 \text{ mg/L})(56/44)$

   $= 6 \text{ mg/L}$

$B = (\text{Alkalinity, mg/L})(56/100)$

   $= (135 \text{ mg/L})(56/100)$

   $= 76 \text{ mg/L}$

$C = 0 \text{ mg/L}$

$D = (Mg^{+2}, \text{mg/L})(56/24.3)$

   $= (28 \text{ mg/L})(56/24.3)$

   $= 65 \text{ mg/L}$

Then calculate the estimated quicklime dosage:

$$\text{Quicklime Dosage,} \atop \text{mg/L} = \frac{(6 \text{ mg/L} + 76 \text{ mg/L} + 0 + 65 \text{ mg/L})\,(1.15)}{0.90}$$

$$= \boxed{188 \text{ mg/L CaO}}$$

Example 2: (Lime Dosage)
❑ A water sample has the following characteristics:
8 mg/L CO_2 as CO_2 , 120 mg/L total alkalinity as $CaCO_3$,
and 18 mg/L magnesium as Mg^{+2}. What is the estimated
hydrated lime ($Ca(OH)_2$) (90% purity) dosage in mg/L
required for softening? (Assume 15% excess lime.)

First determine the A – D factors:

A = $(CO_2 , mg/L)(74/44)$

  = (8 mg/L)(74/44)

  = 13 mg/L

B = (Alkalinity, mg/L)(74/100)

  = (120 mg/L)(74/100)

  = 89 mg/L

C = 0 mg/L

D = $(Mg^{+2}, mg/L)(74/24.3)$

  = (18 mg/L)(74/24.3)

  = 55 mg/L

Then calculate the estimated hydrated lime dosage:

$$\text{Hydrated Lime Dosage, mg/L} = \frac{(13\ mg/L + 89\ mg/L + 0 + 55\ mg/L)\ (1.15)}{0.90}$$

$$= \boxed{201\ mg/L}$$

Example 3: (Lime Dosage)
❑ The characteristics of a water are as follows: 5 mg/L CO_2
as CO_2 , 178 mg/L total alkalinity as $CaCO_3$, and 21 mg/L
magnesium as Mg^{+2} What is the estimated hydrated lime
($Ca(OH)_2$) (90% pure) dosage in mg/L required for
softening? (Assume 15% excess lime.)

First determine the A – D factors:

A = $(CO_2 , mg/L)(74/44)$

  = (5 mg/L)(74/44)

  = 8 mg/L

B = (Alkalinity, mg/L)(74/100)

  = (178 mg/L)(74/100)

  = 132 mg/L

C = 0 mg/L

D = $(Mg^{+2}, mg/L)(74/24.3)$

  = (21 mg/L)(74/24.3)

  = 64 mg/L

Then calculate the estimated hydrated lime dosage:

$$\text{Hydrated Lime Dosage, mg/L} = \frac{(8\ mg/L + 132\ mg/L + 0 + 64\ mg/L)\ (1.15)}{0.90}$$

$$= \boxed{261\ mg/L\ Ca(OH)_2}$$

14.7 SODA ASH DOSAGE FOR REMOVAL OF NONCARBONATE HARDNESS

Soda ash is used to remove noncarbonate hardness from water. To calculate the soda ash dosage required, therefore, you must first determine the noncarbonate hardness, then calculate soda ash dosage, as shown in the equations to the right.

Examples 1-3 illustrate these calculations.

TO CALCULATE SODA ASH DOSAGE, mg/L

1. First calculate the noncarbonate hardness:*

$$\begin{array}{ccc} \text{Total Hardness,} & = & \text{Carbonate Hardness,} & + & \text{Noncarbonate} \\ \text{mg/}L \text{ as CaCO}_3 & & \text{mg/}L \text{ as CaCO}_3 & & \text{Hardness} \\ & & & & \text{mg/}L \text{ as CaCO}_3 \end{array}$$

When the alkalinity (as $CaCO_3$) is less than total hardness, then the <u>alkalinity represents the carbonate hardness</u>.

2. Then calculate the soda ash required:

$$\begin{array}{ccc} \text{Soda Ash} & = & \text{(Noncarbonate)} & \dfrac{(106)}{100} \\ \text{(Na}_2\text{CO}_3) & & \text{Hardness} & \\ \text{Feed, mg/}L & & \text{mg/}L \text{ as CaCO}_3 & \end{array}$$

Example 1: (Soda Ash Dosage)

❏ A water has a total hardness of 270 mg/L as CaCO$_3$ and a total alkalinity of 185 mg/L. What soda ash dosage (mg/L) will be required to remove the noncarbonate hardness?

First calculate the noncarbonate hardness:

$$\begin{array}{ccc} \text{Total Hardness,} & = & \text{Carbonate Hardness,} & + & \text{Noncarbonate} \\ \text{mg/}L \text{ as CaCO}_3 & & \text{mg/}L \text{ as CaCO}_3 & & \text{Hardness} \\ & & & & \text{mg/}L \text{ as CaCO}_3 \end{array}$$

$$270 \text{ mg/}L = 185 \text{ mg/}L + x \text{ mg/}L$$

$$270 \text{ mg/}L - 185 \text{ mg/}L = x$$

$$85 \text{ mg/}L = x$$

Then calculate the soda ash required:

$$\begin{array}{ccc} \text{Soda Ash,} & = & \text{(Noncarbonate)} & \dfrac{(106)}{100} \\ \text{mg/}L & & \text{Hardness,} & \\ & & \text{mg/}L \text{ as CaCO}_3 & \end{array}$$

$$= (85 \text{ mg/}L) \frac{(106)}{100}$$

$$= \boxed{90 \text{ mg/}L \text{ soda ash}}$$

---

* For a review of total/carbonate/noncarbonate hardness refer to Section 14.3.

Example 2: (Soda Ash Dosage)
❑ The alkalinity of a water is 118 mg/L as $CaCO_3$ and the total hardness is 210 mg/L as $CaCO_3$. What soda ash dosage (in mg/L) is required to remove the noncarbonate hardness?

The noncarbonate hardness is calculated as:

$$210 \text{ mg/}L = 118 \text{ mg/}L + x \text{ mg/}L$$

$$210 \text{ mg/}L - 118 \text{ mg/}L = x$$

$$92 \text{ mg/}L = x$$

The soda ash requirement can now be determined:

$$\text{Soda Ash, mg/}L = \frac{\text{(Noncarbonate) Hardness mg/}L \text{ as } CaCO_3 \times (106)}{100}$$

$$= \frac{(92 \text{ mg/}L) (106)}{100}$$

$$= \boxed{98 \text{ mg/}L \text{ soda ash}}$$

Example 3: (Soda Ash Dosage)
❑ Calculate the soda ash required (in mg/L) to soften a water if the water has a total hardness of 198 mg/L and a total alkalinity of 101 mg/L.

First determine the noncarbonate hardness:

$$198 \text{ mg/}L = 101 \text{ mg/}L + x \text{ mg/}L$$

$$198 \text{ mg/}L - 101 \text{ mg/}L = x$$

$$97 \text{ mg/}L = x$$

Then calculate soda ash required:

$$\text{Soda Ash, mg/}L = \frac{\text{(Noncarbonate) Hardness, mg/}L \text{ as } CaCO_3 \times (106)}{100}$$

$$= \frac{(97 \text{ mg/}L) (106)}{100}$$

$$= \boxed{103 \text{ mg/}L \text{ soda ash}}$$

14.8 CARBON DIOXIDE REQUIRED FOR RECARBONATION

Carbon dioxide is generally pumped into a water (recarbonated) after the addition of excess lime. This lowers the pH of the water to about 10.4, thus promoting better precipitation of calcium carbonate and magnesium hydroxide.

Carbon dioxide is sometimes also added <u>after</u> the addition of soda ash (or caustic soda). This lowers the pH even further to about 9.8 and encourages precipitation.

The equations used to estimate carbon dioxide dosage are shown to the right.

Examples 1-3 illustrate these calculations.

CARBON DIOXIDE (CO_2) REQUIRED FOR RECARBONATION

1. First calculate mg/L excess lime:

$$\text{Excess Lime, mg/L} = (A + B + C + D)(0.15)$$

Where A: CO_2 in Source water— (mg/L as CO_2)(56/44)

B: Bicarbonate alkalinity removed in softening (mg/L as $CaCO_3$)(56/100)

C: Hydroxide alkalinity in softener effluent (mg/L as $CaCO_3$)(56/100)

D: Magnesium removed in softening (mg/L as Mg^{+2})(56/24.3)

0.15: Excess lime dosage (using 15 percent excess)

Note: If hydrated lime, ($Ca(OH)_2$) is used instead of quicklime, replace the 56 factor with 74 in A, B, C and D.

2. Then calculate total carbon dioxide dosage (mg/L) required:

$$\text{Total } CO_2 \text{ Dosage, mg/L} = \left[Ca(OH)_2 \text{ Excess, mg/L} \right] \frac{(44)}{74} + (Mg^{+2}) \text{ Residual mg/L} \frac{(44)}{24.3}$$

Example 1: (Carbon Dioxide Dosage)

❏ The A, B, C and D factors of the excess lime equation have been calculated as follows: A = 12 mg/L; B = 129 mg/L; C = O; D = 68 mg/L. If the residual magnesium is 4 mg/L, what is the carbon dioxide dosage (in mg/L) required for recarbonation?

First calculate the excess lime concentration:

$$\text{Excess Lime, mg/L} = (A + B + C + D)(0.15)$$
$$= (12 \text{ mg/L} + 129 \text{ mg/L} + 0 + 68 \text{ mg/L})(0.15)$$
$$= (209 \text{ mg/L})(0.15)$$
$$= 31 \text{ mg/L}$$

Then determine the required carbon dioxide dosage:

$$\text{Total } CO_2 \text{ Dosage, mg/L} = (31 \text{ mg/L})\frac{(44)}{74} + (4 \text{ mg/L})\frac{(44)}{24.3}$$
$$= 18 \text{ mg/L} + 7 \text{ mg/L}$$
$$= \boxed{25 \text{ mg/L } CO_2}$$

Example 2: (Carbon Dioxide Dosage)
❏ The A, B, C and D factors of the excess lime equation have been calculated as: A = 7 mg/L; B = 108 mg/L; C = O; D = 51 mg/L. If the magnesium residual is 3 mg/L, what is the carbon dioxide dosage required for recarbonation?

The excess lime is calculated as:

$$\text{Excess Lime, mg/}L = (A + B + C + D)\,(0.15)$$

$$= (7 \text{ mg/}L + 108 \text{ mg/}L + 0 + 51 \text{ mg/}L)\,(0.15)$$

$$= (166 \text{ mg/}L)\,(0.15)$$

$$= 25 \text{ mg/}L$$

And the required carbon dioxide dosage is therefore:

$$\text{Total CO}_2 \text{ Dosage, mg/}L = (25 \text{ mg/}L)\,\frac{(44)}{74} + (3 \text{ mg/}L)\,\frac{(44)}{24.3}$$

$$= 15 \text{ mg/}L + 5 \text{ mg/}L$$

$$= \boxed{20 \text{ mg/}L \text{ CO}_2}$$

Example 3: (Carbon Dioxide Dosage)
❏ The A, B, C and D factors of the excess lime equation have been calculated as: A = 11 mg/L; B = 89 mg/L; C = O; D = 115 mg/L. If the residual magnesium is 6 mg/L, what carbon dioxide dosage would be required for recarbonation?

The excess lime is:

$$\text{Excess Lime, mg/}L = (A + B + C + D)\,(0.15)$$

$$= (11 \text{ mg/}L + 89 \text{ mg/}L + 0 + 115 \text{ mg/}L)\,(0.15)$$

$$= (215 \text{ mg/}L)\,(0.15)$$

$$= 32 \text{ mg/}L$$

And the required carbon dioxide dosage for recarbonation is:

$$\text{Total CO}_2 \text{ Dosage, mg/}L = (32 \text{ mg/}L)\,\frac{(44)}{74} + (6 \text{ mg/}L)\,\frac{(44)}{24.3}$$

$$= 19 \text{ mg/}L + 11 \text{ mg/}L$$

$$= \boxed{30 \text{ mg/}L \text{ CO}_2}$$

14.9 CHEMICAL FEEDER SETTINGS

Once the mg/L lime and soda ash dosage has been determined by calculation (see Sections 14.6 and 14.7), or by jar testing data, the lbs/day or lbs/min feed rates can be calculated. Examples 1-3 illustrate this calculation.

CALCULATING LBS/DAY OR LBS/MIN DOSAGE

To calculate lbs/day chemical required:*

$$(\text{mg/}L \text{ Chemical}) \ (\text{MGD Flow}) (8.34 \text{ lbs/gal}) \ = \ \frac{\text{lbs/day}}{\text{Chemical}}$$

To calculate the lbs/min chemical required:**

$$\frac{\text{Chemical, lbs/day}}{1440 \text{ min/day}} \ = \ \text{Chemical, lbs/min}$$

Example 1: (Chemical Feeder Settings)

❑ Jar tests indicate that the optimum lime dosage is 210 mg/L. If the flow to be treated is 4.1 MGD, what should be the chemical feeder setting in lbs/day and lbs/min?

First calculate the lbs/day feed rate using the mg/L to lbs/day equation:*

$$(\text{mg/}L \text{ Chemical}) \ (\text{MGD flow}) \ (8.34 \text{ lbs/gal}) \ = \ \text{lbs/day Chemical}$$

$$(210 \text{ mg/}L) \ (4.1 \text{ MGD}) \ (8.34 \text{ lbs/gal}) \ = \boxed{7181 \text{ lbs/day}}$$

This feed rate can be converted to lbs/min feed rate:

$$\frac{7181 \text{ lbs/day}}{1440 \text{ min/day}} \ = \boxed{5.0 \text{ lbs/min}}$$

---

* For a review of mg/L to lbs/day calculations, refer to Chapter 3.

** Flow conversions are described in Chapter 8 of *Basic Math Concepts*.

Example 2: (Chemical Feeder Settings)
❏ A total of 55 mg/*L* soda ash is required to remove noncarbonate hardness from a water. What should be the chemical feeder setting in lbs/day and lbs/min if the flow to be treated is 3.6 MGD?

The lbs/day soda ash feed rate is:

(mg/*L* Chemical) (MGD flow) (8.34 lbs/gal) = lbs/day Chemical

(55 mg/*L*) (3.6 MGD) (8.34 lbs/gal) = $\boxed{\text{1651 lbs/day}}$

This can be converted to lbs/min feed rate, as follows:

$$\frac{1651 \text{ lbs/day}}{1440 \text{ min/day}} = \boxed{1.1 \text{ lbs/min}}$$

Example 3: (Chemical Feeder Settings)
❏ What should the lime chemical feeder setting be, in lbs/day and lbs/hr, if the optimum lime dosage has been determined to be 145 mg/*L* and the flow to be treated is 1.8 MGD?

The lbs/day feed rate for lime is:

(mg/*L* Chemical) (MGD flow) (8.34 lbs/gal) = lbs/day Chemical

(145 mg/*L*) (1.8 MGD) (8.34 lbs/gal) = $\boxed{\text{2177 lbs/day}}$

This can be converted to lbs/min feed rate, as follows:

$$\frac{2177 \text{ lbs/day}}{24 \text{ hrs/day}} = \boxed{91 \text{ lbs/hr}}$$

CONVERTING TO LBS/HR

Depending on the particular chemical feeder used, the feed rate may need to be expressed as lbs/hr. The equation to be used in this conversion is:

$$\boxed{\frac{\text{Chemical, lbs/day}}{24 \text{ hrs/day}} = \text{Chemical, lbs/hr}}$$

14.10 EXPRESSIONS OF HARDNESS—mg/*L* and gpg

In most water and wastewater calculations, concentration is expressed as mg/*L* or percent. As an example, in Section 14.2 hardness is expressed as mg/*L* as CaCO₃.

For ion exchange softening, however, another expression of concentration is used—grains per gallon (gpg). A **grain** is a unit of weight equivalent to 0.0648 grams.

In making ion exchange calculations, (such as described in Sections 14.11-14.14), it is sometimes necessary to convert data from mg/*L* to gpg. The equation used in making these conversions is shown to the right. The "box method" of conversions* may be used to help determine whether to multiply or divide by the 17.12 conversion factor.

Examples 1-3 illustrate these conversions.

WHEN CONVERTING BETWEEN mg/*L* AND gpg, REMEMBER THAT mg/*L* IS ALWAYS THE LARGER NUMBER

$$1 \text{ gpg} = 17.12 \text{ mg/}L$$

When using the "box method" of conversion, the following diagram is used:

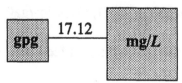

- **To convert from gpg to mg/*L***, you are moving from a smaller box to a larger box. Therefore, <u>multiplication is indicated.</u>

- **To convert from mg/*L* to gpg**, you are moving from a larger box to a smaller box. Thus, <u>division is indicated.</u>

Example 1: (mg/*L* and gpg)
❑ The total hardness of a water is 280 mg/*L*. What is this hardness expressed as grains per gallon?

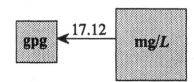

Converting from mg/*L* to gpg, you are moving from a larger box to a smaller box. Therefore, <u>division is indicated</u>:

$$\frac{280 \text{ mg/}L}{17.12 \text{ mg/}L\text{/gpg}} = \boxed{16.4 \text{ gpg}}$$

This answer makes sense as gpg must be a <u>smaller number</u> than mg/*L*.

---

* For a review of the "box method" of conversions, refer to Chapter 8 in *Basic Math Concepts*.

Example 2: (mg/*L* and gpg)
❑ The total hardness of a water is reported as 255 mg/*L*. What is this hardness expressed in terms of grains per gallon?

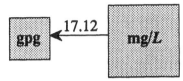

Converting from mg/*L* to gpg requires <u>division</u> by the conversion factor:

$$\frac{255 \text{ mg/}L}{17.12 \text{ mg/}L/\text{gpg}} = \boxed{14.9 \text{ gpg}}$$

Check to be sure that the gpg hardness is a <u>smaller number</u> than the mg/*L* hardness: 14.9 gpg vs. 255 mg/*L*.

Example 3: (mg/*L* and gpg)
❑ The total hardness of a water is 13.6 gpg. What is this concentration expressed as mg/*L*?

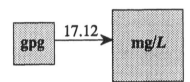

Converting from gpg to mg/*L*, you are moving from a smaller box to a larger box. Thus, <u>multiplication is indicated:</u>

$$(13.6 \text{ gpg}) (17.12 \frac{\text{mg/}L}{\text{gpg}}) = \boxed{233 \text{ mg/}L}$$

Check : mg/*L* should be a larger number than gpg: 13.6 gpg vs. 233 mg/*L*.

14.11 ION EXCHANGE CAPACITY

An ion exchange softener is a treatment unit designed to remove the primary hardness constituents of water—calcium and magnesium. As illustrated in the diagram, the exchange resin contains sodium ions. When water is passed through the softener, the sodium ions are then exchanged for (replaced by) calcium and magnesium ions. The sodium ions flow out of the unit with the treated water, while the calcium and magnesium ions are retained on the resin. As a result, the total hardness of the treated water is significantly reduced.

The removal capacity of an exchange resin is generally reported as grains of hardness removal per cubic foot of resin. To calculate the removal capacity of the softener, you must multiply the removal capacity for each cubic foot of resin (grains/cu ft) by the entire cu ft of resin in the softener, as shown in the equation to the right.

Examples 1-2 illustrate the calculation of softener exchange capacity.

THE RESIN OF AN ION EXCHANGE SOFTENER EXCHANGES SODIUM IONS FOR HARDNESS IONS (CALCIUM AND MAGNESIUM)

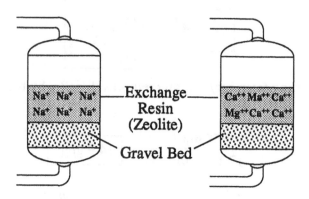

Exchange Resin (Zeolite)

Gravel Bed

SOFTENER AFTER REGENERATION
(Exchange resin contains all sodium ions)

SOFTENER WHEN RESIN IS EXHAUSTED
(Exchange resin contains calcium and magnesium ions)

To calculate the exchange capacity of a softener, use the following equation:

$$\text{Exchange Capacity, grains} = \text{(Removal Capacity, grains/cu ft)}\text{(Media Vol., cu ft)}$$

Example 1: (Ion Exchange Capacity)
❑ The hardness removal capacity of an exchange resin is 22,000 grains/cu ft. If the softener contains a total of 80 cu ft of resin, what is the total exchange capacity (grains) of the softener?

$$\text{Exchange Capacity, grains} = \text{(Removal Capacity, grains/cu ft)}\text{(Media Volume, cu ft)}$$

$$= (22,000 \text{ grains/cu ft})\ (80 \text{ cu ft})$$

$$= \boxed{1,760,000 \text{ grains}}$$

---

* This is true of sodium ion exchange resins only. There are other exchange media available.

Example 2: (Ion Exchange Capacity)

❑ The hardness removal capacity of an exchange resin is 20 kilograins/cu ft. If the softener contains a total of 250 cu ft of resin, what is the total exchange capacity of the softener (in grains)?

First convert kilograins to grains:

$$20 \text{ kilograins} = 20{,}000 \text{ grains}$$

Then calculate the exchange capacity of the softener:

$$\begin{matrix} \text{Exchange} \\ \text{Capacity,} \\ \text{grains} \end{matrix} = \begin{matrix} \text{(Removal Capacity,)(Media Volume,)} \\ \text{grains/cu ft} \qquad \text{cu ft} \end{matrix}$$

$$= (20{,}000 \text{ grains/cu ft}) \ (250 \text{ cu ft})$$

$$= \boxed{5{,}000{,}000 \text{ grains}}$$

Example 3: (Ion Exchange Capacity)

❑ An ion exchange water softener has a diameter of 6 ft. The depth of resin is 4 ft. If the resin has a removal capacity of 21 kilograins/cu ft, what is the total exchange capacity of the softener (in grains)?

Before the exchange capacity of a softener can be calculated, the cu ft resin volume must be known:

$$\begin{matrix} \text{Vol.,} \\ \text{cu ft} \end{matrix} = (0.785) \ (D^2) \ (\text{Depth, ft})$$

$$= (0.785) \ (6 \text{ ft}) \ (6 \text{ ft}) \ (4 \text{ ft})$$

$$= 113 \text{ cu ft}$$

Now calculate the exchange capacity of the softener:

$$\begin{matrix} \text{Exchange} \\ \text{Capacity,} \\ \text{grains} \end{matrix} = \begin{matrix} \text{(Removal Capacity,)(Media Volume,)} \\ \text{grains/cu ft} \qquad \text{cu ft} \end{matrix}$$

$$= (21{,}000 \text{ grains/cu ft}) \ (113 \text{ cu ft})$$

$$= \boxed{2{,}373{,}000 \text{ grains}}$$

KILOGRAINS REMOVAL CAPACITY

The removal capacity of many softeners is expressed in terms of kilograins per cu ft rather than grains per cu ft. In this case, express the kilograins as grains before continuing with the exchange capacity calculation.

Since a kilograin is 1000 grains (the prefix "kilo" means "1000"), **to convert from kilograins to grains, simply add a comma and 3 zeros to the right,** as follows:

$$15 \text{ kilograins} = 15{,}000 \text{ grains}$$

Add a comma and 3 zeros

Examples 2 and 3 illustrate the conversion from kilograins to grains.

WHEN RESIN VOLUME MUST BE CALCULATED

Occasionally you will need to calculate the cu ft resin volume in the softener. To calculate the cu ft volume, use the following volume equation:*

$$\boxed{\begin{matrix} \text{Vol.,} \\ \text{cu ft} \end{matrix} = (0.785) \ (D^2) \ (\text{Depth, ft})}$$

where D = diameter, ft

---

* For a review of volume calculations, refer to Chapter 1 in this text or Chapter 11 in *Basic Math Concepts*.

14.12 WATER TREATMENT CAPACITY

To calculate the volume of water that can be softened before the resin must be regenerated, you must know two factors:

• The exchange capacity of the softener (grains), and

• The hardness of the water, grains/gallon.

The equation used for this calculation is:

$$\text{Water Treatment Capacity, gal} = \frac{\text{Exch. Capac., grains}}{\text{Hardness, gpg}}$$

The exchange capacity indicates the total number of grains hardness that can be removed by the resin before regeneration of the resin is required. Hardness indicates the number of grains hardness to be removed from each gallon of water (gpg). By dividing the total grains by the number of grains representing each gallon, you can determine the total gallons that can be treated before regeneration.

For example, suppose a resin can remove a total of 500,000 grains hardness before it is exhausted and that each gallon of water to be treated contains 50 grains hardness. Since each gallon of water is represented by 50 grains, by dividing 500,000 by 50 grains, in effect you are saying "how many gallons are represented by 500,000 grains":

$$\frac{500,000 \text{ grains}}{50 \text{ grains for each gallon}} = 10,000 \text{ gal}$$

Examples 1-4 illustrate this calculation.

Example 1: (Water Treatment Capacity)

❏ An ion exchange softener has an exchange capacity of 2,373,000 grains. If the hardness of the water to be treated is 18.4 grains/gallon, how many gallons of water can be treated before regeneration of the resin is required?

$$\text{Water Treatment Capacity, gal} = \frac{\text{Exchange Capacity, grains}}{\text{Hardness, grains/gallon}}$$

$$= \frac{2,373,000 \text{ grains}}{18.4 \text{ gpg}}$$

$$= \boxed{128,967 \text{ gallons} \atop \text{water treated}}$$

Example 2: (Water Treatment Capacity)

❏ An ion exchange softener has an exchange capacity of 5,000,000 grains. If the hardness of the water to be treated is 15.1 grains/gallon, how many gallons of water can be treated before regeneration of the resin is required?

$$\text{Water Treatment Capacity, gal} = \frac{\text{Exchange Capacity, grains}}{\text{Hardness, grains/gallon}}$$

$$= \frac{5,000,000 \text{ grains}}{15.1 \text{ gpg}}$$

$$= \boxed{331,126 \text{ gallons} \atop \text{water treated}}$$

Example 3: (Water Treatment Capacity)
❑ An ion exchange softener has an exchange capacity of 1,540,000 grains. If the hardness of the water is 290 mg/L, how many gallons of water can be treated before regeneration of the resin is required?

The hardness of the water must first be expressed as grains/gallon:*

$$\frac{290 \text{ mg}/L}{17.12 \text{ mg}/L \text{ gpg}} = 16.9 \text{ gpg}$$

Now the gallons water treated can be calculated:

$$\frac{\text{Water Treatment}}{\text{Capacity, gal}} = \frac{\text{Exchange Capacity, grains}}{\text{Hardness, grains/gallon}}$$

$$= \frac{1,540,000 \text{ grains}}{16.9 \text{ gpg}}$$

$$= \boxed{\begin{array}{c} 91,124 \text{ gallons} \\ \text{water treated} \end{array}}$$

Example 4: (Water Treatment Capacity)
❑ The hardness removal capacity of an ion exchange resin is 23 kilograins/cu ft. The softener contains a total of 150 cu ft of resin. If the water to be treated contains 14.4 gpg hardness, how many gallons of water can be treated before regeneration of the resin is required?

Both the water hardness and the exchange capacity of the softener must be determined before the gallons water can be calculated.

$$\begin{array}{ccc} \text{Exchange Capacity,} \\ \text{grains} \end{array} = \begin{array}{cc} (\text{Removal Capac.,}) (\text{Media Vol.,}) \\ \text{grains/cu ft} \quad\quad \text{cu ft} \end{array}$$

$$= (23,000 \text{ grains/cu ft}) (150 \text{ cu ft})$$

$$= 3,450,000 \text{ grains}$$

Now calculate the gallons water treated:

$$\frac{\text{Water Treatment}}{\text{Capacity, gal}} = \frac{3,450,000 \text{ grains}}{14.4 \text{ gpg}}$$

$$= \boxed{\begin{array}{c} 239,583 \text{ gallons} \\ \text{water treated} \end{array}}$$

---

* For a review of these calculations, refer to Section 14.10.

14.13 OPERATING TIME (UNTIL REGENERATION REQUIRED)

Once you have calculated the total number of gallons water to be treated before regeneration (see Section 14.12), you can also calculate the operating time required to treat that much water. Additional data required to make this calculation is:

- The flow rate treated (gpm or gph)

If you desire to calculate the operating time **in hours** then the flow rate should also be expressed in terms of hours (gph):

$$\text{Operating Time, hrs} = \frac{\text{Water Treated, gal}}{\text{Flow Rate, gph}}$$

Note that this equation is similar in structure to detention time calculations.*

Examples 1-4 illustrate the operating time calculation.

Example 1: (Operating Time)

❑ An ion exchange softener can treat a total of 628,967 gallons before regeneration is required. If the flow rate treated is 24,000 gph how many hours of operation are there before regeneration is required?

$$\text{Operating Time, hrs} = \frac{\text{Water Treated, gal}}{\text{Flow Rate, gph}}$$

$$= \frac{628,967 \text{ gal}}{24,000 \text{ gph}}$$

$$= \boxed{26.2 \text{ hrs of operation before regeneration}}$$

Example 2: (Operating Time)

❑ An ion exchange softener can treat a total of 850,000 gallons of water before regeneration of the resin is required. If the water is to be treated at a rate of 30,000 gph, how many hours of operation are there until regeneration is required?

$$\text{Operating Time, hrs} = \frac{\text{Water Treated, gal}}{\text{Flow Rate, gph}}$$

$$= \frac{850,000 \text{ gal}}{30,000 \text{ gph}}$$

$$= \boxed{28.3 \text{ hrs of operation before regeneration}}$$

---

* For a discussion of detention time problems, refer to Chapter 5.

Example 3: (Operating Time)

❑ An ion exchange softener can treat a total of 331,126 gallons before regeneration is required. If the flow rate to be treated is 250 gpm, what is the operating time (in hrs) until regeneration of the resin will be required?

Since the operating time is desired in hours, the flow rate should be expressed in terms of hours as well:

$$(250 \text{ gpm}) (60 \text{ min/hr}) = 15,000 \text{ gph}$$

The operating time required is:

$$\text{Operating Time, hrs} = \frac{\text{Water Treated, gal}}{\text{Flow Rate, gph}}$$

$$= \frac{331,126 \text{ gal}}{15,000 \text{ gph}}$$

$$= \boxed{22.1 \text{ hrs of operation}}$$

CONVERTING FLOW RATE TO GPH

As described previously, the flow rate should be expressed in gal/hour if the operating time is desired in hours.

To convert from gpd or gpm flow rate to gph, use the following equations:*

$$\boxed{\frac{(\text{Flow, gpd})}{24 \text{ hrs/day}} = \text{Flow, gph}}$$

Or

$$\boxed{(\text{Flow, gpm})(60 \frac{\text{min}}{\text{hr}}) = \text{Flow gph}}$$

Example 4: (Operating Time)

❑ The exchange capacity of an ion exchange softener is 3,580,000 grains. The water to be treated contains 12 gpg total hardness. If the flow to be treated is 200 gpm, how many hours of operation until regeneration is required?

First calculate the total gallons treated before regeneration: (refer to Section 14.12)

$$\text{Water Treatment Capacity, gal} = \frac{\text{Exchange Capacity, grains}}{\text{Hardness, gpg}}$$

$$= \frac{3,580,000 \text{ grains}}{12 \text{ gpg}}$$

$$= 298,333 \text{ gal treated}$$

Then calculate the operating time required:

$$\text{Operating Time, hrs} = \frac{\text{Water Treated, gal}}{\text{Flow Rate, gph}}$$

$$= \frac{298,333 \text{ gal}}{(200 \text{ gpm}) (60 \text{ min/hr})}$$

$$= \boxed{24.9 \text{ hrs of operation}}$$

---

* For a review of flow conversions, refer to Chapter 8 in *Basic Math Concepts*.

14.14 SALT AND BRINE REQUIRED FOR REGENERATION

When all the sodium ions in the exchange resin have been replaced by calcium and magnesium ions, the resin is exhausted and can no longer remove the hardness ions from the water. When this occurs, the resin must be regenerated by pumping a concentrated brine solution (10-14% sodium chloride solution) onto the resin. When complete, the regeneration process results in a resin completely recharged with sodium ions and ready for softening again.

The salt dosage required to prepare the brine solution ranges from 5-15 lbs salt/cu ft resin. In Examples 1 and 2, the total pounds of salt required are calculated:

$$\begin{array}{ccc} \text{Salt} \\ \text{Req'd,} & = & \text{(Salt Req'd,) (Hardness)} \\ \text{lbs} & & \text{lbs/kilograins Removed,} \\ & & \text{removed kilograins} \end{array}$$

And in Examples 3 and 4, the total gallons brine is determined:

$$\begin{array}{cc} \text{Brine,} \\ \text{gal} & = & \dfrac{\text{Salt Required, lbs}}{\text{Brine Sol'n, lbs salt/gal brine}} \end{array}$$

Example 1: (Salt and Brine Required)

❑ An ion exchange softener will remove 2,178,000 grains hardness from the water until the resin must be regenerated. If 0.4 lbs salt are required for each kilograin removed, how many pounds of salt will be required for preparing the brine to be used in resin regeneration?

$$\begin{array}{ccl} \text{Salt Required,} & = & \text{(Salt Req'd,) (Hardness Removed,)} \\ \text{lbs} & & \text{lbs/1000 grains} \quad \text{kilograins} \\[2mm] & = & \dfrac{\text{(0.4 lbs salt)} \quad \text{(2178 kilograins)}}{\text{kilograins rem.}} \\[2mm] & = & \boxed{871.2 \text{ lbs salt required}} \end{array}$$

Example 2: (Salt and Brine Required)

❑ An ion exchange softener removes 1,280,000 grains hardness from the water before the resin must be regenerated. If 0.3 lbs salt are required for each kilograin removed, how many pounds of salt will be required for preparing the brine to be used in resin regeneration?

$$\begin{array}{ccl} \text{Salt Required,} & = & \text{(Salt Req'd,) (Hardness Removed,)} \\ \text{lbs} & & \text{lbs/1000 grains} \quad \text{kilograins} \\[2mm] & = & \dfrac{\text{(0.3 lbs salt)} \quad \text{(1280 kilograins)}}{\text{kilograins rem.}} \\[2mm] & = & \boxed{384 \text{ lbs salt required}} \end{array}$$

---

* For a discussion of detention time problems, refer to Chapter 5.

Example 3: (Salt and Brine Required)
❑ In Example 2 it was calculated that 384 lbs salt would be required in making up the brine solution for regeneration. If the brine solution is to be a 14% solution of salt, how many gallons of brine will be required for regeneration of the softener? (Use the table to the right to determine the salt content, lbs/gal, of a 14% brine solution.)

A 14% brine solution contains 1.29 lbs salt in each gallon of brine. This means that each 1.29 lbs of salt represents 1 gallon of brine required. When calculating the gallons brine required you're essentially asking, "how many units of 1.29 lbs salt are required?"

Salt Solutions		
% NaCl	$\dfrac{\text{lbs NaCl}}{\text{gal}}$	$\dfrac{\text{lbs NaCl}}{\text{cu ft}}$
10	0.874	6.69
11	0.990	7.41
12	1.09	8.14
13	1.19	8.83
14	1.29	9.63
15	1.39	10.4

$$\frac{\text{Brine,}}{\text{gal}} = \frac{\text{Salt Required, lbs}}{\text{Brine Solution, lbs salt/gal brine}}$$

$$= \frac{384 \text{ lbs salt}}{1.29 \text{ lbs salt/gal brine}}$$

$$= \boxed{298 \text{ gal of 14\% brine}}$$

Example 4: (Salt and Brine Required)
❑ A total of 420 lbs salt will be required to regenerate an ion exchange softener. If the brine solution is to be a 12% brine solution, how many gallons brine will be required? (Use the Salt Solutions table to determine the lbs salt/gal brine for a 12% brine solution.)

$$\frac{\text{Brine,}}{\text{gal}} = \frac{\text{Salt Required, lbs}}{\text{Brine Solution, lbs salt/gal brine}}$$

$$= \frac{420 \text{ lbs salt}}{1.09 \text{ lbs salt/gal brine}}$$

$$= \boxed{385 \text{ gal of 12\% brine}}$$

Therefore, 420 lbs salt used to make up a total of 385 gallons brine will result in the desired 12% brine solution.

NOTES:

15 *Laboratory*

SUMMARY

1. **Estimating Flow From A Faucet**

$$\text{Flow, gpm} = \frac{\text{Volume, gal}}{\text{Time, min}}$$

2. **Service Line Flushing Time**

$$\text{Flushing Time, min} = \frac{(\text{Pipe Volume, gal}) (2)}{\text{Flow Rate, gpm}}$$

Or

$$\text{Flushing Time, min} = \frac{(0.785) (D^2) (\text{Length, ft}) (7.48 \text{ gal/cu ft}) (2)}{\text{Flow Rate, gpm}}$$

3. **Solution Concentration**

   Three common methods of expressing solution concentration are:

   - **Percent***
   - **mg/*L***
   - **Normality**

$$\text{Normality} = \frac{\text{No. of Equivalents of Solute}}{\text{Liter of Solution}}$$

   Note: "Equivalents" are described in Chapter 14, Section 1.

   When preparing **standard solutions**, the following proportion procedure may be used:

$$\underbrace{\frac{\text{Desired Wt., g}}{\text{Desired Sol'n Vol., m}L}}_{\text{Desired Mixture}} = \underbrace{\frac{\text{Actual Wt, g}}{\text{Actual Sol'n Vol., m}L}}_{\text{Actual Mixture}}$$

* Described in Chapter 13, Section 1 and Chapter 8 of *Basic Math Concepts*.

SUMMARY—Cont'd

4. Estimated Chlorine Residual

$$\text{Estimated Chlorine Residual, mg/L} = \frac{(\text{Sample Chlorine Residual, mg/L})(\text{Distilled Water, m}L)}{(\text{Sample Volume, drops})(0.05 \text{ m}L/\text{drop})}$$

5. Temperature

Fahrenheit to Celsius : (Conventional Equation)

$$^\circ C = \frac{5}{9} \, (^\circ F - 32^\circ)$$

Celsius to Fahrenheit : (Conventional Equation)

$$^\circ F = \frac{9}{5} \, (^\circ C) + 32^\circ$$

3-Step Method: (Converting either direction)

1. Add 40°.

2. Multiply by 5/9 or 9/5.
   (Depending on the direction of conversion.)

3. Subtract 40°.

SUMMARY—Cont'd

The following calculations have been described in other chapters, as indicated.

• **Chlorine Demand** (Refer to Chapter 12, Section 2)

Chlorine Demand, mg/L	=	Chlorine Added, mg/L	−	Free Residual Chlorine, mg/L

This is the chlorine dose

• **Alkalinity**

(Refer to Chapter 14, Softening, for a description of alkalinity calculations.)

• **Hardness**

(Refer to Chapter 14, Softening, for a discussion of hardness calculations.)

• **Coliform Bacteria**

Because the various coliform bacteria tests require extensive explanation of the laboratory procedures, it has not been included in this text. For a description of the tests and calculations, refer to the following sources:

Standard Methods, 17th Edition, pp. 9-66 through 9-99.

Water Treatment Plant Operation, Volume 1, California State University, Sacramento, School of Engineering, 1989, p. 500.

Introduction to Water Quality Analyses, American Water Works Association, 1982, p. 72.

15.1 ESTIMATING FLOW FROM A FAUCET

Many small water systems do not have special sampling taps from which to take water samples. The water samples must therefore be taken from a customer's front yard faucet.

To estimate the flow from a faucet, use a one-gallon container and record the time it takes to fill the container:

$$\text{Flow, gpm} = \frac{\text{Volume, gal}}{\text{Time, min}}$$

Practically speaking, at a flow rate of 1 gpm, the one-gallon container will be full in one minute. At a flow rate of 0.5 gpm, the container will be half full at the end of the one-minute period. A convenient flow rate for taking water samples is about 0.5 gpm.

Example 1: (Flow from a Faucet)
❏ The flow from a faucet filled up the gallon container in 45 seconds. What was the gpm flow rate from the faucet?

Since flow rate is desired in <u>minutes</u> the time should also be expressed as <u>minutes</u>:

$$\frac{45 \text{ seconds}}{60 \text{ sec/min}} = 0.75 \text{ min}$$

Now calculate flow rate from the faucet:

$$\text{Flow, gpm} = \frac{\text{Volume, gal}}{\text{Time, min}}$$

$$= \frac{1 \text{ gal}}{0.75 \text{ min}}$$

$$= \boxed{1.3 \text{ gpm}}$$

Example 2: (Flow from a Faucet)
❏ The flow from a faucet filled up the gallon container in 52 seconds. What was the gpm flow rate from the faucet?

First express time in terms of minutes:

$$\frac{52 \text{ seconds}}{60 \text{ sec/min}} = 0.87 \text{ min}$$

The flow rate may now be calculated:

$$\text{Flow, gpm} = \frac{\text{Volume, gal}}{\text{Time, min}}$$

$$= \frac{1 \text{ gal}}{0.87 \text{ min}}$$

$$= \boxed{1.1 \text{ gpm}}$$

Example 3: (Flow from a Faucet)
❑ The flow from a faucet filled up the gallon container in 1 minute 12 seconds. What was the gpm flow rate from the faucet?

The time must <u>all</u> be expressed in terms of minutes:

$$\frac{12 \text{ seconds}}{60 \text{ sec/min}} = 0.2 \text{ min}$$

Therefore the total time is 1.2 min. The flow rate can now be determined:

$$\begin{aligned} \text{Flow,} \atop \text{gpm} &= \frac{\text{Volume, gal}}{\text{Time, min}} \\[2mm] &= \frac{1 \text{ gal}}{1.2 \text{ min}} \\[2mm] &= \boxed{0.8 \text{ gpm}} \end{aligned}$$

Example 4: (Flow from a Faucet)
❑ At a flow rate of 1.25 gpm, how long (minutes) should it take to fill the one-gallon container?

Use the same equation, fill in the given data, then solve for the unknown value*.

$$\begin{aligned} \text{Flow,} \atop \text{gpm} &= \frac{\text{Volume, gal}}{\text{Time, min}} \\[2mm] 1.25 \text{ gpm} &= \frac{1 \text{ gal}}{x \text{ min}} \\[2mm] (x)(1.25) &= 1 \\[2mm] x &= \frac{1}{1.25} \\[2mm] x &= \boxed{0.8 \text{ min}} \end{aligned}$$

(This could be expressed in seconds, if desired: 0.8 min x 60 sec/min = 48 seconds.)

SOLVING FOR OTHER UNKNOWN VALUES

These flow rate problems have three variables:

- Flow Rate, gpm

- Volume, gallons

- Time, min

In Examples 1-3, the unknown value was the gpm flow rate. However, any of the three variables can be the unknown factor. Use the same equation, fill in the given two factors and then solve for the unknown factor. Example 4 illustrates one such calculation.

---

* For a review of solving for the unknown value, refer to Chapter 2 in *Basic Math Concepts*.

15.2 SERVICE LINE FLUSHING TIME

In order to obtain an accurate indication of the system water quality, it is essential that the samples are representative of the water delivered to the consumer.

Therefore, to ensure that a sample taken from a consumer's outside faucet is typical of the water delivered, the service line is flushed before taking a water sample. **Flushing is considered adequate when the water in the service line has been replaced twice.**

The equations to be used in flushing time calculations are shown to the right. Notice that these equations are similar to detention time equations.*

FLUSH A SERVICE LINE <u>TWICE</u> BEFORE SAMPLING

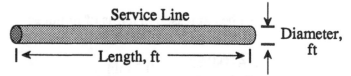

Simplified Equation:

$$\text{Flushing Time, min} = \frac{(\text{Pipe Volume, gal}) (2)}{\text{Flow Rate, gpm}}$$

Expanded Equation:

$$\text{Flushing Time, min} = \frac{(0.785) (D^2) (\text{Length, ft}) (7.48 \text{ gal/cu ft}) (2)}{\text{Flow Rate, gpm}}$$

Example 1: (Flushing Time)
❏ How long (minutes) will it take to flush a 50-ft length of 3/4-inch diameter service line if the flow through the line is 0.5 gpm?

Before pipe volume and flushing time can be calculated, the diameter of the pipe must be expressed in terms of feet:

$$\frac{(0.75 \text{ inches})}{12 \text{ inches/ft}} = 0.06 \text{ ft}$$

Now flushing time can be calculated:

$$\text{Flushing Time, min} = \frac{(0.785) (D^2) (\text{Length, ft}) (7.48 \text{ gal/cu ft}) (2)}{\text{Flow Rate, gpm}}$$

$$= \frac{(0.785) (0.06 \text{ ft}) (0.06 \text{ ft}) (50 \text{ ft}) (7.48 \text{ gal/cu ft}) (2)}{0.5 \text{ gpm}}$$

$$= \boxed{4.2 \text{ min}}$$

---

* For a review of detention time calculations, refer to Chapter 5.

Example 2: (Flushing Time)
❏ How many minutes will it take to flush a 40-ft length of 3/4-inch diameter service line if the flow through the line is 0.7 gpm?

As calculated in Example 1, the 3/4-inch diameter is equal to 0.06 ft.

The flushing time is calculated as follows:

$$\frac{\text{Flushing}}{\text{Time,}\atop\text{min}} = \frac{(0.785) \, (D^2) \, (\text{Length, ft}) \, (7.48 \text{ gal/cu ft}) \, (2)}{\text{Flow Rate, gpm}}$$

$$= \frac{(0.785) \, (0.06 \text{ ft}) \, (0.06 \text{ ft}) \, (40 \text{ ft}) \, (7.48 \text{ gal/cu ft}) \, (2)}{0.7 \text{ gpm}}$$

$$= \boxed{2.4 \text{ min}}$$

Example 3: (Flushing Time)
❏ At a flow rate of 0.5 gpm, how long (minutes and seconds) will it take to flush a 65-ft length of 3/4-inch service line?

The 3/4-inch diameter is equal to 0.06 ft, as calculated in Example 1.

The flushing time required is:

$$\frac{\text{Flushing}}{\text{Time,}\atop\text{min}} = \frac{(0.785) \, (D^2) \, (\text{Length, ft}) \, (7.48 \text{ gal/cu ft}) \, (2)}{\text{Flow Rate, gpm}}$$

$$= \frac{(0.785) \, (0.06 \text{ ft}) \, (0.06 \text{ ft}) \, (65 \text{ ft}) \, (7.48 \text{ gal/cu ft}) \, (2)}{0.5 \text{ gpm}}$$

$$= 5.5 \text{ minutes}$$

Convert the fractional part of a minute (0.5 min) to seconds:

$$(0.5 \text{ min}) \, (60 \text{ sec/min}) = 30 \text{ sec}$$

Therefore the total flushing time required is:

$$\boxed{5 \text{ min } 30 \text{ sec}}$$

15.3 SOLUTION CONCENTRATION

There are three common methods of expressing solution concentration:

- Percent
- Milligrams per liter (mg/L), and
- Normality.

Percent and mg/L concentration have been described in various places in this text including Chapter 1, 3, and 13.

NORMALITY

The **molarity** of a solution refers to its **concentration** (the solute dissolved in the solution), the **normality** of the solution refers more specifically to the reacting characteristics of the solution. One of the first concepts to be learned in a basic chemistry course is that the sharing or transfer of electrons is responsible for chemical activity or the reacting characteristics of an element or compound.

The concept of **equivalents** parallels this concept by relating the number of electrons available to be transferred (valence) and the atomic weight associated with each of these valence electrons.*

Since the concept of equivalents is based upon the "reacting power" of an element or compound, it follows that **a specific number of equivalents of one substance will react with the same number of equivalents of another substance.** For example, two equivalents of a substance will react with two equivalents of another substance. If, however, one equivalent of Substance A is mixed with two equivalents of Substance B, only one equivalent of each substance will react, leaving an excess of one equivalent of Substance B.

NORMALITY IS A MEASURE OF THE "REACTING POWER" OF A SOLUTION

| 1 Equivalent of a Substance | ←→   *Reacts With* | 1 Equivalent of another Substance |

$$\text{Normality} = \frac{\text{No. of Equivalents of Solute}}{\text{Liters of Solution}}$$

This equation may be rearranged as:

$$\begin{array}{ccc} (\text{Normality}) & (\text{Liters}) & = & \text{Equivalents} \\ \text{of Sol'n} & \text{Sol'n} & & \text{in Solution} \end{array}$$

Example 1: (Solution Concentration)

❑ If 2.5 equivalents of a chemical are dissolved in 1.5 liters solution, what is the normality of the solution?

$$\text{Normality} = \frac{\text{No. of Equivalents of Solute}}{\text{Liters of Solution}}$$

$$= \frac{2.5 \text{ Equivalents}}{1.5 \text{ liters}}$$

$$= \boxed{1.67 \, N}$$

---

* The valence number may be either positive or negative, depending upon whether electrons were added to or taken from the particular element or compound.

Example 2: (Solution Concentration)
❏ A 600-m*L* solution contains 1.8 equivalents of a chemical. What is the normality of the solution?

First convert 600 m*L* to liters:

$$\frac{600 \text{ m}L}{1000 \text{ m}L/L} = 0.6 L$$

Then calculate the normality of the solution:

$$\text{Normality} = \frac{\text{No. of Equivalents of Solute}}{\text{Liters of Solution}}$$

$$= \frac{1.8 \text{ Equivalents}}{0.6 \text{ Liters}}$$

$$= \boxed{3 N}$$

Example 3: (Solution Concentration)
❏ How many milliliters of 0.5 *N* NaOH will react with 500 m*L* of 0.01 *N* HCl?

Set the normality and volume of the first solution equal to the normality and volume of the second solution:

$$N_A V_A = N_B V_B$$

$$(0.5)(x \text{ m}L) = (0.01)(500 \text{ m}L)$$

$$x = \frac{(0.01)(500)}{0.5}$$

$$= \boxed{10 \text{ m}L \text{ NaOH}}$$

Practically speaking, if the concept of equivalents is ignored when making up solutions, most likely chemicals will be wasted as excess amounts.

WHEN MILLILITERS VOLUME IS GIVEN

Many times the volume of solution is given as milliliters. To calculate normality, the volume must be expressed in liters. Therefore, convert the milliliters volume to liters:

$$\boxed{\frac{\text{m}L}{1000 \text{ m}L/L}}$$

NORMALITY AND TITRATIONS

The second equation, given on the previous page, indicates that the normality of a solution times the volume of the solution is equal to the number of equivalents of the solution:

$$\boxed{\begin{array}{ccc}(\text{Normality})(\text{Volume}) = & \text{No. of} \\ \text{of Sol'n} \quad L & \text{Equiv.}\end{array}}$$

Because chemicals react on the basis of equivalents, this relationship is of great importance in understanding titrations (such as used in the COD test) or acid/base neutralizations. In general, where N = normality of the solution and V = volume of the solution:

$$\boxed{N_A V_A = N_B V_B}$$

When using this equation, the solution volume may be expressed as liters or milliliters. However, whichever term is used (m*L* or *L*) on one side of the equation, the same term must also be used on the other side of the equation.

STANDARD SOLUTIONS

A **standard solution** is a solution whose exact concentration is known. Laboratory procedures texts indicate the amount of chemical and distilled water that must be combined to produce a solution with the desired concentration.

Occasionally a <u>different</u> quantity of reagent is used. To maintain the same concentration of solution, therefore, the corresponding quantity of distilled water to be added must be calculated, as this will be different as well. The equation to be used in these calculations is shown to the right.

IN MAKING A STANDARD SOLUTION, IF THE QUANTITY OF REAGENT IS CHANGED, THE QUANTITY OF DISTILLED WATER MUST ALSO BE CHANGED

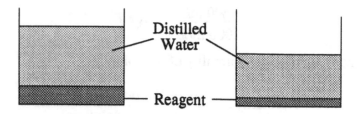

Use a proportion:*

Desired Mixture   Actual Mixture

$$\frac{\text{Desired Wt., grams}}{\text{Desired Sol'n Vol., mL}} = \frac{\text{Actual Wt, grams}}{\text{Actual Sol'n Vol., mL}}$$

Example 4: (Solution Concentration)

❑ To prepare a standard solution, the directions indicate that 7.6992 grams of chemical are to be weighed out and diluted to one liter. If 7.3521 grams of chemical are used in making the solution, how many milliliters solution should be prepared?

$$\frac{\text{Desired Wt., grams}}{\text{Desired Sol'n Vol., mL}} = \frac{\text{Actual Wt, grams}}{\text{Actual Sol'n Vol., mL}}$$

$$\frac{7.6992 \text{ g}}{1000 \text{ mL}} = \frac{7.3521 \text{ g}}{x \text{ mL}}$$

$$(x)(7.6992) = (7.3521)(1000)$$

$$x = \frac{(7.3521)(1000)}{7.6992}$$

$$x = \boxed{955 \text{ mL}}$$

---

* For a review of proportions, refer to Chapter 5 in *Basic Math Concepts.*

Example 5: (Solution Concentration)
❑ To prepare a particular standard solution of sodium hydroxide (NaOH), 50 grams NaOH are to be dissolved in water and diluted to 1 liter. If 46.42 grams are weighed out, how many milliliters solution should be prepared?

$$\frac{\text{Desired Wt., grams}}{\text{Desired Sol'n Vol., m}L} = \frac{\text{Actual Wt, grams}}{\text{Actual Sol'n Vol., m}L}$$

$$\frac{50 \text{ grams}}{1000 \text{ m}L} = \frac{46.42 \text{ grams}}{x \text{ m}L}$$

$$(x)(50) = (46.42)(1000)$$

$$x = \frac{(46.42)(1000)}{50}$$

$$x = \boxed{928 \text{ m}L}$$

Example 6: (Solution Concentration)
❑ *Standard Methods* indicates that a 0.0192N solution of silver nitrate ($AgNO_3$) is to be prepared using 3.27 grams $AgNO_3$ dissolved in 1 liter distilled water. If 3.14 grams of $AgNO_3$ are used to prepare the solution, how many milliliters distilled water should be used in making the solution?

$$\frac{\text{Desired Wt., grams}}{\text{Desired Dist. H}_2\text{O Vol., m}L} = \frac{\text{Actual Wt, grams}}{\text{Actual Dist. H}_2\text{O Vol., m}L}$$

$$\frac{3.27 \text{ grams}}{1000 \text{ m}L} = \frac{3.14 \text{ grams}}{x \text{ m}L}$$

$$(3.27)(x) = (3.14)(1000)$$

$$x = \frac{(3.14)(1000}{3.27}$$

$$x = \boxed{960 \text{ m}L}$$

DIFFERENT DENOMINATOR

In some instructions for preparing solutions you will be asked to dissolve chemical in some distilled water and dilute to one liter. (Such as in Examples 1 and 2.) However, other times you will be asked to dissolve the chemical in liter distilled water. This is slightly different. In the first case, the entire solution (chemical and water) is brought up to one liter. In the second case, the distilled water alone constitutes the one liter then the chemical is added to the distilled water. The denominator of the proportion should reflect this difference. (Compare the denominator of the equation used in Examples 1 and 2 with the denominator of the equation in Example 3.)

15.4 ESTIMATED CHLORINE RESIDUAL

When disinfecting clear wells, distribution reservoirs, or mains, high chlorine residuals must be measured. Instead of using the amperometric titration or DPD colorimetric method to determine chlorine residual, **a drop-dilution technique may be used to estimate the chlorine residual.**

To use the method, 10 mL distilled water and 0.5 mL DPD solution are added to a sample tube. The water to be tested is then added to the sample tube, a drop at a time until a color is produced. The following equation is then used to estimate the chlorine residual in the sample:

$$\text{Estim.} \atop \text{Cl}_2 \text{ Resid.,} \atop \text{mg/}L} = \frac{(\text{Cl}_2 \text{ Resid.,}) (\text{Dist. H}_2\text{O,})}{\underset{\text{Vol., drops}}{(\text{Sample})} \underset{\text{drop}}{(0.05 \text{ m}L)}}$$

Examples 1-4 illustrate this calculation.

Example 1: (Chlorine Residual)

❏ The chlorine residual indicated by the color change is 0.4 mg/L. If 2 drops of sample water were used until the color was produced in the 10 mL distilled water, what is the estimated actual chlorine residual of the sample?

$$\text{Estim. Cl}_2 \text{ Resid.,} \atop \text{mg/}L} = \frac{\underset{\text{mg/}L}{(\text{Cl}_2 \text{ Resid.,})} \underset{\text{m}L}{(\text{Distilled H}_2\text{O,})}}{\underset{\text{drops}}{(\text{Sample Vol.,})} \underset{\text{drop}}{(0.05 \text{ m}L)}}$$

$$= \frac{(0.4 \text{ mg/}L) (10 \text{ m}L)}{(2 \text{ drops}) (0.05 \text{ m}L/\text{drop})}$$

$$= \boxed{40 \text{ mg/}L}$$

Example 2: (Chlorine Residual)

❏ The chlorine residual indicated by the color change produced during the drop-dilution method is 0.3 mg/L. If 3 drops of sample water were used until the color was produced in the 10 mL distilled water, what is the estimated actual chlorine residual of the sample?

$$\text{Estim. Cl}_2 \text{ Resid.,} \atop \text{mg/}L} = \frac{\underset{\text{mg/}L}{(\text{Cl}_2 \text{ Resid.,})} \underset{\text{m}L}{(\text{Distilled H}_2\text{O,})}}{\underset{\text{drops}}{(\text{Sample Vol.,})} \underset{\text{drop}}{(0.05 \text{ m}L)}}$$

$$= \frac{(0.3 \text{ mg/}L) (10 \text{ m}L)}{(3 \text{ drops}) (0.05 \text{ m}L/\text{drop})}$$

$$= \boxed{20 \text{ mg/}L}$$

Example 3: (Chlorine Residual)

❏ The drop-dilution method is used to estimate the actual chlorine residual of a sample. The chlorine residual indicated by the test is 0.5 mg/L. If 4 drops of sample water were required to produce the color change in the 10 mL distilled water, what is the estimated actual chlorine residual of the sample?

$$
\begin{aligned}
\text{Estim. Cl}_2\text{ Resid.,}\atop\text{mg/L} \; &= \; \frac{(\text{Cl}_2\text{ Resid.,}\atop\text{mg/L})(\text{Distilled H}_2\text{O, }\atop\text{mL})}{(\text{Sample Vol.,}\atop\text{drops})\left(\dfrac{0.05\text{ mL}}{\text{drop}}\right)} \\[2mm]
&= \; \frac{(0.5\text{ mg/L})\,(10\text{ mL})}{(4\text{ drops})\,(0.05\text{ mL/drop})} \\[2mm]
&= \; \boxed{25\text{ mg/L}}
\end{aligned}
$$

Example 4: (Chlorine Residual)

❏ Five drops of sample water are required to produce a color change in the 10 mL distilled water during the drop-dilution method of estimating chlorine residual. The chlorine residual indicated by the test is 0.4 mg/L. What is the estimated actual chlorine residual of the water tested?

$$
\begin{aligned}
\text{Estim. Cl}_2\text{ Resid.,}\atop\text{mg/L} \; &= \; \frac{(\text{Cl}_2\text{ Resid.,}\atop\text{mg/L})(\text{Distilled H}_2\text{O, }\atop\text{mL})}{(\text{Sample Vol.,}\atop\text{drops})\left(\dfrac{0.05\text{ mL}}{\text{drop}}\right)} \\[2mm]
&= \; \frac{(0.4\text{ mg/L})\,(10\text{ mL})}{(5\text{ drops})\,(0.05\text{ mL/drop})} \\[2mm]
&= \; \boxed{16\text{ mg/L}}
\end{aligned}
$$

15.5 TEMPERATURE

Since the temperature of the wastewater affects its general characteristics, this test is one of the most frequently performed tests. The formulas normally used to convert Fahrenheit (°F) to Celsius (°C), are as follows:

Fahrenheit to Celsius:
°C = 5/9 (°F-32°)

Celsius to Fahrenheit:
°F = 9/5 (°C) + 32°

Because these formulas have parentheses around different terms, and one requires the addition of 32° while the other requires the subtraction of 32°, they are difficult conversions to remember unless used constantly.

Another method of conversion is more easily remembered and can be used regardless of whether the conversion is from Fahrenheit to Celsius or vice versa. This method consists of three steps, as illustrated at the top of this page.

Step 2 involves the only variable in these conversions. The decision of whether to multiply by 5/9 or 9/5 is dependent whether a larger or smaller number is desired.

The Celsius temperature scale is a lower range scale than that of Fahrenheit. Therefore, **the Celsius reading should always be a smaller number.**

THE THREE-STEP METHOD OF TEMPERATURE CONVERSION

Step 1: Add 40°

Step 2: Multiply by 5/9 or 9/5
(Depending on direction of conversion.)

°F → x $\frac{5}{9}$ → = °C

°F = ← $\frac{9}{5}$ x ← °C

Step 3: Subtract 40°

THE EFFECT OF MULTIPLYING BY 5/9 OR 9/5

When **multiplying by 5/9** (approximately 1/2), the answer will be **less than the original number**. For example:

100 x $\frac{5}{9}$ = $\frac{500}{9}$ = 56

Compare

When **multiplying by 9/5** (approximately 2), the answer will be **greater than the original number**. For example:

100 x $\frac{9}{5}$ = $\frac{900}{5}$ = 180

Compare

Example 1: (Temperature)
❑ The influent to a treatment plant has a temperature of 75° F. What is this temperature expressed in degrees Celsius?

Step 1: (Add 40°)

$$75°$$
$$\underline{+\,40°}$$
$$115°$$

Step 2: (Multiply by 5/9 or 9/5)

In this example the conversion is from Fahrenheit to Celsius. Since the answer should be a **smaller number**, multiply by 5/9:

$$\frac{(5)}{9}\,\frac{(115°)}{1} = \frac{575°}{9}$$
$$= 64°$$

Step 3: (Subtract 40°)

$$64°$$
$$\underline{-\,40°}$$
$$24°$$

$$\boxed{75°\ F = 24°\ C}$$

Example 2: (Temperature)
❑ The effluent of a treatment plant is 23° C. What is this expressed in degrees Fahrenheit?

Step 1: (Add 40°)

$$23°$$
$$\underline{+\,40°}$$
$$63°$$

Step 2: (Multiply by 5/9 or 9/5)

In this example the conversion is from Celsius to Fahrenheit. Since the answer should be a **larger number**, multiply by 9/5:

$$\frac{(9)}{5}\,\frac{(63°)}{1} = \frac{567°}{5}$$
$$= 113°$$

Step 3: (Subtract 40°)

$$113°$$
$$\underline{-\,40°}$$
$$73°$$

$$\boxed{23°\ C = 73°\ F}$$

NOTES:

Appendix 1

APPENDIX 1

Thrust Load Calculations

THRUST LOADS

(The total downthrust produced is the sum of the hydraulic thrust
plus the static thrust (dead weight) of the shaft and impellers.)

SHAFT WEIGHTS AND AREAS														
DIA. (IN's)	¾	1	1¼	1½	1¹¹⁄₁₆	1¹⁵⁄₁₆	2¼	2⁷⁄₁₆	2¹¹⁄₁₆	2¹⁵⁄₁₆	3³⁄₁₆	3⁷⁄₁₆	3¹¹⁄₁₆	3¹⁵⁄₁₆
LBS. / FT.	1.50	2.67	4.17	6.01	7.60	10.02	13.52	15.87	19.29	23.04	27.13	31.55	36.31	41.40
AREA	0.44	0.78	1.23	1.77	2.24	2.95	3.97	4.67	5.67	6.77	7.98	9.28	10.67	12.17

TOTAL THRUST FORMULA

TOTAL THRUST = (K x H x SG) + (W x S) + (Imp. Weight x No. of Stages)

K = Thrust Factor for Pump
H = Bowl Head (Total Head + Column Friction Loss) in Feet
W = Weight of Shaft in Pounds
S = Setting (Total Column Length) in Feet
SG = Specific Gravity of Liquid

EXAMPLE: The total thrust for a 12DKH 5-stage pump with a 1½" line shaft with a total
head of 250 feet and a setting of 230 feet, would be calculated as follows:

TOTAL THRUST = (K x H x SG) + (W x S) + (Imp. Wt. x No. of Stages)

K = 12.0   H = 250   W = 6.01   S = 230   SG = 1.0 for Water

TOTAL THRUST = (12.0 x 250 x 1.0) + (6.01 x 230) = 3000 + 1382.3 + (15 x 5) = 4457.3 Lbs.

NOTES:

The driver selected must have thrust capacity greater than the total thrust value.

Thrust factors (K) and impeller weights can be found on the performance curve page for the specific curve.

THRUST BEARING HORSEPOWER

Convert total thrust to loss in HP by the following formula:

$$\text{Thrust Bearing HP} = .0075 \times \frac{\text{RPM}}{100} \times \frac{\text{THRUST}}{1000}$$

Example: Total Thrust of 4457.3

$$\text{Thrust Bearing HP} = .0075 \times \frac{1770}{100} \times \frac{4457.3}{1000}$$

$$= 0.6 \text{ Thrust Bearing HP Loss.}$$

The published efficiencies of drivers do not include any
external thrust on rotor. The thrust load as a unit operating
loss must be added to the brake horsepower along with
mechanical friction in BHP to arrive at the actual pump brake
horsepower requirements of pump (field BHP).

Source: *Turbine Data Handbook*, Peabody Floway, Fresno, CA

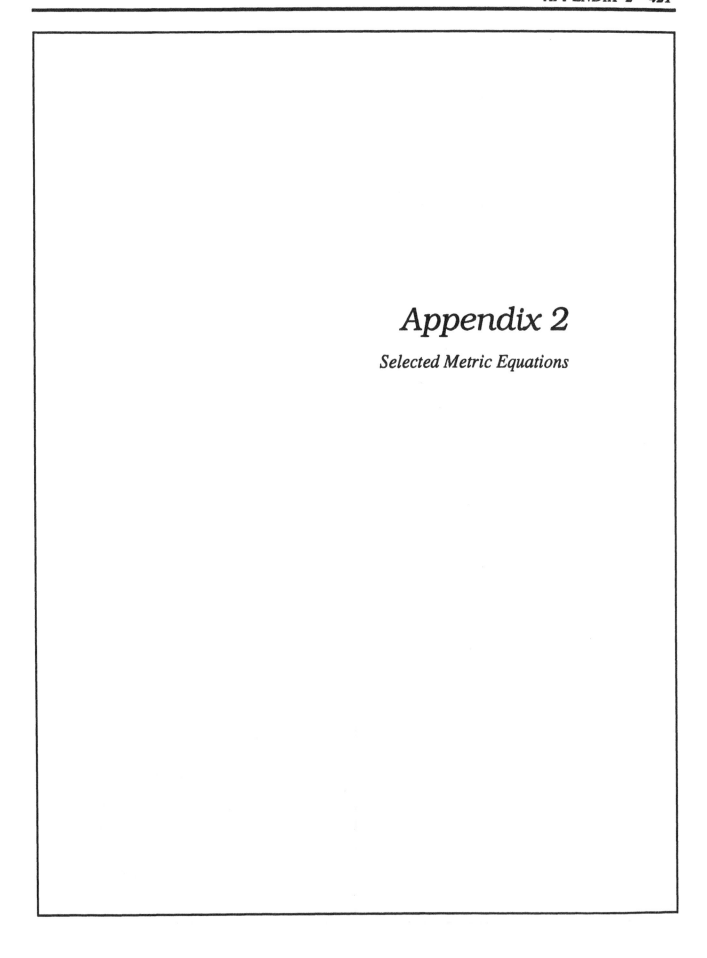

Appendix 2

Selected Metric Equations

Area and Volume Equations—Metric System

1. Areas

The equations for the four basic shapes most often used in area calculations are as follows:

Rectangle

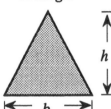

$$A = lw$$

where:

A = area, m^2
l = length, m
w = width, m

Triangle

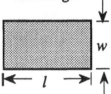

$$A = \frac{bh}{2}$$

where:

A = area, m^2
b = base, m
h = height, m

Trapezoid

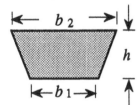

$$A = \frac{(b_1 + b_2)h}{2}$$

where:

A = area, m^2
b_1 = smaller base, m
b_2 = larger base, m
h = height, m

Circle

$$A = \frac{\pi D^2}{4}$$

or $A = (0.785)(D^2)$

where:

A = area, m^2
D = diameter, m

2. Volumes

The equations for the five basic shapes most often used in volume calculations are given below. The equation for oxidation ditch volume is also given.

Rectangular Prism

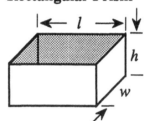

$$V = lwh$$

where:

V = volume, m^3
l = length, m
w = width, m
h = height, m

Triangular Prism

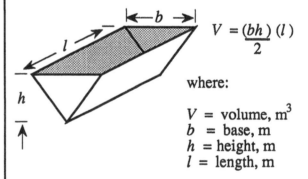

$$V = \frac{(bh)}{2}(l)$$

where:

V = volume, m^3
b = base, m
h = height, m
l = length, m

Trapezoidal Prism

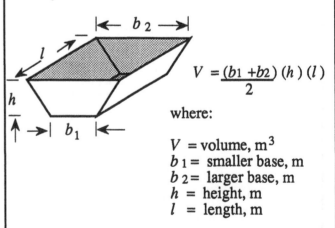

$$V = \frac{(b_1 + b_2)(h)(l)}{2}$$

where:

V = volume, m^3
b_1 = smaller base, m
b_2 = larger base, m
h = height, m
l = length, m

SUMMARY OF EQUATIONS

Cylinder

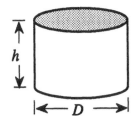

$V = (0.785)(D^2)(h.)$

where:

V = volume, m^3
D = diameter, m
h = height, m

Cone

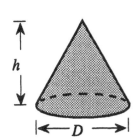

$V = \frac{1}{3}(0.785)(D^2)(h)$

where:

V = volume, m^3
D = diameter, m
h = height, m

Oxidation Ditch

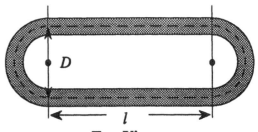

Top View

Dashed line represents total ditch length (L). This is equal to 2 half circumferences + 2 lengths

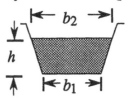

Cross-Section

$$\frac{\text{Volume}}{\text{cu ft}} = \frac{\text{(Trapezoidal)}}{\text{Area}}\text{(Total Length)}$$

$$V = \left[\frac{(b_1 + b_2)(h)}{2}\right]\left[\begin{array}{c}(\text{Length}) \\ \text{of} \\ 2 \text{ Sides}\end{array} + \begin{array}{c}(\text{Length}) \\ \text{Around} \\ 2 \text{ Half Circles}\end{array}\right]$$

$$V = \left[\frac{(b_1 + b_2)(h)}{2}\right]\left[(2l + \pi D)\right]$$

where:

V = volume, m^3
b_1 = smaller base, m
b_2 = larger base, m
h = height, m
l = length, m

Water Sources and Storage—Metric System

1. Well Drawdown

$$\text{Drawdown, m} = \frac{\text{Pumping}}{\text{Water}} - \frac{\text{Static}}{\text{Water}}$$
$$\text{Drawdown, m} = \text{Water Level, m} - \text{Water Level, m}$$

2. Well Yield

$$\text{Well Yield, } L\text{/sec} = \frac{\text{Flow, liters}}{\text{Duration of Test, sec}}$$

3. Specific Yield

$$\text{Specific Yield, } L\text{/sec/m} = \frac{\text{Well Yield, } L\text{/sec}}{\text{Drawdown, m}}$$

4. Well Casing Disinfection

First calculate chlorine required, kg:

$$\frac{(\text{mg/}L \text{ Cl}_2)(\text{Water-filled Casing Volume, m}^3)}{1000} = \text{kg Cl}_2$$

Then calculate chlorine compound required, if applicable:

$$\text{Chlorine Compound, kg} = \frac{\text{Chlorine, kg}}{\dfrac{\% \text{ Avail. Cl}_2}{100}}$$

5. Pond or Small Lake Storage Capacity

$$\text{Storage Capac., m}^3 = \text{(Aver.) Length, m} \times \text{(Aver.) Width, m} \times \text{(Aver.) Depth, m}$$

6. Copper Sulfate Dosing

The desired copper sulfate dosage may be expressed in three different ways, as indicated below:

• Given mg/L Copper

First calculate the kg copper (Cu) required:

$$\frac{(\text{mg/}L \text{ Cu})(\text{Vol. of Reser., m}^3)}{1000} = \text{kg Cu}$$

Then determine the kg of copper sulfate pentahydrate ($\text{CuSO}_4 \cdot 5\text{H}_2\text{O}$): (This compound contains 25% copper.)

$$\frac{\text{kg Cu}}{\dfrac{\% \text{ Copper}}{100}} = \text{kg CuSO}_4 \cdot 5\text{H}_2\text{O}$$

SUMMARY OF EQUATIONS

• <u>Given g Copper Sulfate/m^3</u>
(Using 0.3 g CuSO$_4$/m^3 as an example)

$$\text{Copper Sulfate, g} = \frac{(0.3 \text{ g CuSO}_4)(\text{Actual m}^3)}{1 \text{ m}^3}$$

• <u>Given kg Copper Sulfate/ha</u>
(Using 6 kg CuSO$_4$/ha as an example)

$$\text{Copper Sulfate, kg} = \frac{(6 \text{ lbs CuSO}_4)(\text{Actual ha})}{1 \text{ ha}}$$

Coagulation and Flocculation—Metric System

SUMMARY OF EQUATIONS

1. Chamber or Basin Volume

$$\text{Vol, m}^3 = (\text{length, m}) (\text{width, m}) (\text{depth, m})$$

2. Detention Time

For Flash Mix Chambers:

$$\frac{\text{Detention}}{\text{Time, sec}} = \frac{\text{Volume of Chamber, m}^3}{\text{Flow, m}^3/\text{s}}$$

For Flocculation Basins:

$$\frac{\text{Detention}}{\text{Time, min}} = \frac{\text{Volume of Basin, m}^3}{\text{Flow, m}^3/\text{min}}$$

3. Determining Chemical Feeder Setting— Dry Chemical Feeder (kg/day)

$$\frac{(\text{mg}/L \text{ Chem.}) (\text{m}^3/\text{d flow})}{1000} = \frac{\text{Chemical}}{\text{kg/d}}$$

4. Determining Chemical Feeder Setting— Solution Chemical Feeder, mL/min

$$\frac{\begin{array}{c}(\text{m}^3/\text{d}) (1000 \, L/\text{m}^3) (1000 \text{ m}L/L)\\ \text{flow}\end{array}}{1440 \text{ min/day}} = \frac{\text{Chem.,}}{\text{m}L/\text{min}}$$

5. Percent Strength of Solutions

Percent strength using dry chemicals:

$$\% \text{ Strength} = \frac{\text{Chemical, g}}{\text{Solution, g}} \times 100$$

Or

$$\frac{\%}{\text{Strength}} = \frac{\text{Chemical, g}}{\text{Water, g} + \text{Chem., g}} \times 100$$

Percent strength using liquid chemicals:

$$\begin{array}{c}(\text{Liq.})\\ \text{Poly.,}\\ \text{g}\end{array} \frac{(\% \text{ Strength})}{\text{of Liq. Poly.}}{100} = \begin{array}{c}(\text{Poly.})\\ \text{Sol'n,}\\ \text{g}\end{array} \frac{(\%\text{Strength})}{\text{of Poly. Sol'n.}}{100}$$

6. Mixing Solutions of Different Strength

$$\frac{\% \text{ Strength}}{\text{of Mixture}} = \frac{\text{Chem. in Mixture, g}}{\text{Sol'n Mixture, g}} \times 100$$

7. Dry Chemical Feeder Calibration

$$\frac{\text{Chemical}}{\text{Feed}}{\text{Rate, g/min}} = \frac{\text{Chemical Used, g}}{\text{Length of Application, min}}$$

SUMMARY OF EQUATIONS

8. Solution Chemical Feeder Calibration
(Given mL/min flow rate)

First calculate the m^3/d flow rate:

$$\frac{(mL/min)\ (1440\ min/day)}{(1000\ mL/L)\ (1000\ L/m^3)} = m^3/d$$

Then calculate chemical dosage, kg/day:

$$\frac{(mg/L\ Chem.)\ (m^3/d\ flow)}{1000} = \frac{Chemical,}{kg/d}$$

9. Solution Chemical Feeder Calibration
(Given Drop in Solution Tank Level)

Diameter, m

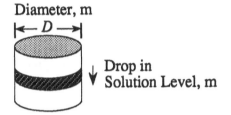

Drop in
Solution Level, m

$$\frac{Flow,}{mL/min} = \frac{Volume\ Pumped,\ mL}{Duration\ of\ Test,\ min}$$

Or

$$\frac{Flow}{mL/min} = \frac{(0.785)(D^2)(Drop,\ m)(1000\ L/m^3)(1000\ mL/L)}{Duration\ of\ Test,\ min}$$

10. Chemical Use Calculations

First determine the average chemical use:

$$\frac{Average\ Use}{kg/day} = \frac{Total\ Chem.\ Used,\ kg}{Number\ of\ Days}$$

Or

$$\frac{Average\ Use}{L/d} = \frac{Total\ Chem.\ Used,\ L}{Number\ of\ Days}$$

Then calculate days' supply in inventory

$$\frac{Days'}{Supply\ in\ Inventory} = \frac{Total\ Chem.\ in\ Inventory,\ kg}{Average\ Use,\ kg/day}$$

Or

$$\frac{Days'}{Supply\ in\ Inventory} = \frac{Total\ Chem.\ in\ Inventory,\ L}{Average\ Use,\ L/d}$$

Sedimentation—Metric System

SUMMARY OF EQUATIONS

**1. Tank Volume—
Rectangular Sedimentation Basin**

$$\text{Vol, m}^3 = (\text{length, m}) \, (\text{width, m}) \, (\text{depth, m})$$

Tank Volume—Circular Clarifier

$$\text{Vol, m}^3 = (0.785) \, (D^2) \, (\text{depth, m})$$

Where D = Diameter, m

2. Detention Time

$$\frac{\text{Detention}}{\text{Time, hrs}} = \frac{\text{Tank Volume, m}^3}{\text{Flow Rate, m}^3/\text{h}}$$

3. Surface Overflow Rate

$$\begin{array}{l}\text{Surface} \\ \text{Overflow} \\ \text{Rate,} \\ \text{mm/sec}\end{array} = \frac{(\text{Flow, m}^3/\text{d}) \, (1000 \text{ mm/m})}{\underset{\text{m}^2 \quad \text{min/day} \quad \text{sec/min}}{(\text{Area,}) \, (1440) \, (60)}}$$

4. Mean Flow Velocity

$$\underset{\text{m}^3/\text{min}}{Q} = \underset{(\text{m})(\text{m})}{A} \; \underset{(\text{m/min})}{V}$$

Or, if the flow rate is given as L/sec:

$$\underset{L/\text{sec}}{Q} = \frac{\underset{(\text{m})(\text{m})}{A} \; \underset{(\text{m/min})}{V} \, (1000 \text{ L/m}^3)}{60 \text{ sec/min}}$$

5. Weir Loading Rate

$$\begin{array}{l}\text{Weir Loading} \\ \text{Rate, m}^3/\text{d/m}\end{array} = \frac{\text{Flow, m}^3/\text{d}}{\text{Length of weir, m}}$$

Calculations associated with the solids contact clarification process include:

- Detention Time,

- Percent Settled Sludge, ("volume over volume" test)

- Lime Dose Required, mg/L, and

- Lime Dose Required, lbs/day or grams/min.

The detention time calculation has been described above (see calculation #2). Equations for the remaining calculations are given below.

6. Percent Settled Sludge (V/V Test)

$$\begin{array}{l}\text{\% Settled} \\ \text{Sludge}\end{array} = \frac{\text{Settled Sludge Volume, m}L}{\text{Total Sample Volume, m}L} \times 100$$

SUMMARY OF EQUATIONS

7. Lime Dose Required, mg/*L*

The lime dose requirement (in mg/*L*) for the solids contact clarification process is calculated in three steps:

- **Step 1**—Determine the <u>total alkalinity</u> (HCO_3^-) required to react with the alum to be added and provide proper precipitation: (Use 1 mg/*L* alum reacts with 0.45 mg/*L* alkalinity, HCO_3^-.)

$$
\begin{array}{c}
\text{Total Alk.} \\
\text{Required,} \\
\text{mg/}L
\end{array}
=
\begin{array}{c}
\text{Alk. Required to} \\
\text{React with} \\
\text{the Alum,} \\
\text{mg/}L
\end{array}
+
\begin{array}{c}
\text{Alk. Required to} \\
\text{Assure Proper} \\
\text{Precipitation of} \\
\text{Alum,} \\
\text{mg/}L
\end{array}
$$

- **Step 2**—Determine alkalinity (HCO_3^-) to be added to the water:

$$
\begin{array}{c}
\text{Total Alk.} \\
\text{Required,} \\
\text{mg/}L
\end{array}
-
\begin{array}{c}
\text{Alk.} \\
\text{Present in} \\
\text{Water, mg/}L
\end{array}
=
\begin{array}{c}
\text{Alk. to be} \\
\text{Added to the} \\
\text{Water, mg/}L
\end{array}
$$

- **Step 3**—Determine the lime required to meet this alkalinity need: (Using the relationships 1 mg/*L* alum reacts with 0.45 mg/*L* alkalinity, HCO_3^-, and 1 mg/*L* alum reacts with 0.35 mg/*L* lime, the following proportion can be constructed.)

From Step 2 above ↓

$$
\frac{0.45 \text{ mg/}L \text{ } HCO_3^-}{0.35 \text{ mg/}L \text{ } Ca(OH)_2} = \frac{\square \text{ mg/}L \text{ } HCO_3^-}{x \text{ mg/}L \text{ } Ca(OH)_2}
$$

8. Lime Dose Required, lbs/day

$$
\frac{(\text{mg/}L \text{ Lime}) (\text{Flow, m}^3\text{/d})}{1000} = \begin{array}{c} \text{Lime,} \\ \text{kg/day} \end{array}
$$

9. Lime Dose Required, g/min

$$
\frac{(\text{Lime, kg/day}) (1000 \text{ g/kg})}{1440 \text{ min/day}} = \begin{array}{c} \text{Lime,} \\ \text{g/min} \end{array}
$$

Filtration—Metric System

SUMMARY OF EQUATIONS

1. Flow Rate Through A Filter, L/sec

Using Flow Meter

$$\frac{(\text{Flow Rate, m}^3/\text{d}) (1000 \, L/\text{m}^3)}{(1440 \text{ min/day}) (60 \text{ sec/min})} = \frac{\text{Flow Rate,}}{L/\text{sec}}$$

Using Total Liters Produced

$$\frac{\text{Flow Rate,}}{L/\text{sec}} = \frac{\text{Total Liters Produced, } L}{(\text{Filter Run, min}) (60 \text{ sec/min})}$$

Using Water Drop Data
(This is a Q = AV problem)

$$Q_{L/\text{sec}} = \qquad (A) \quad (V_{\text{cm/min}})$$

$$Q_{L/\text{sec}} = \frac{(\text{Length,}_{\text{m}})(\text{Width,}_{\text{m}})(\text{Drop})(1000 \, L/\text{m}^3) \, \text{Veloc.,}_{\text{cm/min}}}{(100 \text{ cm/m}) (60 \text{ sec/min})}$$

2. Filtration Rate

$$\frac{\text{Filtration Rate,}}{L/\text{sec/m}^2} = \frac{\text{Flow Rate, } L/\text{sec}}{\text{Filter Surface Area, m}^2}$$

Or

$$\frac{\text{Filtration Rate,}}{\text{mm/sec}} = \frac{(\text{Flow Rate, } L/\text{sec}) (1000 \text{ mm/m})}{(\text{Filter Surface Area, m}^2) (1000 \text{ L/m}^3)}$$

3. Unit Filter Run Volume (UFRV), L/m^2

$$\frac{\text{UFRV,}}{L/\text{m}^2} = \frac{\text{Total Water Filtered, } L}{\text{Filter Surface Area, m}^2}$$

Filtration rate data can also be used to calculate UFRV:

$$\frac{\text{UFRV}}{L/\text{m}^2} = \frac{\underset{\text{mm/sec}}{(\text{Filtr. Rate})} \; \underset{\text{Time, min}}{(\text{Filter Run})} \; (60) \; (1000) \atop \text{sec/min} \; L/\text{m}^3}{1000 \text{ mm/m}}$$

4. Backwash Rate,

$$\frac{\text{Backwash Rate,}}{\text{m}^3/\text{d/m}^2} = \frac{\text{Flow Rate, m}^3/\text{d}}{\text{Filter Surface Area, m}^2}$$

Or

$$\frac{\text{Backwash Rate,}}{L/\text{min/m}^2} = \frac{\text{Flow Rate, } L/\text{min}}{\text{Filter Surface Area, m}^2}$$

Backwash rate is sometimes expressed as mm/min:

$$\frac{(\text{Backwash Rate,}) (1000 \text{ mm/m}) \atop L/\text{min/m}^2}{1000 \, L/\text{m}^3} = \frac{\text{Backwash Rate,}}{\text{mm/min}}$$

SUMMARY OF EQUATIONS

5. Volume of Backwash Water Required, L

$$\begin{array}{l} \text{Backwash} \\ \text{Water} \\ \text{Required, } L \end{array} = \begin{array}{c} \text{(Backwash)} \\ \text{Flow,} \\ L/\text{sec} \end{array} \begin{array}{c} \text{(Duration of)} \\ \text{Backwash,} \\ \text{min} \end{array} \begin{array}{c} (\,60\,) \\ \text{sec/min} \end{array}$$

6. Required Depth of Backwash Water Tank, m

<u>If Tank is Cylindrical</u>

$$\begin{array}{l} \text{Tank Vol.} \\ \text{Req'd, m}^3 \end{array} = (0.785)\,(D^2)\,(\text{Depth, m})$$

Where $D =$ Diameter, m

<u>If Tank is Rectangular</u>

$$\begin{array}{l} \text{Tank Vol.} \\ \text{Req'd, m}^3 \end{array} = \begin{array}{c} \text{(Length,)} \\ \text{m} \end{array} \begin{array}{c} \text{(Width,)} \\ \text{m} \end{array} \begin{array}{c} \text{(Depth,)} \\ \text{m} \end{array}$$

7. Backwash Pumping Rate, L/sec

$$\begin{array}{l} \text{Backwash} \\ \text{Pumping} \\ \text{Rate, } L/\text{sec} \end{array} = \frac{\begin{array}{c} \text{(Desired Backwash)} \\ \text{Rate, } L/\text{min/m}^2 \end{array} \begin{array}{c} \text{(Filter Area,)} \\ \text{m}^2 \end{array}}{60 \text{ sec/min}}$$

8. Percent of Product Water Used for Backwashing

$$\begin{array}{l} \% \text{ Backwash} \\ \text{Water} \end{array} = \frac{\text{Backwash Water, } L}{\text{Water Filtered, } L} \times 100$$

9. Percent Mud Ball Volume

$$\begin{array}{l} \% \text{ Mud Ball} \\ \text{Volume} \end{array} = \frac{\text{Mud Ball Vol., m}L}{\text{Total Sample Vol., m}L} \times 100$$

Chlorination—Metric System

SUMMARY OF EQUATIONS

1. Chlorine Feed Rate kg/d

$$\frac{(mg/L\ Cl_2)\ (m^3/d\ flow)}{1000} = \text{Chlorine, kg/d}$$

2. Chlorine Dose, Demand and Residual

$$Cl_2\ Dose = Cl_2\ Demand + Cl_2\ Residual$$

To determine if chlorination is above the **breakpoint**, compare the expected increase in residual with the actual increase in residual:

Expected Increase in Residual

$$\frac{(mg/L\ Expected\ Inc.)\ (m/d\ flow)}{1000} = \text{Increase in } Cl_2\ \text{Dose, kg/d}$$

Actual Increase in Residual

$$\begin{array}{l}\text{Actual Increase in Resid., mg/L} = \text{New Residual, mg/L} - \text{Old Residual, mg/L}\end{array}$$

3. Dry Hypochlorite Feed Rate

Simplified Equation:

$$\text{Hypochlorite, kg/day} = \frac{kg/day\ Cl_2}{\dfrac{\%\ Available\ Cl_2}{100}}$$

3. Dry Hypochlorite Feed Rate—Cont'd

Expanded Equations:

$$\text{Hypochlorite, kg/day} = \frac{\dfrac{(mg/L\ Cl_2)\ (m^3/d\ flow)}{1000}}{\dfrac{\%\ Strength\ of\ Hypochl.}{100}}$$

Or

$$\text{Hypochlorite, kg} = \frac{\dfrac{(mg/L\ Cl_2)\ (Tank\ Vol.,\ m^3)}{1000}}{\dfrac{\%\ Strength\ of\ Hypochl.}{100}}$$

4. Hypochlorite Solution Feed Rate

Simplified Equation:

$$\begin{array}{l}\text{Actual Dose kg/day} = \text{Solution Feeder Dose kg/day}\end{array}$$

Expanded Equation:

$$\frac{(mg/L)\ (m^3/d\ Flow)}{1000}\ Dose\ Treated = \frac{(mg/L)\ (m^3/d)}{1000}\ Dose\ Sol'n$$

5. Percent Strength of Solutions

Percent strength using dry chlorine:

$$\%\ Cl_2\ Strength = \frac{Chlorine,\ kg}{Water,\ kg + Chlorine,\ kg} \times 100$$

SUMMARY OF EQUATIONS

5. Percent Strength of Solutions—Cont'd

Percent strength using chlorine solution:

$$\% \; Cl_2 \; \text{Strength} = \frac{\text{Chlorine, kg}}{\text{Water, kg} + \text{Chlor. Cmpd., kg}} \times 100$$

Or

$$\% \; Cl \; \text{Strength} = \frac{(\text{Liquid}) \; (\% \; \text{Strength}) \; \text{Hypo., kg} \; \dfrac{\text{of Liq. Hypo.}}{100}}{\text{Water, kg} + \text{Liq. Hypo., kg}} \times 100$$

6. Mixing Hypochlorite Solutions

Simplified Equation:

$$\% \; Cl_2 \; \text{Strength of Mixture} = \frac{Cl_2 \; \text{in Mixture, kg}}{\text{Sol'n Mixture, kg}} \times 100$$

Expanded Equations:

$$\% \; Cl_2 \; \text{Strength of Mixture} = \frac{\text{kg } Cl_2 \text{ from Sol'n 1} + \text{kg } Cl_2 \text{ from Sol'n 2}}{\text{kg Sol'n 1} + \text{kg Sol'n 2}} \times 100$$

7. Chemical Use Calculations

Average Chemical Use:

$$\text{Average Use kg/day} = \frac{\text{Total Chem. Used, kg}}{\text{Number of Days}}$$

Or

$$\text{Average Use } L/\text{d} = \frac{\text{Total Chem. Used, } L}{\text{Number of Days}}$$

7. Chemical Use Calculations—Cont'd

Days' Supply In Inventory:

$$\text{Days' Supply in Inventory} = \frac{\text{Total Chem. in Inventory, kg}}{\text{Average Use, kg/day}}$$

Or

$$\text{Days' Supply in Inventory} = \frac{\text{Total Chem. in Inventory, } L}{\text{Average Use, } L/\text{d}}$$

Total Chemical Use, lbs:

$$\text{Total Chemical Used, kg} = (\text{Chem. Use,}) \; (\text{Days Use}) \; \text{kg/d}$$

Or

$$\text{Total Chemical Used, kg} = (\text{Chem. Use,}) \; (\text{Hrs Use}) \; \text{kg/h}$$

Fluoridation—Metric System

SUMMARY OF EQUATIONS

1. Percent Fluoride Ion in a Compound

$$\text{\% Fluoride Ion in a Compound} = \frac{\text{Molec. Wt. of Fluoride}}{\text{Molec. Wt. of Compound}} \times 100$$

2 Calculating Dry Feed Rate of a Fluoride Compound, lbs/day

$$\frac{\dfrac{(\text{mg/L F})\,(\text{m}^3/\text{d flow})}{1000}}{\dfrac{(\text{\% Comm. Purity})}{100}\,\dfrac{(\text{\% F in Comp.})}{100}} = \begin{array}{l}\text{kg/d} \\ \text{Fluoride} \\ \text{Comp.}\end{array}$$

3. Percent Strength of a Solution

Simplified Equation:

$$\text{\% Strength of a Solution} = \frac{\text{kg Chemical}}{\text{kg Solution}} \times 100$$

Expanded Equations:

When the commercial chemical is 100% pure:

$$\text{\% Strength} = \frac{\text{kg Chemical}}{\text{kg Water} + \text{kg Chemical}} \times 100$$

When the commercial chemical is less than 100% pure:

$$\text{\% Strength} = \frac{\dfrac{(\text{Chem.,})\,(\text{\% Comm. Purity})}{\text{kg}}{100}}{\text{kg Water} + \text{kg Chem.}} \times 100$$

4. Fluoride Solution Feed Rate Calculations

There are two related methods of calculating feed rates:

• **The mg/L to lbs/day method:**

Simplified Equation:

$$\begin{array}{l}\text{Actual Feed,} \\ \text{kg F/d}\end{array} = \begin{array}{l}\text{Solution Feed,} \\ \text{kg F/d}\end{array}$$

Expanded Equation:

$$\frac{(\text{mg/L})\,(\text{m}^3/\text{d})}{1000}\,\begin{array}{c}\text{F}\\\text{flow}\end{array} = \frac{(\text{mg/L})\,(\text{m}^3/\text{d})}{1000}\,\begin{array}{c}\text{Comp. flow}\end{array}\,\frac{(\text{sp. gr.})\,(\text{\% F})}{100}\,\begin{array}{c}\text{in Comp.}\end{array}$$

When applicable

• **The rate/concentration method:**

$$R_1 \times C_1 = R_2 \times C_2$$

where:
R_1 = feed rate of first solution
C_1 = concentration of first solution
R_2 = feed rate of second solution
C_2 = concentration of second solution

5. Calculating mg/L Fluoride Dosage

Given kg/d Fluoride:

$$\frac{(\text{mg/L F})\,(\text{m}^3/\text{d Flow})}{1000} = \text{kg/day F}$$

SUMMARY OF EQUATIONS

5. Calculating mg/L Fluoride Dosage —Cont'd

For Dry Feed Rate Calculations:

$$\frac{\dfrac{(mg/L)\,(m^3/d)}{1000}}{\dfrac{(\%\ Comm.\ Pur.)}{100}\,\dfrac{(\%\ F\ in\ Comp.)}{100}} = Compound,\ kg/d$$

For Solution Feed Rate Calculations:

$$\underset{F\ flow}{\frac{(mg/L)\,(m^3/d)}{1000}} = \underset{Comp.\ flow}{\frac{(mg/L)\,(m^3/d)}{1000}}\ (sp.\ gr.)\ \underset{in\ Comp.}{\frac{(\%\ F)}{100}}$$

When applicable

6. Solution Mixtures

<u>Simplified Equation:</u>

$$\begin{array}{c} \%\ Strength \\ of\ Mixture \end{array} = \frac{kg\ Chemical}{kg\ Solution} \times 100$$

<u>Expanded Equation:</u>

$$\begin{array}{c} \% \\ Strength \\ of \\ Mixture \end{array} = \frac{\dfrac{kg\ Chem.\ in}{Sol'n\ 1} + \dfrac{kg\ Chem.\ in}{Sol'n\ 2}}{kg\ Sol'n\ 1 + kg\ Sol'n\ 2} \times 100$$

By moving the denominator to the left side of the equation, the equations above can be stated generally as:

$$R_3\,C_3\ =\ R_1\,C_1\ +\ R_2\,C_2$$

where: R_3 = kg (or L) of Sol'n Mixture
C_3 = % Strength of Mixture
R_1 = kg (or L) of Sol'n 1
C_1 = % Strength of Sol'n 1
R_2 = kg (or L)) of Sol'n 2
C_2 = % Strength of Sol'n 2

Softening—Metric System

1. Equivalent Weight and Equivalents

$$\text{Equivalent Weight} = \frac{\text{Atomic Weight}}{\text{Valence}}$$

Or

$$\text{Equivalent Weight} = \frac{\text{Molecular Weight}}{\text{Net Positive Valence}}$$

Or

$$\text{Equivalent Weight} = \frac{\text{Molecular Weight}}{\text{Number of Equivalents}}$$

The number of equivalents or milliequivalents of a chemical may be calculated as:

$$\text{Number of Equivalents} = \frac{\text{Chemical, grams}}{\text{Equivalent Wt.}}$$

And

$$\text{Number of Milliequivalents} = \frac{\text{Chemical, mg}}{\text{Equivalent Wt.}}$$

2. Hardness, as $CaCO_3$

Calcium Hardness, as $CaCO_3$:

$$\frac{\text{Ca Hardness, mg/}L \text{ as } CaCO_3}{\text{Equivalent Wt. of } CaCO_3} = \frac{\text{Ca, mg/}L}{\text{Equivalent Wt. of Calcium}}$$

Magnesium Hardness, as $CaCO_3$:

$$\frac{\text{Mg Hardness, mg/}L \text{ as } CaCO_3}{\text{Equivalent Wt. of } CaCO_3} = \frac{\text{Mg, mg/}L}{\text{Equivalent Wt. of Magnesium}}$$

2. Hardness, as $CaCO_3$ —Cont'd

Total Hardness, as $CaCO_3$:

$$\begin{array}{c}\text{Total Hardness,} \\ \text{mg/}L \\ \text{as } CaCO_3\end{array} = \begin{array}{c}\text{Ca Hardness,} \\ \text{mg/}L \\ \text{as } CaCO_3\end{array} + \begin{array}{c}\text{Mg Hardness,} \\ \text{mg/}L \\ \text{as } CaCO_3\end{array}$$

3. Carbonate and Noncarbonate Hardness

The alkalinity of the water dictates the type of hardness present:

If the alkalinity (as $CaCO_3$) is greater than the total hardness, then all the hardness is carbonate hardness:

$$\begin{array}{c}\text{Total Hardness,} \\ \text{mg/}L \text{ as } CaCO_3\end{array} = \begin{array}{c}\text{Carbonate Hardness,} \\ \text{mg/}L \text{ as } CaCO_3\end{array}$$

If the alkalinity (as $CaCO_3$) is less than the total hardness, the alkalinity is carbonate hardness and noncarbonate hardness is present as well:

$$\begin{array}{c}\text{Total Hardness,} \\ \text{mg/}L \text{ as } CaCO_3\end{array} = \begin{array}{c}\text{Carbonate} \\ \text{Hardness,} \\ \text{as } CaCO_3\end{array} + \begin{array}{c}\text{Noncarbonate} \\ \text{Hardness,} \\ \text{mg/}L \text{ as } CaCO_3\end{array}$$

These terms represent the same number

$$\begin{array}{c}\text{Total Hardness,} \\ \text{mg/}L \text{ as } CaCO_3\end{array} = \begin{array}{c}\text{Alkalinity,} \\ \text{mg/}L \text{ as } CaCO_3\end{array} + \begin{array}{c}\text{Noncarbonate} \\ \text{Hardness,} \\ \text{mg/}L \text{ as } CaCO_3\end{array}$$

SUMMARY OF EQUATIONS

4. Phenolphthalein and Total Alkalinity

$$\text{Phenolphthalein Alkalinity } mg/L \text{ as } CaCO_3 = \frac{(A)(N)(50,000)}{mL \text{ of Sample}}$$

$$\text{Total Alkalinity } mg/L \text{ as } CaCO_3 = \frac{(B)(N)(50,000)}{mL \text{ of Sample}}$$

Where: A = mL titrant used to pH 8.3
$\quad\quad\quad B$ = Total mL of titrant used to pH 4.5
$\quad\quad\quad N$ = normality of the acid

5. Lime Dosage for Softening

Using Quicklime, CaO:

$$\text{Quicklime (CaO) Feed, } mg/L = \frac{(A+B+C+D)\,1.15}{\dfrac{\% \text{ Purity of Lime}}{100}}$$

Where:

A: CO_2 in Source water—
$\quad$ (mg/L as CO_2)(56/44)

B: Bicarbonate alkalinity removed in softening
$\quad$ (mg/L as $CaCO_3$)(56/100)

C: Hydroxide alkalinity in softener effluent
$\quad$ (mg/L as $CaCO_3$)(56/100)

D: Magnesium removed in softening
$\quad$ (mg/L as Mg^{+2})(56/24.3)

1.15: Excess lime dosage
$\quad\quad$ (using 15 percent excess)

Using hydrated Lime, Ca(OH)$_2$:

To calculate hydrate lime required, use the same equation as given for quicklime except substitute 74 for 56 in A, B, C & D.

6. Soda Ash Dosage for Removal of Noncarbonate Hardness

1. First calculate noncarbonate hardness. (Review Section 14.3)

2. Then calculate soda ash dosage required:

$$\text{Soda Ash Feed, } mg/L = \text{(Noncarbonate) Hardness } mg/L \text{ as } CaCO_3 \; \frac{(106)}{100}$$

7. Carbon Dioxide Required for Recarbonation

1. First calculate mg/L excess lime:

$$\text{Excess Lime, } mg/L = (A+B+C+D)(0.15)$$

Where A, B, C and D are the same factors described in Section 14.6.

2. Then calculate total carbon dioxide dosage (mg/L) required:

$$\text{Total } CO_2 \text{ Feed, } mg/L = \left[\begin{array}{c} Ca(OH)_2 \\ \text{Excess,} \\ mg/L \end{array}\right]\frac{(44)}{74} + \begin{array}{c}(Mg^{+2}) \\ \text{Residual} \\ mg/L\end{array}\frac{(44)}{24.3}$$

8. Chemical Feeder Settings

To calculate kg/day chemical required:

$$\frac{(mg/L \text{ Chemical})(m^3/d \text{ Flow})}{1000} = \frac{kg/day}{\text{Chemical}}$$

To express this feed rate as kg/min:

$$\frac{\text{Chemical, kg/day}}{1440 \text{ min/day}} = \frac{\text{Chemical,}}{kg/min}$$

Softening Cont'd—Metric System

SUMMARY OF EQUATIONS

9. Expressions of Hardensss – mg/L and gpg

$$1 \text{ gpg} = 17.12 \text{ mg}/L$$

When using the "box method" of conversion, the following diagram is used:

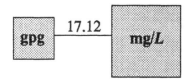

10. Ion Exchange Capacity

$$\begin{array}{l}\text{Exchange} \\ \text{Capacity,} \\ \text{mg}\end{array} = \begin{array}{l}\text{(Removal)} \\ \text{Capacity,} \\ \text{mg/m}^3\end{array} \begin{array}{l}\text{(Media Vol.,)} \\ \text{m}^3\end{array}$$

11. Water Treatment Capacity (before Regeneration)

$$\begin{array}{l}\text{Water Treatment} \\ \text{Capacity,} L\end{array} = \frac{\text{Exchange Capacity, mg}}{\text{Hardness Removed, mg}/L}$$

12. Operating Time (Until Regeneration Required)

$$\begin{array}{l}\text{Operating} \\ \text{Time, hrs}\end{array} = \frac{\text{Water Treated, } L}{(\text{Flow Rate, } L/\text{min}) (60 \text{ min/hr})}$$

13. Salt and Brine Required for Regeneration

$$\begin{array}{l}\text{Salt Required,} \\ \text{kg}\end{array} = \begin{array}{l}(\text{Salt Req'd,}) \\ \text{kg/1000 mg}\end{array} \begin{array}{l}(\text{Hardness}) \\ \text{Removed,} \\ \text{mg}\end{array}$$

$$\begin{array}{l}\text{Brine} \\ \text{Required,} \\ L\end{array} = \frac{\text{Salt Required, kg}}{\text{Brine Solution, kg Salt/}L \text{ Brine}}$$

Laboratory—Metric System

SUMMARY OF EQUATIONS

1. Estimating Flow From A Faucet

$$\text{Flow, } L/\text{min} = \frac{\text{Volume, } L}{\text{Time, min}}$$

2. Service Line Flushing Time

$$\text{Flushing Time, min} = \frac{(\text{Pipe Vol., m}^3)\ (1000\ L/\text{m}^3)\ (2)}{\text{Flow Rate, } L/\text{min}}$$

3. Solution Concentration—Normality

$$\text{Normality} = \frac{\text{No. of Equivalents of Solute}}{\text{Liter of Solution}}$$

Note: "Equivalents" are described in Chapter 14, Section 1.

When preparing **standard solutions**, the following proportion procedure may be used:

Desired Mixture   Actual Mixture

$$\frac{\text{Desired Wt., g}}{\text{Desired Sol'n Vol., m}L} = \frac{\text{Actual Wt, g}}{\text{Actual Sol'n Vol., m}L}$$

4. Estimated Chlorine Residual

$$\text{Estimated Cl}_2 \text{ Resid., mg/}L = \frac{(\text{Sample Cl}_2)\ (\text{Distilled})}{(\text{Sample Vol.,})\ (0.05\ \text{m}L/\text{drop})\ \text{drops}}$$

5. Temperature

Fahrenheit to Celsius :
(Conventional Equation)

$$°C = \frac{5}{9}\ (°F - 32°)$$

Celsius to Fahrenheit :
(Conventional Equation)

$$°F = \frac{9}{5}\ (°C) + 32°$$

3-Step Method:
(Converting either direction)

1. Add 40°.

2. Multiply by 5/9 or 9/5. (Depending on the direction of conversion.)

3. Subtract 40°.

Index

Index

Index—Cont'd

Index—Cont'd

NOTES: